Studies in Systems, Decision and Control

Volume 355

Series Editor

Janusz Kacprzyk, Systems Research Institute, Polish Academy of Sciences, Warsaw, Poland

The series "Studies in Systems, Decision and Control" (SSDC) covers both new developments and advances, as well as the state of the art, in the various areas of broadly perceived systems, decision making and control–quickly, up to date and with a high quality. The intent is to cover the theory, applications, and perspectives on the state of the art and future developments relevant to systems, decision making, control, complex processes and related areas, as embedded in the fields of engineering, computer science, physics, economics, social and life sciences, as well as the paradigms and methodologies behind them. The series contains monographs, textbooks, lecture notes and edited volumes in systems, decision making and control spanning the areas of Cyber-Physical Systems, Autonomous Systems, Sensor Networks, Control Systems, Energy Systems, Automotive Systems, Biological Systems, Vehicular Networking and Connected Vehicles, Aerospace Systems, Automation, Manufacturing, Smart Grids, Nonlinear Systems, Power Systems, Robotics, Social Systems, Economic Systems and other. Of particular value to both the contributors and the readership are the short publication timeframe and the world-wide distribution and exposure which enable both a wide and rapid dissemination of research output.

Indexed by SCOPUS, DBLP, WTI Frankfurt eG, zbMATH, SCImago.

All books published in the series are submitted for consideration in Web of Science.

More information about this series at http://www.springer.com/series/13304

El Hassan Zerrik · Oscar Castillo

Stabilization of Infinite Dimensional Systems

With Contribution by Lahcen Ezzaki
and Abderrahmane Ait Aadi

Springer

El Hassan Zerrik
Faculty of Sciences
Department of Mathematics
Moulay Ismail University
Meknes, Morocco

Oscar Castillo
Division of Graduate Studies
and Research
Tijuana Institute of Technology
Tijuana, Mexico

ISSN 2198-4182 ISSN 2198-4190 (electronic)
Studies in Systems, Decision and Control
ISBN 978-3-030-68602-4 ISBN 978-3-030-68600-0 (eBook)
https://doi.org/10.1007/978-3-030-68600-0

This Springer imprint is published by the registered company Springer Nature Switzerland AG
The registered company address is: Gewerbestrasse 11, 6330 Cham, Switzerland

Preface

Systems theory is a branch of applied mathematics, it is interdisciplinary and develops activities in fundamental research which are at the frontier of mathematics, automation and engineering sciences. It is everywhere, innumerable and daily, moreover is there something which is not system: it is present in medicine, in commerce, economy, in psychology, in biological sciences, in finance, in architecture (construction of towers, bridges, etc ...), weather forecast, robotics, automobile, aeronautics, localization systems and so on. These are the few fields of application that are useful or even essential for our society. It is a question of studying the behavior of systems and acting on their evolution. Among the most important notions in system theory which attracted the most attention is stability.

The notion of stability is crucial. Indeed, the evolution of a dynamic system is described by a system of differential equations, partial differential equations... These systems generally have an infinite number of solutions, unless initial conditions are fixed. Depending on the choice of these initial values, there can be an infinity of operating regimes. To illustrate this idea, we cite the example of a clock. This one works with a well-determined amplitude of the pendulum although at the starting time, the pendulum can deviate more or less from its vertical position. If at the starting time of the clock one does not deviate sufficiently the pendulum, it will stop after a few oscillations. On the other hand, if the deviation is large enough, the amplitude of the oscillations of the pendulum will become fully determined after a short time and the clock will operate with this amplitude, for an infinite time. So the equations describing the clock functioning have two stationary solutions: an equilibrium position corresponding to the pendulum at rest and a periodic solution corresponding to the normal operation of the clock. So to understand the evolution of any dynamic system, it is important to know all the stable solutions of its equations, these are equilibrium states.

The existing literature on systems stability is quite important but disparate and the purpose of this book is to bring together in one document the essential results on the stability of infinite dimensional dynamical systems. In addition and as such systems evolve in time and space, explorations and research on their stability were mainly focused on the whole domain in which the system evolved. We have strongly felt that, in this sense, important considerations were missing: those which are consist in considering that the system of interest may be unstable on the whole domain but stable in a certain region of the whole domain. This is the case in many applications ranging from engineering sciences to living science. For this reason, we have dedicated this work to extension of classical results on stability to the regional case.

This book has its origins in the book I wrote with Professor El Jai [1]. An important consideration was that it should be accessible to mathematicians and to graduate engineering with a minimal background in functional analysis. Moreover, for the majority of the students this would be their only acquaintance with infinite dimensional system. It is organized by following increasing difficulty. The two first chapters deal with stability and stabilization of infinite dimensional linear systems described by partial differential equations.

The following chapters concern original and innovative aspects of stability and stabilization of certain classes of systems motivated by real applications, that is to say bilinear and semilinear systems. The stability of these systems has been considered from a global and regional point of view.

A particular aspect concerning the stability of the gradient has also been considered for various classes of systems.

This book is for students of master's degrees, engineering students and researchers interested in the stability of infinite dimensional dynamical systems, in various aspects.

Our thanks to Prof. Abdelhaq El Jai the president of the "Systems Theory" network who initiated research on applied mathematic in Morocco in 1978. He has greatly contributed to the development of research in systems theory in most Moroccan universities. To all the researchers of MACS Laboratory particularly M. Ouzahra and Y. Benslimane with whom I initiated the concept of regional stabilization. This new concept has led to developments of important results and to openings on other lines of research that can enrich approaches in systems theory, as well as all the colleagues who were part of their constructive remarks.

Finally we would like to thank the Hassan II Academy of Sciences and Techniques for its support, through the network *Systems Theory*, which made it possible to finalize this work.

Meknes, Morocco El Hassan Zerrik
August 2020

Reference

El Jai, A. and Zerrik, E. *Stabilité des systèmes dynamiques*, Presses Universitaires De Perpignan, 2014.

Contents

Notations

Ω	open of $\mathbb{R}^n$
$\partial\Omega$	boundary of Ω
$f(.,t)$	function : $x \in \Omega \mapsto f(x,t)$
$\frac{\partial f}{\partial x_i}$	partial derivative with respect to x_i
ω	subregion of Ω
χ_ω	characteristic function of $\omega \subset \Omega$
γ_0	trace application of order zero
∇y	gradient of y : $\nabla y = (\frac{\partial y}{\partial x_1}, ..., \frac{\partial y}{\partial x_n})$
Δy	Laplacien of y : $\Delta y = \sum_{i=1}^n \frac{\partial^2 y}{\partial x_i^2}$
$\Delta^2 y$	bilaplacien of y : $\Delta^2 y = \sum_{i=1}^n \frac{\partial^4 y}{\partial x_i^4}$
$D(A)$	domain of A linear operator
$N(A)$	kernel set of A
$Im(A)$	images set of A
A^{-1}	Inverse of A
A^*	adjoint operator of A
$\rho(A)$	resolvent set of A
$\sigma(A)$	pointwise spectrum of A
$r(A)$	spectral radius of A
$L^2(\Omega)$	space of square integrable functions on Ω
$H^1(\Omega)$	$\{y \in L^2(\Omega) \mid y' \in L^2(\Omega)\}$
$H_0^1(\Omega)$	$\{y \in H^1(\Omega) \mid y = 0 \text{ on } \partial\Omega\}$
$H^{\frac{1}{2}}(\partial\Omega) = Im(\gamma_0)$	Sobolev space of order $\frac{1}{2}$ on $\partial\Omega$
$H^s(\Omega)$	Sobolev space of order s
$\vert\vert\vert.\vert\vert\vert$	uniform norm on $\mathcal{L}(X)$
$<,>$	inner product defined on X
$\vert\vert.\vert\vert$	norm defined on X
$\vert\vert.\vert\vert_A$	graph norm
X_w	X endowed with its weak topology

$V \hookrightarrow X$	V is dense in X and the injection is continous
$y(t)$	solution of a system
$u(t), v(t)$	controls that act on the system
$\frac{dy(t)}{dt}$	derivative of the function y with respect to the variable t
$\frac{\partial y}{\partial x}$	partial derivative of the function y with respect to the variable x
$\frac{\partial^2 y}{\partial x^2}$	second partial derivative of the function y with respect to the variable x
X	the state space
X'	dual space of X
$\mathcal{L}(X, Y)$	bounded linear operators from X to Y
$\mathcal{L}(X)$	bounded linear operators from X to X
$\mathcal{R}(\lambda, A)$	resolvent of the operator A
$\rightarrow$	strong convergence
$\rightharpoonup$	weak convergence

Chapter 1
Introduction

Systems theory has been extensively developed over the last few decades. It is present in various fields of research, and extends its applications in the development of many sectors of life such as medicine, economy or industry. Real systems, which have become more and more complex, are made up of a large number of components interconnected to each other following a determined structure. To modelize a system amounts to determine a virtual model which represents it by symbols and signs. We are looking for a model because we cannot repeat an experiment endlessly. In order to design and automatize these systems, it is obviously suitable to give a mathematical representation, i.e. a set of equations that can be joined to initial conditions (for evolution phenomena), and to boundary conditions (when speaking on a spatial domain where the phenomena is studied), whose the resolution makes possible to define the state of system. This state can be represented by a vector formed of a finite number of components, in this case we speak of localized systems, as it can be seen in a space of infinite dimension, we then speak of infinite dimensional systems.

Modeling becomes an important area of scientific research in different disciplines. The goal is obviously to understand as best as possible the phenomenon we wants to study in view to predicting the behaviour of the state of the considered system.

Many phenomena in different area, physical, biological, economic, ecological, ... can be modeled by linear or non-linear partial differential equations, and the study of these equations partly goes through a better understanding of the properties of their solutions.

Thus, most physical problems are represented by non-linear models. As an example, we quote Boltzmanns equation in statistical mechanics, Navier Stokes equations in fluid mechanics, the Von Karman equations of flat plates in large displacement, etc. ... However, these models are generally difficult to study both theoretically or numerically. In this case we use other types of systems such as semi linear systems, which constitute a transition class between linear systems and non-linear ones. In

E. H. Zerrik and O. Castillo, *Stabilization of Infinite Dimensional Systems*,
Studies in Systems, Decision and Control 355,
https://doi.org/10.1007/978-3-030-68600-0_1

addition, a large number of industrial or natural processes inherently have a semi linear structure. For examples; the transfers of heat by conduction-convection, the kinetics of neutrons in a nuclear reactor, the dynamics of sense organs, the mass action law in chemistry, the diffusion of a semi-permeable membrane, cell multiplication by cell division, electric power production, direct current motors in mechanics, the economy of certain countries, etc. ...

The modeling of some physical problems can be directly obtained (without approximations) by Linear equations: this is the case of the transport equation of neutrons, the diffusion of heat in a metal bar, the propagation of an electromagnetic wave in a conductive material, certain chemical reactions, ... Other linear models can be derived from non linear systems either by ignoring certain terms, which is valid in certain situations: small displacements, slow movements, ..., either by linearizing around a special solution. In addition, the methods developed for solving linear systems are often useful for the study of non linear ones.

It is classic, in any phase of modeling to describe dynamic systems following two approaches:

- The first one, said external representation, and consists in modeling a system from an experimental study linking inputs and outputs magnitudes, which can be either controls or disturbances that the system undergoes, i.e. measurements, which correspond to the observations that can be made on the behavior of the system.
- The other said, internal representation and allows to write, from the fundamental laws, the evolution equations and to analyze some intrinsic properties. Thus, a dynamical system can then be designed as a dynamic transformation of inputs into outputs.

However, the modeling phase is still linked to an identification phase which consists of choosing from a class of specific models; the one that fits best with achieved experiences. It is a very broad notion that can be defined as the induction of mathematical models starting from experiences and a priori knowledge on systems.

It is possible today, to modify the traditional scheme: modeling—identification—command, by the scheme

$$\text{Modeling} \longrightarrow \text{Analysis} \longrightarrow \text{Control}$$

where the analysis step consists in studying a set of notions and concepts that allows a better knowledge of the properties of the system. Among these notions, we note that of controllability, observability, stability, detectability, etalability ... This analysis can be done through operators appearing in the model and also through the spatial distribution of actions or measures in the geometrical evolution domain representing the system.

Any study concerning the analysis of a dynamical system is usually followed by a control step, which consists in determining a command which allows to drive the studied system to a certain objective. For example, the possibility of bringing a system from an initial state to a desired one or to reach its neighbors at a finite time, which respectively defines the notion of exact and weak controllability. In the case of infinite time, we obtain the notion of asymptotic controllability.

By duality, the notion of exact or weak observability consists in reconstructing the state in a finite time, based on knowledge of dynamics and the output of the system. In the case of infinite time, we speak of the concept of detection.

These problems have been the subject of numerous studies. Lions [1–3] considered many more or less standard situations. These results have been followed and completed by numerous developments concerning the different concepts of analysis and control of a system. In finite dimension, the controllability has been characterized by a rank condition which relates the dynamics to the controllability matrix, this algebraic condition has extensions to the case of infinite dimension, notably a temporal version involving the semigroup generated by the dynamic of the system (see for example [4] for the linear case, and [5] for the bilinear case).

For some applications, we can be brought to design a control system that achieves an objective while minimizing a performance criterion. This kind of problem has dominated the literature: In the finite time case Pritchard [6, 7], used a dynamic programming to reduce the minimization problem from a quadratic cost to a resolution of a Riccati differential equation, and to the resolution of an algebraic Riccati equation in the case of infinite time. This study was completed by Zabczyk [8], then developed by other authors: Krasovski [9], Lions [2, 10], Curtain and Pritchard [11–13], Lukes and Russell [14], Datko [15, 16], Quinn [17], Staffans [18], Weiss and Weiss [19].

In addition, control of a system is often sought by a feedback control requiring continuous knowledge of the state of the system, according to the used description. As an example, the problem of regulation which consists in finding a way to maintain the state of a system as close as possible to a desired state one. It is obviously possible to regulate in open loop, but such control is very sensitive to disturbances, so it is better to regulate by a closed loop, adjusting the control so that the state of the system be close to the desired one.

We explore in this book the concept of stability which is one of the most important concepts in systems theory. And an unstable system is useless and potentially dangerous. Qualitatively a system is stable if whenever it is disturbed from its equilibrium state, it remains around this state thereafter.

Analysis of the stability of a system induces two methods: the linearization method and the direct method of Lyapunov. The linearization method makes it possible to derive conclusions regarding the local stability of a dynamic system around an equilibrium state, from the properties of stability of its linear approximation. As for the direct method, introduced in the 19th century by Lyapunov, it is not restricted to a local character and makes it possible to determine the stability properties of a nonlinear system using an energy function.

A dynamic system is always affected by disturbances, and for a good functioning of the system, it is therefore essential to remedy these disturbances, by considering the notion of stabilization, which consists in bringing the disturbed system back to its equilibrium state.

By duality and by need to reconstruct the state of the system from measurements, we define a second system called observer which estimates, at each time, a part of the state, (or the whole state). It is about to determine an estimate of the state, in infinite time from the data of the equation defining the system, the control and the output.

In finite dimension, the problem of the stabilization of the linear and nonlinear systems is today perfectly established, and has given rise to a rich literature (Whonham [20], Azzo-Houpis [21], Lasalle [22], Jurdjevic-Quin in [5], Quin [17], Slemrod [23], Gauthier-Bornard [24] and Outbib-Sallet [25], ...).

In the case of infinite dimensional systems, work on strong and or weak stabilization of such systems are mainly due to Slemrod [23, 26], Triggiani [27], Benchimol [28], Russell [29, 30], Balakrishnan [31], Ball [32], Ball-Slemrod [33], ...

The link between the asymptotic behavior of a system, the spectral properties of its dynamic and the existence of a Lyapunov function are explored in [34]. The exponential stabilization was studied in [27, 35] via an appropriate decomposition of the state space. Asymptotic and exponential stabilization were studied in [31] and [34] respectively, by means of the Riccati equation.

Stabilization results for finite dimensional systems have always guided the study of the infinite dimensional case, and the transition from the first class to the second, leads to difficulties mainly due to the presence of spatial variable describing the geometrical domain of the system. The enhancement of this variable, conventionally hidden in spaces of infinite dimension, allows however the right understanding of the reality of the phenomenon.

The stabilization of infinite dimensional semilinear systems has been the subject of many works. Stabilization of such systems was considered in [36], where one shows that a feedback control weakly stabilizes a bilinear system using the compactness of control operator and a weak controllability condition. In [37], the author proves exponential stabilization under the observability inequality. The same condition allows strong stabilization for a larger class of bilinear systems (see [38]).

At the beginning of the 90s, and with motivations linked to real applications, El Jai and Zerrik [39–42] introduced the notion of regional analysis, it is a question of analyzing and controlling a system defined on a geometrical domain Ω in order to achieve a goal on a given region ω of Ω. A new line of research was opened, and the analysis of the infinite dimensional systems can be treated in other way. The usual notions of controllability and of observability have been reconsidered with another point of view, indeed if we consider a region $\omega \subset \Omega$, regional control over ω consists in driving the considered system from its initial state to any desired state on ω. It has been shown that a system can be controlled just on a region without being on its whole evolution domain. In addition, the cost of regional controllability is lower [42].

What has been done for an internal region of the spacial domain, has been extended thereafter in case where the region of interest is located on the boundary of Ω [43–45]. Then all these notions have been also studied through the input-output parametric structure, that is to say again, through the sensors and actuators structure [40, 42, 46, 47].

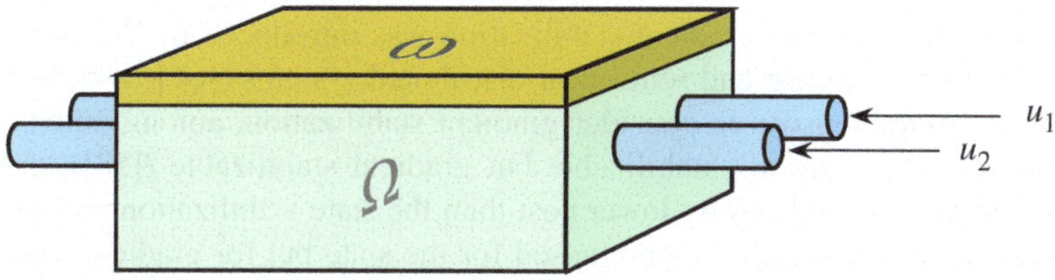

Fig. 1.1 Stabilization on the boundary region ω

Thus, in the presence of a system evolving on a spatial domain, it became possible to analyze it or to control it only on a privileged region of the system evoling domain. As examples, we cite:

- The example of certain oven tunnel in the ceramic industry, modeled by thermal equations, where the problem is to excite the system so to maintain, over a region of the oven, a prescribed temperature.
- The problem of water purification of a river where the question is naturally to reach a minimum pollution degree in a given region.
- The example of a thermal exchanger: It is about a heat exchanger (see Fig. 1.1) with parallel flow where two liquids are sent in the exchanger with speeds $v_1(t)$ and $v_2(t)$ with input temperatures $T_1(t)$ and $T_2(t)$ respectively. We want the output temperature $z(t) = T_s(t)$ of the first fluid be as close as possible to a desired temperature T_c on Γ and for that it is possible to act on the flow rate of the second fluid by choosing $v_2(t)$.

The established results and the various examples prove that the regional approach in the study of infinite dimensional systems is a natural extension of the classical notions in systems theory. The question is naturally what becomes the notion of stability in this new design.

Classically, any system is considered stable or else unstable. However, there are systems that are unstable on their geometrical domains Ω, but do not be the same on all Ω, indeed, they can be stable on certain regions ω of Ω. In addition, you may need to stabilize a system or improve its degree of stability only on a part of its evolution domain, which will require a lower cost [48, 49].

The problem of regional stabilization consists in studying the asymptotic behavior of an infinite dimensional system only in a region ω interior to the evolution domain Ω or a part of the boundary $\partial\Omega$ of Ω.

Regional weak, strong and exponential stabilization for bilinear systems was studied in [49, 50] using various controls. In [51], authors proved regional strong and weak stabilization of bilinear systems with unbounded control operator. In [52], authors considered regional weak, strong and exponential stabilization of bilinear systems with control operator assumed to be bounded with respect to the graph norm of the dynamics operator. In [53, 54], authors considered exponential, strong and weak output stabilization of semilinear systems.

Recently, the notion of gradient stabilization was introduced by Zerrik and co-workers for linear, bilinear and semilinear distributed systems (see [55–58]).

There are many reasons to consider gradient stabilization, among others, there exist systems which are not stabilizable but gradient stabilizable ([58]), gradient stabilization may be realized by lower cost than the state stabilization and in many real problems the stabilization is not asked for the state but for gradient. As in the thermal insulation problem where the purpose is to keep a constant temperature in a local with regards to the outside environment, thus one has to regulate the local temperature in order to vanish the exchange thermal flux.

On the other hand the strong and exponential stabilization for a class of bilinear and semilinear second order systems was proved in [59, 60] using the observability inequality, furthermore, the established results have been extended to regional case in [61].

We now give a general overview of the content of chapters.

- In the first chapter, we recall the most important results on the theory of linear semigroups and their asymptotic behavior, it also concerns characterizations of the stability and stabilization of linear systems.
- Chapter 2 concerns some results on the stabilization of semilinear systems, we then discuss the existence and the uniqueness of the solutions of bilinear and semilinear systems and then we give the main results of stabilization developed for such systems.
- In chapter three, we introduce the notion of regional stabilization for linear systems. We give definition and characterizations of regional stability. And we present some motivating examples we give characterizations of the controls achieving regional stabilization, stabilizing controls that maintain the system state bounded over the whole domain and minimizes a performance criterion. The used techniques come from the classical case and are essentially based on spectral properties, Lyapunov function or Riccati equation [27, 31, 34, 62, 63]. Numerical examples illustrate the obtained results and extension to boundary region is also discussed.

Chapter four focus on studying regional stabilization of infinite dimensional bilinear systems. Under sufficient conditions, we obtain exponential, strong and weak stabilization, also we discuss a regional optimal stabilization problem.

The fifth chapter is devoted to regional stabilization of semilinear systems. So we give results about exponential, strong and weak stabilization. The strong stabilization is also obtained minimizing a functional cost.

Chapter 6 deals with the output stabilization of bilinear and semilinear systems, then gives results on the exponential, strong and weak stabilisation. It is an extension of regional stabilization of such systems. Illustration by numerical example are also given.

Chapter 7 discuss stabilization of infinite dimensional second order semilinear systems, so we establish results on strong and exponential stabilization for bilinear system, that we extend to semi linear ones. Moreover, we study regional stabilization of bilinear and semi linear second order systems.

In Chap. 8, we introduce the notion of gradient stabilization of infinite dimensional linear systems, this requires a specific approach and some technical precautions. Thus we give definitions and some properties of gradient stability. Then we establish results on the gradient stabilization by three approaches: the state space decomposition, the Riccati method and minimizing a functional cost. We also give numerical algorithm that we illustrate by simulations.

Chapter 9 discuss regional gradient stabilization of infinite dimensional linear systems. And stabilization results are established using stabilizing controls, state space decomposition method and by a control that minimizes a specific functional cost.

In Chap. 10 we deal with the gradient stabilization of infinite dimensional bilinear systems, so we establish exponential, strong and weak gradient stabilization. Then we extend the obtained results to regional gradient stabilization.

Chapter 11 is about regional gradient stabilization of infinite dimensional semi-linear systems. Then we characterize gradient stabilizing controls, and the one that stabilizes the gradient and minimizes a given functional cost. The developed results are successfully illustrated by simulations.

The last chapter presents possible perspectives to this work where we we give open questions useful for future developments.

References

1. Lions, J.L.: Contrôlabilité exacte. Perturbations et stabilisation des systèmes distribués, Tome 1, Contrôlabilité Exacte, Masson (1988)
2. Lions, J.L.: Contrôle optimal des systèmes gouvernés par des équations aux dérivées partielles. Dunod, Paris (1968)
3. Lions, J.L., et Magénes, E.: Problèmes aux limites non homogènes et application, vol. 1. Dunod, Paris (1968)
4. Dunford, N., Schwartz, J.T.: Linear Operators. Part II. Willy Classics Library, New York (1963)
5. Jurjevic, V., Quinn, J.P.: Controllability and stability. J. Differ. Equ. **28**, 381–389 (1978)
6. Prichard, A.J.: Stability and control of distributed systems. Proc. IEEE., 1433–1438 (1969)
7. Prichard, A.J.: Stability and control of distributed systems governed by wave equations. In: Proceedings of IFAC. Conference on Distributed Parameter Systems, Banf, Canada (1971)
8. Zabczyk, J.: Remarks on the algebraic Riccati equation in Hilbert space. Appl. Math. Opt **3**, 251–258 (1976)
9. Krasovskii, N.N.: On analytical design of optimum regulators in time-delay systems. Prikl. Mat. Mekh. **1**, 39–52 (1962)
10. Lions, J.L.: Sur le contrôle optimal des systèmes decrits par des équations aux dérivées partielles linéaires. C.R. Acad. Sc. Paris **263**, 661–663, 713–715, 776–779 (1966)
11. Curtain, R.F., Pritchard, A.J.: Infinite Dimensional Linear Systems Theory. Springer, Berlin (1978)
12. Curtain, R.F., Pritchard, A.J.: The infinite dimensional Riccati equation. J. Math. Anal. Appl. **47**, 43–57 (1974)
13. Curtain, R.F., Pritchard, A.J.: The infinite dimensional Riccati equation for systems defined by evolution operators. SIAM J. Control and Optim. **14**, 955–469 (1989)
14. Lukes, D.L., Russel, D.L.: The quadratic criterion for distributed systems. SIAM J. Control **7**, 101–121 (1969)

15. Datko, R.: A linear control problem in abstrat Hilbert space. J. Diff. Eq. **9**, 346–359 (1971)
16. Datko, R.: unconstrained control problem with quadratic cost. SIAM J. Control **11**, 32–52 (1973)
17. Quinn, J.P.: Stabilization of bilinear systems by quadratic feedback control. J. Math. Anal. Appl. **75**, 66–80 (1980)
18. Staffans, O.: Quadratic optimal control of stable systems through spectral factorization. Math. Control Signals Syst. **8**, 167–197 (1995)
19. Weiss, M., Weiss, G.: Optimal control of stable weakly regular linear systems. Math. Control Signals Syst. **10**, 287–330 (1997)
20. Whonham, W.M.: Linear Multivariable Control : A Geometric Approach. Springer, Berlin (1979)
21. D'azzo, J.J., Houpis, C.H.: Linear Control System: Analysis and Design, 2nd edn. Mc Graw-Hill, New York (1981)
22. Lasalle, J.P.: Stability theory for ordinary differntial equations. J. Diff. Equ. **4**, 307–324 (1968)
23. Slemrod, M.: A note on complete controllability and stability for linear control systems in Hilbert space. SIAM. J. Control **12**, 500–508 (1974)
24. Gauthier, J.P., Bornard, G.: Stabilisation des systèmes nonlinéaires. Outils et modèles mathématiques pour l'automatique, l'analyse des systèmes et le traitement du signal. CNRS Edit (1981)
25. Outbib, R., Sallet, G.: Stabilizability of the angular velocity body revised. Syst. Control Lett. **18**, 93–98 (1992)
26. Slemrod, M.: Feedback stabilisation of a linear control system in Hilbert space with a priori bounded control. Math. Control Sign. Syst. **2**, 265–285 (1989)
27. Triggiani, R.: On the stabilizability problem in Banach space. J. Math. Anal. Appl. **52**, 383–403 (1975)
28. Benchimol, C.: A note on weak stabilizability of contraction semi-groups. SIAM J. Control Optim. **16**, 373–379 (1978)
29. Russell, D.L.: Controllability and stabilizability theory for linear partial differential equations: recent progress and open questions. SIAM J. Control Optim. **24**, 797–816 (1986)
30. Russell, D.L. : Linear stabilization of the linear oscillator in Hilbert space. J. Math. Anal. Appl. **25**, 663–675 (1969)
31. Balakrishnan, A.V.: Strong stability and the steady state Riccati equation. Appl. Math. Optim. **7**, 335–345 (1981)
32. Ball, J.M.: On the asymptotic behavior of generalized processes with application to nonlinear evolutions equations. J. Diff. Eqs .**27**, 224–265 (1978)
33. Ball, J.M.: On the asymptotic behavior of generalized processes with application to nonlinear evolutions equations. J. Diff. Equ. **27**, 224–265 (1978)
34. Ball, J.M., Slemrod, M.: Feedback stabilization of distributed semilinear control systems. Appl. Math. Optim. **5**, 169–179 (1979)
35. Prichard, A.J., Zabczyk, J.: Stability and stabilizability of infinite dimensional systems. SIAM Rev. **23**, 25–51 (1981)
36. Rabah, R., Ionescu, D.: Stabilization problem in Hilbert space. Int. J. Control **46**, 2035–2042 (1987)
37. Ouzahra, M.: Global stabilization of semilinear systems using switching controls. Automatica **48**(5), 837–843 (2012)
38. Ouzahra, M.: Strong stabilization with decay estimate of semilinear systems. Syst. Control Lett. **57**(10), 813–815 (2008)
39. Amouroux, M., El Jai, A., Zerrik, E.: Regional observability of distributed systems. Int. J. Syst. Sci. **25**, 301–313 (1994)
40. El Jai, A., Pritchard, A.J.: Sensors and actuators in distributed systems analysis. Ellis Horwood Series in Applied Mathematics, J. Wiley, Hoboken (1988)
41. El Jai, A., Simon, M.C., Zerrik, E., Pritchard, A.J.: Regional controllability of distributed parameter systems. Int. J. Control **62**(6), 1351–1365 (1995)

42. Zerrik, E.: Analyse régionale des systèmes distribués. Université Mohammed V, Rabat, Thèse d'état (1993)
43. Zerrik, E., Boutoulout, A., El Jai, A.: Actuators and regional boundary controllability of parabolic system. Int. J. Syst. Sci. **31**(1), 73–82 (2000)
44. Zerrik, E., Badraoui, L.El, Jai, A.: Sensors and regional boundary reconstruction for parabolic systems. Sens. Actuators J. **75**, 102–117 (1999)
45. Zerrik, E., Boutoulout, A., Bourray, H.: Boundary strategic actuators. Sens. Actuators J. A **94**, 197–203 (2001)
46. Berrahmoune, L.: Localisation d'actionneurs pour la contrôlabilité des systèmes paraboliques et hyperboliques. Université Med V, Application par dualité à la localisation de capteurs. Diplôme d'étude supérieures (1984)
47. Berrahmoune, L.: Actionneurs et capteurs dans la contrôlabilité et la stabilisation des systèmes distribués flexibles. Thèse d'état. Université Med V, Rabat (1990)
48. Zerrik, E., Ouzahra, M.: Regional stabilization for infinite-dimensional systems. Int. J. Control **76**(1), 73–81 (2003)
49. Zerrik, E., Ouzahra, M., Ztot, K.: Regional stabilization for infinite bilinear systems. IEEE Control Theory. Appl. **151**(1), 109–116 (2004)
50. Zerrik, E., Ezzaki, L.: Output stabilization of distributed bilinear systems. Control Theory Technol. **16**(1), 58–71 (2018)
51. Zerrik, E., Ait Aadi, A., Larhrissi, R.: On the stabilization of infinite dimensional bilinear systems with unbounded control operator. J. Nonlinear Dyn. Syst. Theory **18**, 418–425 (2018)
52. Zerrik, E., Ait Aadi, A., Larhrissi, R.: Regional stabilization for a class of bilinear systems. IFAC-PapersOnLine **50** 4540–4545 (2017)
53. Zerrik, E., Ezzaki, L.: An output stabilization of infinite dimensional semilinear systems. IMA J. Math. Control. Inf. **36**(1), 101–123 (2019)
54. Zerrik, E., Ait Aadi, A., Larhrissi, R.: On the output feedback stabilization for distributed semilinear systems. Asian J. Control (2019). https://doi.org/10.1002/asjc.2081
55. Zerrik, E., Ezzaki, L.: Stabilisation of the gradient of distributed bilinear systems. International Journal of Control (2019). https://doi.org/10.1080/00207179.2019.1652767
56. Zerrik, E., Ezzaki, L.: Regional gradient stabilization of semilinear distributed systems. J. Dyn. Control Syst. **23**(2), 405–420 (2016)
57. Zerrik, E., Ouzahra, M.: Output stabilization for infinite bilinear systems. Int. J. Appl. Math. Comput. Sci. **15**(2), 187–196 (2005)
58. Zerrik, E., Benslimane, Y., El Jai, A.: Regional gradient stabilization of distributed linear systems. Int. Rev. Autom. Control (IREACO) **4**(5), 755–765 (2011)
59. Ezzaki, L., Zerrik, E.: Stabilization of second order bilinear and semilinear systems. Int. J. Control **2020**, 1770335 (2020)
60. Zerrik, E., Ezzaki, L.: Strong and exponential stabilization for a class of second order semilinear systems. Recent Advances in Modeling, Analysis and Systems Control: Theoretical Aspects and Applications. Springer Nature Switzerland AG (2019). https://doi.org/10.1007/978-3-030-26149-8
61. Ezzaki, L., Zerrik, E.: Regional gradient stabilization of semilinear distributed systems (2020)
62. Datko, R.: Extending a theorem of A. M. Liapunov to Hilbert Space. J. of Math. Anal. Appl. **32**, 610–616 (1970)
63. Pazy, A.: On the applicability of Lyapunov's theorem in Hilbert space. SIAM J. Math. Anal. **3**(2), 291–294 (1972)

Chapter 2
Stabilization of Infinite Dimensional Linear Systems

In this chapter, we present the main results concerning the stabilization of infinite dimensional linear systems. The considered approach uses the semigroup theory.

2.1 Linear Semigroups

In this section we recall the main results of semigroups that we will use in the sequel.

Let us consider a linear system described by the equation:

$$\begin{cases} \dfrac{dy(t)}{dt} = Ay(t) \\ \\ y(.,0) = y_0 \in X \end{cases} \tag{2.1}$$

where A is a linear operator defined on a complex Hilbert space X (state space) provided with a scalar product $<,>$ and the corresponding norm $||.||$.

To interpret (2.1), we assume the state $y(t)$ depends only on the previous one. We can then define a linear operator of two parameters $S(t,s)$ such that

$$y(t) = S(t,s)y(s) \tag{2.2}$$

If we assume that the state $y(t)$ depends only on the past and that the variation of the state from one moment to another depends only on the time difference, $\tau = t - s$, then (2.2) is autonomous and becomes

$$y(s+\tau) = S(\tau)y(s) \tag{2.3}$$

E. H. Zerrik and O. Castillo, *Stabilization of Infinite Dimensional Systems*,
Studies in Systems, Decision and Control 355,
https://doi.org/10.1007/978-3-030-68600-0_2

It is then clear that $S(0) = Id$, the identity operator of X and $y(s) = S(s)y_0$, $y_0 \in X$. It follows

$$S(s+\tau) = S(\tau)S(s), \quad \tau, s \geq 0$$

Definition 2.1 A strongly continuous semigroup (C_0-semigroup) is an operator $S(.)$ from $\mathbb{R}^+$ to $\mathcal{L}(X)$, that satisfy the following properties:
1. $S(t+s) = S(t)S(s)\ \forall\ t, s \geq 0$;
2. $S(0) = Id$;
3. $||S(t)y - y|| \longrightarrow 0^+$ as $t \longrightarrow 0, \quad \forall y \in X$.
$S(t)$ is a contraction semigroup if it is a C_0-semigroup that satisfy $|||S(t)||| \leq 1$ for all $t \geq 0$, where $|||.|||$ is the uniform norm of $\mathcal{L}(X)$.

The above properties imply that $S(t)$ is strongly continuous for every $t > 0$, and that there are constants M and σ such that

$$|||S(t)||| \leq Me^{\sigma t}$$

If $M = 1$ and $\sigma \leq 0$, the semigroup $S(t)$ is a contraction semigroup, in this case A is said a dissipative operator (see [1]):

$$< Ay, y > + < y, Ay > \leq 0, \forall y \in \mathcal{D}(A)$$

Definition 2.2 The infinitesimal generator A of a C_0-semigroup on a Hilbert space X is defined by

$$Ay = \lim_{t \longrightarrow 0} \frac{S(t)y - y}{t} \tag{2.4}$$

whenever the limit exists, the domain of A, being the set $\mathcal{D}(A)$ of elements in X for which the limit (2.4) exists in X.

We cannot, in general, derive $S(t)y_0$. However, if $y_0 \in \mathcal{D}(A)$ then $S(t)y_0 \in \mathcal{D}(A)$, and we can derive $y(t)$ and we have

$$\frac{d}{dt}(S(t)y_0) = AS(t)y_0 = S(t)Ay_0, \ t \geq 0$$

and therefore we deduce that

$$S(t)y_0 - y_0 = \int_0^t AS(s)y_0 ds$$

and that $y(t) = S(t)y_0$ is the solution to the linear system (2.1).

In addition, $S(t)y_0$ can be defined for every $y_0 \in X$ as a continuous function, and in this case it's said weak solution to the system (2.1).

Note that if an operator A generates a C_0-semigroup, then A is closed and has dense domain in X. There is a full characterization of the operators A that generate strongly continuous semigroups; this is given by the following theorem which remains even valid for a Banach space.

Theorem 2.1 (Hille-Yosida [1]) *A closed linear operator A with dense domain $\mathcal{D}(A)$ in X generates a strongly continuous semigroup $S(t)$ if and only if there exist λ_0 and M such that for every $\lambda > \lambda_0$, the operator $\lambda I - A$ is invertible, $(\lambda I - A)^{-1} \in \mathcal{L}(X)$ and*

$$|||(\lambda I - A)^{-r}||| \leq \frac{M}{(\lambda - \lambda_0)^r}, \quad r = 1, 2, \ldots$$

let us recall the resolvent set of A, $\rho(A) \subset \mathbb{C}$, the set of λ, such that $(\lambda I - A)^{-1}$ exists, $R(\lambda, A) = (\lambda I - A)^{-1}$ is called the resolvent operator of A.

Proposition 2.1 ([1]) *Let A be a closed linear operator with dense domain. If the operators A and A^* are dissipative, then A is an infinitesimal generator of a contraction semigroup.*

For contraction semigroups, we have the characterization (see [1]).

Proposition 2.2 *A closed linear operator A with dense domain $\mathcal{D}(A)$ in X generates a contraction semigroup if and only if the resolvent set $\rho(A)$ of A contains $\mathbb{R}^+$ and for all $\lambda > 0$, we have $|||R(\lambda, A)||| \leq \frac{1}{\lambda}$.*

A valid result in a Hilbert space [2] is as follows.

Proposition 2.3 *1. Let A be a closed operator with dense domain in a Hilbert space. If there exists $\beta > 0$ such that*

$$Re(< Ay, y >) \leq \beta ||y||^2, \quad \forall\, y \in \mathcal{D}(A)$$

and

$$Re(< A^*y, y >) \leq \beta ||y||^2, \quad \forall\, y \in \mathcal{D}(A^*)$$

then A generates a C_0-semigroup $S(t)$ such that

$$|||S(t)||| \leq e^{\beta t}$$

2. If $S(t)$ is a compact C_0-semigroup (i.e. $S(t)$ is compact for every $t > 0$), then $S(t)$ is continuous for the uniform norm $|||.|||$.
3. Let A be an infinitesimal generator of a C_0-semigroup $S(t)$. $S(t)$ is compact if and only if it is continuous for the uniform topology and the resolvent operator $R(\lambda, A)$ is compact for every $\lambda \in \rho(A)$.

Example 2.1 Let $(\varphi_n)_{n\geq 1}$ be an orthonormal basis of a Hilbert space X, and let $(\lambda_n)_{n\geq 1}$ be a sequence complex numbers such that the sequence $(e^{\lambda_n t})_{n\geq 1}$ is bounded (which is true if $\sup_{n\geq 1} Re(\lambda_n) < \infty$).

Let us consider the diagonal operator defined by

$$Ay = \sum_{n=1}^{\infty} \lambda_n < \varphi_n, y > \varphi_n$$

with domain $\mathcal{D}(A) = \{y \in X \ / \ \sum_{n=1}^{\infty} |\lambda_n < \varphi_n, y > |^2 < \infty\}$. If moreover A is self-adjoint and $(\lambda I - A)^{-1}$ is compact for some λ, then we have the following properties:
1. There exists a sequence (λ_n) of distinct eigenvalues of A, such that:

- $|\lambda_n| \longrightarrow \infty$;
- the dimension of the eigenspace associated with λ_n is equal to its multiplicity order r_n;
- the sequence $(\lambda_n)_{n \geq 1}$ is strictly decreasing.

2. A has a complete set of eigenvectors $(\varphi_{jk})_{j,k \geq 1}$.
3. The spectrum $\sigma(A)$ is countable.
4. For every $y \in \mathcal{D}(A)$, we have

$$Ay = \sum_{j=1}^{\infty} \lambda_j \sum_{k=1}^{r_j} < \varphi_{jk}, y > \varphi_{jk}$$

(r_j is the multiplicity order of λ_j).
5. The operator A generates the semigroup given by

$$S(t)y = \sum_{j=1}^{\infty} e^{\lambda_j t} \sum_{k=1}^{r_j} < \varphi_{jk}, y > \varphi_{jk}$$

The following result concerns the regularity of the solution of the system (2.1).

Theorem 2.2 ([3]) *For every $y \in \mathcal{D}(A)$, $S(.)y \in \mathcal{C}(\mathbb{R}^+, \mathcal{D}(A)) \cap \mathcal{C}^1(\mathbb{R}^+, X)$ and*

$$\begin{aligned} \frac{dS(t)y}{dt} &= AS(t)y \\ &= S(t)Ay \end{aligned}$$

Proof Let $y \in \mathcal{D}(A)$, we set $z(t) = S(t)y$, and we have

$$\lim_{s \to 0} \frac{1}{s}(z(t+s) - z(t)) = S(t)Ay$$

As $S(.)Ay \in \mathcal{C}(\mathbb{R}^+, X)$, so $z(t) \in \mathcal{C}^1(\mathbb{R}^+, X)$.

In addition, we have

$$\|z(t+s)-z(t)\|_{\mathcal{D}(A)} = \|z(t+s)-z(t)\| + \|Az(t+s)-Az(t)\|$$

$$= \|S(t)(S(s)y-y)\| + \|S(t)(S(s)Ay-Ay)\|$$

therefore $\lim_{s\to 0}\|z(t+s)-z(t)\|_{\mathcal{D}(A)} = 0$ which completes the proof. □

The following result gives sufficient conditions so that a disturbed operator generate a semigroup [1, 4].

Theorem 2.3 ([3]) *Let A be a generator of C_0-semigroup $S(t)$ and B a bounded linear operator, then the operator $K = A + B$ generates a C_0-semigroup $S^K(t)$ given by*

$$S^K(t) = \sum_{k=0}^{+\infty} S_k^K(t)$$

with $S_0^K(t) = S(t)$ and $S_k^K(t) = \int_0^t S(t-s)BS_{k-1}^K(s)ds$ for all $k \geq 1$.

In addition if M and σ are such that $|||S(t)||| \leq Me^{\sigma t}$, then

$$|||S^K(t)||| \leq Me^{(\sigma+M.|||B|||)t}$$

Some topological properties of $S(t)$ remain true for $S^K(t)$. This is given by the following result.

Theorem 2.4 ([3]) *Let A be an infinitesimal generator of C_0-semigroup $S(t)$ and B a bounded linear operator.*

If the semigroup $S(t)$ is compact, then the operator $K = A + B$ is an infinitesimal generator of a compact C_0-semigroup $S^K(t)$.

We will examine the case of an operator A disturbed by a relatively bounded operator B. Let us first recall the following definition.

Definition 2.3 Let two operators A and B of domains respective $\mathcal{D}(A)$ and $\mathcal{D}(B)$ with $\mathcal{D}(B) \subset \mathcal{D}(A)$ in a Hilbert space X. We say that B is relatively bounded with respect to A if there exist $a > 0$ and $b > 0$ such that

$$\forall y \in \mathcal{D}(A) \quad \| By \| \leq a \| Ay \| + b \| y \|$$

We then have the following result.

Theorem 2.5 ([3]) *Let A be an infinitesimal generator of a contraction semigroup.*
1. If B is a dissipative operator and relatively bounded with respect to A with $a < 1$, then $A + B$ generates a contraction semigroup.
2. If B is an operator relatively bounded with respect to A with $a < 1$ and $A + B$ dissipative, then $A + B$ generates a contraction semigroup.

We give the following definitions.

Definition 2.4 Let $K \in \mathcal{L}(X)$ be a bounded operator and Z a subspace of X.
1. We say that Z is a reduced subspace of K if Z is invariant by K and by its adjoint K^*, i.e. $KZ \subset Z$ and $K^*Z \subset Z$.
2. K is said to be unitary if $KK^* = K^*K = Id$.

The following result, classic and important, concerns the decomposition of the state space into a direct sum of two subspaces.

Theorem 2.6 ([5] *If A generates a contraction semigroup $S(t)$ on X, then X can be decomposed as a direct subspaces sum $X = X_u \oplus X_s$ where X_u and X_s are two reduced subspaces by $S(t)$. In addition, we have*
1. The restriction of $S(t)$ to X_u, noted $S_u(t)$, is a unitary C_0-semigroup for all $t > 0$,
2. The restriction of $S(t)$ to X_s, noted $S_s(t)$, is a completely non-unitary C_0-semigroup for all $t > 0$,
3. The decomposition is unique and X_u can be written in the form

$$X_u = \{y \in X; \|S(t)y\| = \|S^*(t)y\| = \|y\|, \ t \geq 0\}$$

and, if we consider the restriction K_u of K,

$$K_u = X_u \cap \mathcal{D}(A) \ \ and \ \ \overline{K}_u = X_u.$$

Proof Letting $L(t) = Id - S^*(t)S(t)$ and $L^*(t) = Id - S(t)S^*(t)$ we have,

$$\|S(t)y\| = \|S^*(t)y\| = \|y\| \Leftrightarrow \langle L(t)y, y\rangle = \langle y, L^*(t)\rangle = 0 \tag{2.5}$$

The operators $L(t)$ and $L^*(t)$ are self adjoint and positive.
Indeed let y and z two elements of X, then we have

$$\begin{aligned}\langle L(t)y, z\rangle &= \langle y, z\rangle - \langle S^*(t)S(t)y, z\rangle \\ &= \langle y, z\rangle - \langle S(t)y, S(t)z\rangle \\ &= \langle y, z\rangle - \langle y, S^*(t)S(t)z\rangle \\ &= \langle y, L(t)z\rangle\end{aligned}$$

so $L(t)$ is self adjoint.
In addition, as $S(t)$ is a contraction, then $L(t)$ is positive, thus

$$\langle L(t)y, y\rangle = \|y\|^2 - \|S(t)y\|^2 \geq 0$$

With the same techniques, we show that $L^*(t)$ is self adjoint and positive, and therefore the operators $L^{\frac{1}{2}}(t)$ and $L^{\frac{1}{2}*}(t)$ are well defined.
Consequently (2.5) is equivalent to $y \in KerL(t) \cap KerL^*(t)$.
The space

$$X_u = \bigcap_{t\geq 0}[KerL(t) \cap KerL^*(t)]$$

is a closed subspace of X and the space $X_s = X_u^{\perp}$.

The spaces X_u and X_s are invariant by $S(t)$ and $S^*(t)$, for all $t \geq 0$. Indeed,

$$y \in X_u \Leftrightarrow \|S(t)S(s)y\| = \|S(t+s)y\| = \|y\|, \ \forall s, t \geq 0. \tag{2.6}$$

thus X_u is invariant by $S(t)$ for all $t \geq 0$. Since $S(t)$ is a contraction we have

$$\|S(t)y\| = \|y\| \Rightarrow S^*(t)S(t)y = y \tag{2.7}$$

since,

$$\begin{aligned} \|S^*(t)S(t)y - y\|^2 &= \|S^*(t)S(t)y\|^2 - 2Re\langle S^*(t)S(t)y, y\rangle + \|y\|^2 \\ &= \|S^*(t)S(t)y\|^2 - 2\|S(t)y\|^2 + \|y\|^2 \\ &\leq \|S^*(t)\|^2\|S(t)y\|^2 - \|y\|^2 \\ &\leq 0 \end{aligned}$$

We conclude that X_u is invariant by $S^*(t)$ for all $t \geq 0$. This gives

$$\|S^*(s)S(t)y\| = \|S^*(s-t)S^*(t)S(t)y\| = \|S^*(s)S(s)S(t-s)y\|$$

Using now (2.7), we obtain

$$\|S^*(s)S(t)y\| = \begin{cases} \|S^*(s-t)y\| = \|y\| \text{ if } s > t \\ \\ \|S(t-s)y\| = \|y\| \ \text{ if } s \leq t \end{cases}$$

and therefore X_u is invariant by $S^*(t)$.

Let us show that X_s is invariant by $S(t)$ and $S^*(t)$ for all $t \geq 0$.

Let $y \in X_s$ then for all $z \in X_u$ we have $\langle y, z\rangle = 0$ and since X_u is invariant by $S^*(t)$, we also have $\langle S(t)y, z\rangle = \langle y, S^*(t)z\rangle = 0$ then X_s is invariant by $S(t)$. Similarly (X_u is invariant by $S(t)$)

$$\langle S^*(t)y, z\rangle = \langle y, S(t)z\rangle = 0$$

1. With the same techniques used to show (2.7), we have

$$\|S^*(t)y\| = \|y\| \Rightarrow S(t)S^*(t)y = y \tag{2.8}$$

therefore (2.7) and (2.8) imply that

$$S_u(t)S_u^*(t) = S_u^*(t)S_u(t) = Id_{X_u}$$

2. We suppose that there exists a subspace $X_0 \subset X_s$ such that the operator S_{X_0} (the restriction of S to X_0) is unitary. For $y \in X_0$ such that $y \neq 0$ we have

$$\|S(t)y\|^2 = \langle S^*(t)S(t)y, y\rangle = \langle S(t)S^*(t)y, y\rangle = \|y\|^2$$

thus $\|S(t)y\| = \|S^*(t)y\| = \|y\|$ and therefore $y \in X_u$, which is a contradiction.
3. If X admits another orthogonal decomposition

$$X = X_0 \oplus X_1$$

such that X_0 and X_1 verify the same assumptions as X_u and X_s respectively.

Since S_{X_0} is unitary then for all $y \in X_0$ and for all $t \geq 0$, we have $\|S(t)y\| = \|S^*(t)y\| = \|y\|$, hence $X_0 \subset X_u$.

Furthermore, since X_0 and X_u are two reduced spaces of $S(.)$, it will be the same for $X_u \cap X_0^{\perp}$.

From $X_u \cap X_0^{\perp} \subset X \cap X_0^{\perp} = X_1$, the restriction of $S(.)$ to $X_u \cap X_0^{\perp}$ is unitary.

But by hypothesis S_{X_1} is completely non-unitary, which implies that $X_u \cap X_0^{\perp} = \{0\}$ and consequently $X_u = X_0$. Hence the uniqueness of the decomposition.

It remains to show that $X_u = \overline{K}_u$, with $K_u = X_u \cap \mathcal{D}(A)$. Since X_u is closed, then $\overline{K}_u \subset X_u$. On the other hand for all $y \in X_u$, we have

$$R(\lambda, A)y = \int_0^{\infty} e^{-\lambda s} S(s)yds \in X_u$$

indeed,

$$\begin{aligned}
S(t)S^*(t)R(\lambda, A)y &= \int_0^{\infty} S(t)S^*(t)S(s)yds, \ \forall t, s \geq 0 \\
&= \int_0^{\infty} S(s)yds, \ \forall s \geq 0 \\
&= R(\lambda, A)y \\
&= \int_0^{\infty} S^*(t)S(t)S(s)yds \\
&= S^*(t)S(t)R(\lambda, A)y
\end{aligned}$$

(because X_u is invariant by $S(t)$ and S_{X_u} is unitary).

From where $S(t)S^*(t)R(\lambda, A)y = S^*(t)S(t)R(\lambda, A)y = R(\lambda, A)y$.

Finally, we obtain

$$\|S(t)R(\lambda, A)y\|^2 = \langle S^*(t)S(t)R(\lambda, A)y, R(\lambda, A)y\rangle = \langle R(\lambda, A)y, R(\lambda, A)y\rangle$$

Similarly, we have

$$\|S^*(t)R(\lambda, A)y\|^2 = \langle S(t)S^*(t)R(\lambda, A)y, R(\lambda, A)y\rangle = \langle R(\lambda, A)y, R(\lambda, A)y\rangle$$

Thus $R(\lambda, A)y \in X_u$; and since $R(\lambda, A)y \in \mathcal{D}(A)$, then for all $y \in X_u$, $\lambda R(\lambda, A)y \in K_u$.

Finally, we know that $\lambda R(\lambda, A)y \to y$, when $Re(\lambda) \to +\infty$ (see [6]), thus $X_u \subset \overline{K}_u$. We conclude that $X_u = \overline{K}_u$. □

Lemma 2.1 *Let $S(t)$ be a contraction semigroup, then*

$$S(t)[Id - S(s)S^*(s)]y \rightharpoonup 0 \;\; when\; t \longrightarrow +\infty \tag{2.9}$$

and

$$S(t)[Id - S^*(s)S(s)]y \rightharpoonup 0 \; when\; t \longrightarrow +\infty \tag{2.10}$$

Proof Let t and s be such that $t \geq s \geq 0$, and as $S^*(t)$ is a contraction we have $\|S^*(t)\|^2 = \|S^*(t-s)S^*(s)\|^2 \leq \|S^*(s)\|^2$.

Hence the function $t \longmapsto |||S^*(t)|||^2$ is positive and decreasing, therefore for fixed $s \geq 0$, $f(t) = \|S^*(t)y\|^2 - \|S^*(t+s)y\|^2 \longrightarrow 0$ when $t \longrightarrow +\infty$. We have

$$f(t) = \langle S^*(t)y, S^*(t)y\rangle - \langle S(s)S^*(s)S^*(t)y, S^*(t)y\rangle$$

$$= \langle (Id - S(s)S^*(s))S^*(t)y, S^*(t)y\rangle$$

Since $(Id - S(s)S^*(s))$ is self adjoint and positive then

$$f(t) = \|(Id - S(s)S^*(s))^{1/2}S^*(t)y\|^2$$

Hence, for all $y \in X$ and $s \geq 0$, we have

$$[Id - S(s)S^*(s)]S^*(t)y \longrightarrow 0 \;\text{ when }\; t \longrightarrow +\infty \tag{2.11}$$

and if we multiply the left term by $S^*(s)$ we get

$$S^*(s)[Id - S(s)S^*(s)]S^*(t)y = [Id - S^*(s)S(s)]S^*(t+s)y$$

Therefore, for all $y \in X$ and $s \geq 0$,

$$[Id - S^*(s)S(s)]S^*(t)y \longrightarrow 0 \text{ when } \; t \longrightarrow +\infty \tag{2.12}$$

We know that if $Y(t)$ is a bounded linear operator, then $Y(t)y \to 0$ when $t \to +\infty$, for all $y \in X$.

If we apply this result to (2.11) and (2.12) we obtain (2.9) and (2.10). □

Theorem 2.7 ([3]) *Let $S(t)$ be a contraction semigroup on X and the set*

$$W = \{y \in X \;\; such\; that \;\; S(t)y \rightharpoonup 0 \; when\; t \to +\infty\},$$

then
1. W is a closed and reduced subspace of $S(t)$, for all $t \geq 0$.
2. $W^{\perp}$ is a reduced subspace of $S(t)$, and the restriction $S_{W^{\perp}}(t)$ is a unitary C_0-semigroup.
3. $W = \{y \in X$ such that $S^*(t)y \rightharpoonup 0$ when $t \to +\infty\}$.

Proof 1. W is a closed subspace of X.
Let $y_n, y \in X$ be such that $\langle y_n, z\rangle \to 0$ when $n \to +\infty$ for all $z \in X^*$ and $\lim_{n\to+\infty} \| y_n - y\| = 0$.

Let $z \in X^*$, $\langle S(t)y, z\rangle = \lim_{n\to+\infty} \langle S(t)y_n, z\rangle = 0$. Which implies that $y \in W$, in other words W is closed in X.

W is invariant by $S(s)$ and by $S^*(s)$, for all $s \geq 0$. Indeed, let $y \in W$, for $s \geq 0$, $S(t)S(s)y = S(t+s)y \rightharpoonup 0$ when $t \to +\infty$ and W is invariant by $S(s)$, for all $s \geq 0$.

Now we will show that for all $y \in W$, then $S^*(s)y \in W$, for all $s \geq 0$.

For $y \in W$, $S(t)y \rightharpoonup 0$ when $t \longrightarrow +\infty$ and with (2.9), we get

$$-S(t)S(s)S^*(s)y = -S(t+s)S^*(s)y \rightharpoonup 0 \text{ when } t \longrightarrow +\infty$$

Which means that $S(t)S^*(s)y \rightharpoonup 0$ when $t \longrightarrow +\infty$.

Consequently W and $W^{\perp}$ are two reduced subspaces of $S(s)$, for all $s \geq 0$.
2. For this point, it suffices to show that $W^{\perp} \subset X_u$.
By (2.9) and (2.10), for all $s \geq 0$ we have

$$ImL(s) \subset W \ \text{ and } ImL^*(s) \subset W \tag{2.13}$$

and since $L(s)$ and $L^*(s)$ are bounded and self adjoint we have

$$(ImL(s))^{\perp} = KerL(s) \quad \text{and} \quad (ImL^*(s))^{\perp} = KerL^*(s)$$

thus (2.13) implies that, for all $s \geq 0$, $W^{\perp} \subset (ImL(s))^{\perp} \cap (ImL^*(s))^{\perp}$.

Which is equivalent to $W^{\perp} \subset \bigcap_{s\geq 0}(KerL(s) \cap KerL^*(s)) = X_u$.

3. Let $z \in X$, $z = z_1 + z_2$ such as $z_1 \in W$ and $z_2 \in W^{\perp}$

$$\langle S^*(t)y, z\rangle = \langle y, S(t)z_1\rangle + \langle \text{ for all } y \in W.y, S(t)z_2\rangle$$

As $(z_1, z_2) \in W \times W^{\perp}$ and W is invariant by $S(t)$ then
$\langle y, S(t)z_1\rangle \longrightarrow 0$ and $\langle y, S(t)z_2\rangle \longrightarrow 0$ when $t \longrightarrow \infty$.
Hence $S^*(t)y \rightharpoonup 0$ when $t \to +\infty$, and by changing the roles of $S(t)$ and $S^*(t)$, and with the same techniques as above we show that $S^*(t)y \rightharpoonup 0$ when $t \longrightarrow +\infty$, hence $S(t)y \rightharpoonup 0$ when $t \longrightarrow +\infty$, which completes the proof. □

Theorem 2.8 *Let A be an operator of domain $\mathcal{D}(A) \subset X$ a Hilbert space. Let $\delta > 0$ and the subsets of the spectrum points of A*

$$\begin{aligned} \sigma_u(A) &= \{\lambda \in \sigma(A) \mid Re(\lambda) \geq -\delta\} \\ \sigma_s(A) &= \{\lambda \in \sigma(A) \mid Re(\lambda) < -\delta\} \end{aligned} \tag{2.14}$$

If the set $\sigma_u(A)$ is bounded and separated from the set $\sigma_s(A)$ by a simple closed curve (which is the case if $\sigma_u(A)$ is finite), then the state space X can be decomposed into two subspaces in the form

$$X = X_u \oplus X_s \quad with \quad X_u = \Pi X, \ X_s = (I - \Pi)X \tag{2.15}$$

where $\Pi \in \mathcal{L}(X)$ is a projection operator given by

$$\Pi = \frac{1}{2\pi i} \int_C (\lambda I - A)^{-1} d\lambda, \tag{2.16}$$

and C is a curve surrounding $\sigma(A)$.

2.2 Stability of Infinite Dimensional Linear Systems

In this section we recall definitions and characterizations of stability of infinite dimensional system following by illustration with many examples.

2.2.1 *Definitions—Examples*

Let us consider the linear system described by the equation

$$\begin{cases} \dfrac{dy(t)}{dt} = Ay(t) \\ y(0) = y_0 \in X \end{cases} \tag{2.17}$$

where $A : \mathcal{D}(A) \subset X \longrightarrow X$ is an infinitesimal generator of a C_0-semigroup $S(t)$, $t \geq 0$ in X.

Definition 2.5 The system (2.17) is said
1. exponentially stable (or A generates a C_0-semigroup $(S(t))$ exponentially stable) if there exist $\exists\, \alpha > 0,\ M \geq 0$ such that $|||S(t)||| \leq Me^{-\alpha t}$.
2. uniformly stable if $\lim\limits_{t\to+\infty} |||S(t)||| = 0$.
3. strongly stable if $\lim\limits_{t\to+\infty} \|S(t)y\| = 0$, for all $y \in X$.
4. weakly stable if $\lim\limits_{t\to+\infty} \langle S(t)y, z\rangle = 0$, for all $(y, z) \in X \times X^*$.

Remark 2.1 1. In the case of finite dimension, the three degrees of stability coincide.
2. Exponential stability $\Longrightarrow$ strong stability $\Longrightarrow$ weak stability.
The converse is not true in the case where $\dim X = +\infty$. This is illustrated by the following examples.

Example 2.2 On $\Omega =]0, +\infty[$, we consider the system described by the equation

$$\begin{cases} \dfrac{\partial y(x,t)}{\partial t} = \dfrac{\partial y(x,t)}{\partial x} \quad \text{on } \Omega\times]0,+\infty[\\ y(x,0) = y_0(x), \quad \text{on }]0,+\infty[\end{cases} \tag{2.18}$$

Here the operator $A = \dfrac{\partial}{\partial x}$ of domain $\mathcal{D}(A) = \{y \in H^1(]0,+\infty[);\ y(0) = 0\}$ generates C_0-semigroup given by

$$(S(t)y)(x) = y(x+t), \quad t \geq 0$$

For all $y \in L^2(]0,+\infty[)$, $||S(t)y||^2 = \displaystyle\int_t^{+\infty} |y(x)|^2 dx \to 0$, when $t \to \infty$, then the system (2.18) is strongly stable.

But $|||S(t)||| = 1$, for all $t \geq 0$, so the system (2.18) is not uniformly neither exponentially stable.

A system can be weakly stable without being strongly stable.

Example 2.3 On $\Omega =]0, +\infty[$ we consider the system

$$\begin{cases} \dfrac{\partial y(x,t)}{\partial t} = -\dfrac{\partial y(x,t)}{\partial x} \quad \text{on } \Omega\times]0,+\infty[\\ y(x,0) = y_0(x) \in L^2(]0,+\infty[) \end{cases} \tag{2.19}$$

The operator $A = -\dfrac{\partial}{\partial x}$ of domain $\mathcal{D}(A) = \{y \in L^2(]0,+\infty[);\ y(0) = 0\}$ generates a C_0-semigroup given by

$$(S(t)y)(x) = \begin{cases} y(x-t) & \text{if } x \geq t \\ 0 & \text{if } x < t \end{cases}$$

We have $||S(t)y|| = ||y||$, for all $t \geq 0$, therefore the system (2.19) is not strongly stable. However it is weakly stable, indeed, for all $y, z \in L^2(]0,+\infty[)$, two functions with compact supports. For t large enough, the supports of $S(t)y$ and z are disjoint, so

$$\langle S(t)y, z\rangle = \int_0^{+\infty} y(x-t)z(x)dx = 0$$

Now let $(y,z) \in (L^2(]0,+\infty[))^2$ then there exist two sequences of functions (y_n) and (z_n) with compact supports such that

$$\|y - y_n\|_{L^2(]0,+\infty[)} < \frac{1}{n} \text{ and } \|z - z_n\|_{L^2(]0,+\infty[)} < \frac{1}{n}$$

So

$$\begin{aligned}|\langle S(t)y, z\rangle| &< |\langle S(t)(y - y_n), z_n\rangle| + |\langle S(t)y, z - z_n\rangle| + |\langle S(t)y_n, z_n\rangle| \\ &< \tfrac{1}{n}(\tfrac{1}{n} + \|z\| + \|y\|) + |\langle S(t)y_n, z_n\rangle|\end{aligned}$$

But for t large enough $|\langle S(t)y_n, z_n\rangle| = 0$, hence $\langle S(t)y, z\rangle \to 0$ when $t \to +\infty$ and therefore the system (2.19) is weakly stable.

2.2.2 Characterizations

In finite dimension, the three degrees of stability coincide and several characterization results have been established, and are based on the spectral properties of A describing the dynamics of the system or on the existence of a Lyapunov function, this turns to establishing the existence of a solution of a Lyapunov equation.

In infinite dimension, if we note by $\sigma(A)$ the point spectrum of A and

$$\sigma_0 = \lim_{t\to+\infty} \frac{1}{t} \ln |||S(t)||| \tag{2.20}$$

and the upper and lower stability indices of the operator A

$$\sigma_- = \sup\{Re\lambda : \lambda \in \sigma(A)\}$$

$$\sigma_+ = \inf\{\mu \mid \exists M > 0;\ |||S(t)||| \le Me^{\mu t}, \forall t \ge 0\}$$

then we have the following result.

Proposition 2.4 *1. We have $\sigma_- \le \sigma_+$, and if the system (2.17) is exponentially stable then $\sigma_- < 0$.*
2. If A generates a semigroup $S(t)$ differentiable for $t > 0$, or if there exists $t_1 > 0$, such that $S(t_1)$ be compact, then $\sigma_- = \sigma_+$. Thus, the inequality $\sigma_- < 0$ imply the exponential stability of the system (2.17).

For the proof we refer to [7–9].

Theorem 2.9 ([3]) *If A generates a C_0-semigroup $S(t)$, then $-\infty \le \sigma_- \le \sigma_0$.*

Definition 2.6 We say that the equality of the spectral decay is verified if $\sigma_0 = \sigma_-$.

The following result gives a necessary and sufficient condition for the exponential stability [10].

Theorem 2.10 ([3]) *The system (2.17) is exponentially stable if and only if*

$$\int_0^{\infty} \|S(t)y\|^2 dt < +\infty \quad \text{for all} \quad y \in X. \tag{2.21}$$

Proof If the system is exponentially stable, there exist $M > 0$ and $\alpha > 0$ such that for $t \geq 0$, $|||S(t)||| \leq Me^{-\alpha t}$, thus for $m \in \mathbb{N}^*$, the operator

$$Q_m(y) = \mathbb{1}_{[0,m]} S(t)y \in \mathcal{L}(X, L^2(\mathbb{R}^+, X))$$

is well defined (see [11]) and then

$$\int_0^{+\infty} ||Q_m S(t)y||^2 \leq a_m ||y||^2$$

where $a_m = \displaystyle\int_0^m |||S(t)|||^2 dt$.

According to Banach Steinhaus theorem, there exists $\gamma > 0$ such that $||S_m|| \leq \gamma, \;\; m \geq 1$.

For $t_0 > 0$, the family $(|||S(t)|||)_{t \leq t_0}$ is bounded and, for $t > t_0$, we have

$$\begin{aligned}
\frac{1 - e^{-2\alpha t}}{2\alpha} ||S(t)y||^2 &= \int_0^t e^{-2\alpha(t-s)} ||S(t)y||^2 ds \\
&\leq \int_0^t e^{-2\alpha(t-s)} ||S(s)y||^2 |||S(t-s)|||^2 ds \\
&\leq M^2 \gamma^2 ||y||^2
\end{aligned}$$

Hence there exists $k > 0$ such that $|||S(t)||| \leq k$, for all $t \geq 0$. In addition, we have

$$\begin{aligned}
t||S(t)y||^2 &\leq \int_0^t |||S(s)|||^2 ||S(t-s)y||^2 ds \\
&\leq k^2 \gamma^2 ||y||^2
\end{aligned}$$

For τ large enough, $|||S(\tau)||| < 1$, thus $(\ln(|||S(\tau)|||) < 0$ and by the Theorem 2.9, there exist $\tilde{M} > 0$ and $\alpha > 0$ such that $|||S(t)||| \leq \tilde{M} e^{-\alpha t}$.

The converse is immediate. □

Corollary 2.1 *If there exists $t_0 > 0$ such that $|||S(t_0)||| < 1$ then the system (2.17) is exponentially stable.*

Proof The proof is immediate since $|||S(t_0)||| < 1$, then $\sigma_0 < 0$. □

Remark 2.2 1. In the case of finite dimension, the exponential stability is examined from the spectrum of the system dynamics so the system (2.17) is exponentially stable if and only if:

$$sup\{Re(\lambda), \lambda \in \sigma(A)\} \leq 0 \tag{2.22}$$

2. In infinite dimension, we always have the inequality

$$sup\{Re(\lambda), \lambda \in \sigma(A)\} \leq \sigma_0.$$

However, we do not need equality to obtain exponential stability as shown in the theorem above.

Example 2.4 We consider the system described by the equation

$$\begin{cases} \frac{\partial y}{\partial t}(x,t) = \frac{\partial y}{\partial x}(x,t) & \text{on }]0,+\infty[^2 \\ y(x,0) \quad = y_0 \in X \end{cases} \tag{2.23}$$

The state space $X = \{y \in L^2(\mathbb{R}^+) | y$ is absolutely continuous over $[a, b]$, with $0 < a < b$ and

$$\int_0^{+\infty} |\frac{dy}{dt}(x)|^2 e^{-2x} < \infty\}$$

endowed with the scalar product

$$\langle y_1, y_2 \rangle = \int_0^{+\infty} y_1(x)\overline{y}_2(x)dx + \int_0^{+\infty} \frac{dy_1}{dx}(x)\overline{\frac{dy_1}{dx}}(x)e^{-2x}dx$$

is a Hilbert space.

The operator $Ay = \frac{\partial y}{\partial}$ with domain $\mathcal{D}(A) = \{y \in X \mid \frac{\partial y}{\partial x} \in X\}$ and generates a C_0-semigroup given by $S(t)y(x) = y(t + x), \forall t > 0$ and $y \in X$. According to [12], any complex $\beta \in \mathbb{C}$ such that $Re(\beta) \geq 0$, is an element of the resolvent of A, and we have $\sigma_0 = 1$ (σ_0 is defined in (2.20)). So the inequality (2.22) is strict, however the system (2.23) is exponentially stable.

The following result characterizes the exponential stability of the system (2.17) [1].

Theorem 2.11 ([3]) *Let A be the infinitesimal generator of a C_0-semigroup $S(t)$ on X.*

The system (2.17) is exponentially stable if and only if there exists $P \in \mathcal{L}(X)$ solution of Lyapunov's equation

$$\langle Ay, Py \rangle + \langle Py, Ay \rangle = -\langle y, y \rangle, \ \forall\, y \in \mathcal{D}(A) \tag{2.24}$$

Proof If the semigroup $S(t)$ is exponentially stable, then the operator

$$Py = \int_0^{+\infty} S^*(s)S(s)y\,ds,\ y \in \mathcal{D}(A)$$

is solution of the Eq. (2.24) and we have

$$\langle y, Py\rangle = \int_0^{\infty} \| S(t)y \|^2\,dt \geq 0,\ \ y \in \mathcal{D}(A)$$

Thus $\langle y, Py\rangle = 0$ implies $\| S(t)y \| = 0$, for all $t \geq 0$ and since $S(t)$ is strongly continuous semigroup, then $y = 0$; so P is a positive definite operator.

Let $y \in \mathcal{D}(A)$ and let $V(y) = \langle Py, y\rangle$. As P is solution of the Eq. (2.24), we have

$$\frac{dV(S(t)y)}{dt} = -\|S(t)y\|^2$$

P is positive defined then $\int_0^{+\infty} \|S(t)y\|^2 \leq \langle Py, y\rangle$.

The conclusion results from the density of $\mathcal{D}(A)$ in X and from the continuity of P. □

Proposition 2.5 *If the semigroup $S(t)$ is differentiable for $t > 0$, then A verify the equality of the spectral decay*

$$\sigma_0 = \sigma_-$$

and it follows that if $\sigma_- < 0$, then the system (2.17) is exponentially stable.

The proof is given in [3].

2.3 Stabilization of Infinite Dimensional Linear Systems

We consider the controlled linear system

$$\begin{cases} \dfrac{dy(t)}{dt} = Ay(t) + Bv(t) \\ y(.,0) = y_0 \in X \end{cases} \tag{2.25}$$

where A is an infinitesimal generator of a C_0-semigroup $S(t), t \geq 0$ in a Hilbert space X and $B \in \mathcal{L}(V, X)$, V the controls space assumed to be a Hilbert space.

We know that for all $K \in \mathcal{L}(X)$, the operator $A + BK$ generates a C_0-semigroup. The weak solution of the system (2.25), is given by

$$y(t) = S(t)y_0 + \int_0^t S(t-s)Bv(s)ds$$

We have the following definition.

Definition 2.7 The system (2.25) is said to be weakly (respectively strongly, exponentially) stabilizable, if there exists a bounded operator $K \in \mathcal{L}(X, V)$ such that the system

$$\begin{cases} \dfrac{dy(t)}{dt} = (A + BK)y(t) \\ \\ y(0) = y_0 \in X \end{cases}$$

be weakly (respectively strongly, exponentially) stable.

2.3.1 *Characterizations*

There are mainly three approaches to tackle the problem of stabilization of the system (2.25). The first is based on the concept of controllability, the second by solving a Riccati equation and the third uses the spectral and structural properties of the operator A.

Theorem 2.12 ([2]) *The system (2.25) is exponentially stabilizable if and only if*

$$\int_0^{\infty} \|T(t)y\|^2 dt < +\infty \quad \text{for all } y \in X, \tag{2.26}$$

where $(T(t))$ is a semigroup generated by the operator $A + BK$.

Proof We use similar techniques to those for the proof of Proposition 2.10. □

Corollary 2.2 ([2]) *If there exists $t_0 > 0$ such that $|||S(t_0)||| < 1$, then the system (2.25) is exponentially stabilizable.*

Proof The proof is similar to that of the Corollary 2.1. □

A characterization of the stability of system (2.25) is given by the following result.

Proposition 2.6 ([2]) *The following properties are equivalent*
1. *The system (2.25) is exponentially stabilizable.*
2. *There exists a control $v(t)$ such that the solution of (2.25) satisfy*

$$\int_0^{\infty} \| y(t) \|^2 + \| v(t) \|^2 \, dt < \infty \quad \text{for all } y_0 \in X$$

3. *There exists an operator P positive defined, solution of the Riccati's equation*

$$\langle Ay, Py\rangle + \langle Py, Ay\rangle - \langle B^*Py, B^*Py\rangle + \langle y, y\rangle \ , \ y \in \mathcal{D}(A) \tag{2.27}$$

4. For all $y_0 \in X$, there exists a control $v(t)$ such that the associated solution $y(t)$ tend to 0 when $t \to \infty$.

Remark 2.3 If the operator P is a solution of (2.27), the control $v(t) = -BPy(t)$ minimizes the cost function

$$q(v) = \int_0^\infty ||y(t)||^2 + ||v(t)||^2 dt < +\infty$$

see [2].

In finite dimension, the stabilization of the system (2.25) is equivalent to the controllability of its unstable modes. In infinite dimension, the spectrum of an operator is not reduced to its punctual part and one cannot in general move the spectrum by a finite rank perturbation. In addition the position of the spectrum is not directly linked to the stability of the system.

Proposition 2.7 ([13]) *The system (2.25) is weakly stabilizable by the control $v(t) = B^* y(t)$ if and only if the unstable states are controllable.*

One of the methods using the spectral properties of A is the decomposition method. Under the hypotheses of the Theorem 2.8, the system (2.25) can be decomposed into a stable system and an unstable one as follows:

$$\begin{cases} \dfrac{dy_u(t)}{dt} = A_u y_u(t) + \Pi B v(t) \\ \\ y_{0u} = \Pi y_0, \quad y_u = \Pi y \end{cases} \tag{2.28}$$

and

$$\begin{cases} \dfrac{dy_s(t)}{dt} = A_s y_s(t) + (I - \Pi) B v(t) \\ \\ y_{0s} = (I - \Pi) y_0, \quad y_s = (I - \Pi) y \end{cases} \tag{2.29}$$

where Π is defined in (2.16), A_s and A_u are the restrictions of A to X_s and X_u respectively, and we have $\sigma(A_s) = \sigma_s(A)$, $\sigma(A_u) = \sigma_u(A)$. In addition, A_u is a bounded operator defined on X_u.

The solutions of (2.28) and (2.29) are given by

$$y_u(t) = S_u(t) y_{0u} + \int_0^t S_u(t - \tau) \Pi B v(\tau) d\tau \tag{2.30}$$

and

$$y_s(t) = S_s(t) y_{0s} + \int_0^t S_s(t - \tau)(I - \Pi) B v(\tau) d\tau \tag{2.31}$$

where $S_u(t)$ and $S_s(t)$ are the restrictions of $S(t)$ on X_u and X_s, which are respectively the C_0- semigroups generated by A_u and A_s.

We know ([7]) that if (2.25) is exactly controllable, then it is stabilizable.

If the operator A_s verifies the following property of spectral decay

$$\lim_{t\to+\infty} \frac{\ln |||S_s(t)|||}{t} = \sup Re(\sigma(A_s)) \tag{2.32}$$

then stabilizing the system (2.25) turns up to stabilizing (2.28).

Proposition 2.8 ([7]) *If there exists $K_u \in \mathcal{L}(X, V)$ such that the control $v = K_u y_u$ stabilizes the system (2.28), then the system (2.25) is exponentially stabilizable using the control operator $K = (K_u, 0)$.*

Proof According to the previous decomposition, we have $\sup Re(\sigma(A_s)) \le -\delta$.

Since A_s verify (2.32), then there exists $a > 0$ and $\mu > 0$ such that $|||S_s(t)||| \le a\, e^{-\mu t}$, for $t \ge 0$.

The solution of system (2.28) is given by

$$y_u(t) = e^{F_u t} y_{0u}, \quad \text{with} \quad F_u = A_u + \Pi B K_u \in \mathcal{L}(X_u)$$

and there exist $0 < \alpha < \mu,\ b > 0$ such that $||y_u(t)|| \le b\, e^{-\alpha t}||y_{0u}||$.

so for $v = K_u z_u$, we have $||v(t)||_V \le b\, e^{-\alpha t}\ |||K_u|||.||y_{0u}||$

It follows

$$\begin{aligned} ||y_s(t)|| &\le a\, e^{-\mu t}||y_{0s}|| + c\,||y_{0u}|| \int_0^t e^{-\mu(t-\tau)} e^{-\alpha\tau} d\tau \\ &\le a\, e^{-\delta' t}||y_{0s}|| + c\,||y_{0u}|| \frac{e^{-\mu t} - e^{-\alpha t}}{\alpha - \mu} \end{aligned}$$

where $c > 0$. Therefore the state of the system (2.25) controlled by $v(t) = Ky(t)$ verifies

$$||y(t)|| \le (a\, e^{-\mu t} + c\, \frac{e^{-\mu t} - e^{-\alpha t}}{\alpha - \mu} + b\, e^{-\alpha t})||y_0||$$

which leads to the exponential stabilization of the system (2.25). □

The following result (see [7]) makes possible to bring back the problem of stabilization to that of the controllability of a finite dimensional system.

Proposition 2.9 *If the condition (2.32) is verified and if, moreover,*
1. the space X_u is of finite dimension,
2. the system (2.28) is controllable on X_u, then the system (2.25) is exponentially stabilizable.

Proof In finite dimension, there is equivalence between controllability and exponential stabilization. This implies that the system (2.28) is exponentially stabilizable. □

One of the results that characterize the weak stabilization of (2.25) is given by the following proposition (see [13]).

Proposition 2.10 *The control $v(t) = -B^* y(t)$ weakly stabilizes the system (2.25) if is only if the unstable states of (2.25) are controllable.*

We recall the definition of weak controllability

Definition 2.8 If A generates a C_0-semigroup $S(t)$ and $y_d \in X$,
1. the system (2.25) is approximately controllable to y_d, if for all $\varepsilon > 0$, there exist $T > 0$ and a control $v \in L^2([0, T], V)$ such that:

$$\|y_d - \int_0^T S(T-s)Bv(s)ds\| < \varepsilon$$

2. If C indicates the set of states approximately controllable, the system (2.25) is said to be approximately controllable if $\overline{C} = X$.

Let the operator $L(t)v = \displaystyle\int_0^t S(t-s)Bv(s)ds$ defined from V in X, we have the following result.

Lemma 2.2 ([3]) *1. C is a closed subspace of X and $C = \overline{\bigcup_{t\geq 0} Im(L(t))}$.*

2. The set of non-controllable states is $C^\perp = \bigcap_{t\geq 0}(Ker[B^* S^*(t)])$.

Proof Let $T > 0$ and $y \in (Im(L(T))^\perp$, then for all $v \in V$, we have

$$\int_0^T \langle v(s), B^* S^*(s)y\rangle ds = 0$$

which implies that $B^* S^*(s)y = 0, \quad 0 < s < T$.
Consequently $y \in \{Im(S(t)B), \quad t \leq T\}^\perp$, which gives

$$L(t)^\perp \subset \{Im(S(t)B), \quad t \leq T\}^\perp$$

Thus

$$Vect\{Im(S(t)B), \quad t \leq T\} \subset (L(T)^\perp)^\perp = \overline{L(T)}$$

where $Vect(E)$ indicates the subspace generated by E. And finally we show that

$$\overline{\left\{\int_a^b S(t)Bydt, \, y \in X\right\}} = X.$$

□

Lemma 2.3 ([3]) *Let $K \in \mathcal{L}(V, X)$ be a bounded operator, and let $T(t)$ be a C_0-semigroup generated by $A + BK$. We have the following equivalence*

$$\left\{\forall t \geq 0, \, B^* S^*(t)y = 0\right\} \Leftrightarrow \left\{\forall \, t \geq 0, \, B^* T^*(t)y = 0\right\}$$

Proof We have

$$T^*(t)y = S^*(t)y + \int_0^t S^*(t-s)K^*B^*T^*(s)yds$$

and

$$S^*(t)y = T^*(t)y + \int_0^t T^*(t-s)K^*B^*S^*(s)yds$$

Therefore $B^*S^*(t)y = 0 \Leftrightarrow B^*T^*(t)y = 0$. □

Proposition 2.11 ([13]) *If the semigroup $S(t)$ is a contraction semigroup, then the two following conditions are equivalent:*
1. The control $v(t) = -B^*y(t)$ weakly stabilizes the system (2.25).
2. The unstable states of (2.25) are controllable.

Proof Let W be the set of weakly stable states of the system (2.25), and C the set of weakly controllable states.

Suppose that the system (2.25) is weakly stabilizable with the control $v(t) = -B^*y(t)$. Property 2. is equivalent to $W^\perp \subset C$, more again $C^\perp \subset W$.

Let $y \in C^\perp$, according to the Lemma 2.2, for all $t \geq 0$, we have $B^*S^*y = 0$. Using the Lemma 2.3, for all $z \in X$ we have

$$\langle S^*(t)y, z\rangle = \langle T^*(t)y, z\rangle = \langle y, T(t)z\rangle \longrightarrow 0 \text{ when } t \longrightarrow +\infty$$

Therefore $S^*(t)y \rightharpoonup 0$ when $t \longrightarrow +\infty$.

As $S^*(t)$ is a contraction semigroup, by Theorem 2.7 we have $S(t)y \rightharpoonup 0$ when $t \longrightarrow +\infty$. Hence $y \in W$ and therefore $C^\perp \subset W$.

Conversely, if we suppose that $C^\perp \subset W$, BB^* is a dissipative bounded operator, then $A - BB^*$ generates a contraction semigroup $(T(t))$ on X.

Using Theorem 2.6, $X = X_u \oplus X_s$ with X_u reduces $T(t)$ to unitary C_0-semigroup, and X_s reduces $T(t)$ into a completely non-unitary C_0-semigroup.

By Theorem 2.7 we have, for all $y \in X_s$:

$$T(t)y \rightharpoonup 0 \text{ when } t \to +\infty$$

It remains to show that, for all $y \in X_u$, $T(t)y \rightharpoonup 0$ when $t \to +\infty$ We note $G_u = X_u \cap \mathcal{D}(A)$. For all $y \in G_u$ and $t \geq 0$, we have

$$\frac{d}{dt}\|T^*(t)\|^2 = \langle (A^* - BB^*)T^*(t)y, T^*(t)y)\rangle + \langle T^*(t)y, (A^* - BB^*)y)\rangle = 0$$

Since A^* and BB^* are dissipative, then for all $t \geq 0$, we have $B^*T^*(t)y = 0$ The Lemma 2.3 implies that for all $t \geq 0$, $B^*S^*(t)y = 0$, but as $C^\perp \subset W$, therefore $G_u \subset W$.

With Lemma 2.3, we have, for all $t \geq 0$ and $y \in G_u$, $T^*(t)y = S^*(t)y$. Thus for $y \in W$, and by Theorem 2.7 we have $S^*(t)y \rightharpoonup 0$ when $t \to 0$.

Consequently $T^*(t)y \rightharpoonup 0$ when $t \to 0$, and with the Theorem 2.7 we have $T(t)y \rightharpoonup 0$ when $t \to 0$.

We deduce that, for all $y \in G_u$, $T(t)y \rightharpoonup 0$ when $t \longrightarrow +\infty$.

Let $y \in X_u = \overline{G}_u$ (with Theorem 2.7), there exists a sequence $(y_n) \in G_u$ such that $\|y_n - y\| \longrightarrow 0$ when $n \longrightarrow +\infty$ and since $T(t)$ is a contraction semigroup, we have

$$|\langle T(t)y, z\rangle| \leq \|y - y_n\|\|z\| + |\langle T(t)y_n, z\rangle| \longrightarrow 0 \text{ when } n \longrightarrow +\infty$$

which ends the proof. □

The following result is very useful in the case where the controls space $V = \mathbb{R}^p$, and the control operator

$$B[(v_1, v_2, \dots, v_p)] = \sum_{j=1}^{p} v_j f_j$$

with $f_j \in X$, for $j = 1, \dots, p$.

We also assume that the dynamics operator A of the system is self adjoint and has a compact resolvent.This implies that A admits a complete set of eigenvectors (Ψ_m), $m \in \mathbb{N}^*$, associated with eigenvalues (λ_{mj}), $j = 1, \dots, r_m$ of multiplicity r_m. If in addition A has a finite number of positive eigenvalues denoted $\lambda_{1d}, \lambda_{2d}, \dots, \lambda_{lj}$, in this case the hypothesis of the decomposition of the state space is verified and $dim(X_u) < +\infty$. So we have the result.

Proposition 2.12 *The system (2.28) is exponentially stabilizable if*

$$\begin{cases} p \geq \sup\limits_{1 \leq m \leq l} r_m \\ rg(G_m) = r_m, \quad m = 1, 2, \dots, l \end{cases} \tag{2.33}$$

where $(Gm)_{ij} = \langle f_i, \Psi_{mj}\rangle$

Proof A_u can be represented by a diagonal matrix such that:

$$A_u = diag(\underbrace{\lambda_1, \dots, \lambda_1}_{r_1 \; times}, \underbrace{\lambda_2, \dots, \lambda_2}_{r_2 \; times}, \dots, \underbrace{\lambda_l, \dots, \lambda_l}_{r_l \; times})$$

We also have $PB = {}^t[{}^tG_1 \, {}^tG_2 \cdots {}^t G_l]$.

The condition (2.33) implies that the system (2.28) is controllable. Consequently it is exponentially stabilizable (because X_u is of finite dimension). Hence the existence of an operator $K_u \in \mathcal{L}(X_u, X)$ which exponentially stabilizes the system (2.28). And Theorem 2.8 allows to conclude. □

2.3.2 Optimal Stabilization Problem

The stabilization at lower cost remains a major problem which has been,among others, the subject of numerous works, see [12, 14]. In this subsection we propose to stabilize the system (2.25) while minimizing the functional

$$q(u) = \int_0^{\infty} (||y(t)||^2 + \langle v(t), Rv(t)\rangle)dt < +\infty. \tag{2.34}$$

where $v \in V_{ad} = \{v \in L^2(\mathbb{R}^+, X) \mid q(v) < +\infty\}$ and $R \in \mathcal{L}(V)$ a self adjoint and positive defined operator.

Theorem 2.13 ([12]) *The system (2.25) is exponentially stabilizable if and only if there exists $P \in \mathcal{L}(X)$ solution of the Riccati equation*

$$\langle Ay, Py\rangle + \langle Py, Ay\rangle - \langle B^*Py, B^*Py\rangle + \langle Ry, y\rangle = 0 \tag{2.35}$$

for all $y \in \mathcal{D}(A)$. In addition the control

$$v^*(t) = -PB^*y^*(t) \tag{2.36}$$

stabilizes the system (2.25) and is the unique control minimizing the functional (2.34).

In reality the stabilizing control is of finite energy, below we assume that the control (2.36) is bounded.

Now we consider a control $v(t) = -f(B^*y(t))$ where the function $f : V \to \mathbb{R}$ to be determined, and the system

$$\begin{cases} \dfrac{\partial y}{\partial t} = Ay(t) - Bf(B^*y(t)) \\ y(0) = y_0 \end{cases} \tag{2.37}$$

The following lemma allows to state the main result.

Lemma 2.4 ([15]) *Let f_0 be a positive and increasing function such that $f_0(0) = 0$.*

We note $f_1(t) = t - (Id + f_0)^{-1}(t)$, and let (s_k) a sequence of positive numbers which verify $f_0(s_{k+1}) + s_{k+1} \leq s_k$, for all $k \geq 0$. If $U(t)\xi$ is solution of the system

$$\begin{cases} \dfrac{dU(t)}{dt} = -f_1(U(t)) \\ U(0) = \xi \geq 0 \end{cases} \tag{2.38}$$

then $s_k \leq U(k)s_0$.

Theorem 2.14 ([16]) *If the function f satisfies the conditions:*

1. there are two positive constants c, d such that for all $v \in V$ with $\|v\|_V \geq 1$, we *have*

$$c\ \|v\|_V^2 \leq \langle f(v), v\rangle_V \text{ and } \|f(v)\|_V^2 \leq d\ \|v\|_V \tag{2.39}$$

2. *there exists a strictly increasing concave function* $\xi : \mathbb{R}^+ \longrightarrow \mathbb{R}$ *such that* $\xi(0) = 0$ *and for all* v *such that* $\|v\| < 1$, *we have*

$$\|f(v)\|_V^2 + \|v\|_V^2 \leq \xi(\langle f(v), v\rangle) \tag{2.40}$$

3. *there exist* $T > 0$, $e > 0$ *such that*

$$\int_0^T \|B^* S(t) y_0\|_V^2 dt \geq e\|y_0\| \tag{2.41}$$

for all $y_0 \in X$.

Then the system (2.37) is strongly stable, moreover we have the estimate

$$\|y(t)\| = O(\frac{1}{\sqrt{t}})$$

More precisely, for all $t > T$, *we have*

$$\frac{1}{2}\|y(t)\|^2 \leq \frac{1}{2}U\left(\frac{t}{T} - 1\right) \tag{2.42}$$

where $U(t) \to 0$ *when* $t \to \infty$, *is a contraction semigroup solution of the differential equation*

$$\begin{cases} \dfrac{dU(t)}{dt} + q(U(t)) = 0 \\ U(0) = \frac{1}{2}\|y_0\|^2 \end{cases} \tag{2.43}$$

where q *is given by*

$$q(s) = s - (I + p)^{-1}(s) \tag{2.44}$$

Proof For all $T > 0$, the solution $y(t)$ of the system (2.37) verify

$$\frac{1}{2}\|y(T)\|^2 + \int_0^T \langle f(B^* y(t)), B^* y(t)\rangle_V dt \leq \frac{1}{2}\|y_0\|^2 \tag{2.45}$$

Let ϕ and ψ be two functions such that $y = \varphi + \psi$, where φ is solution of the system

$$\begin{cases} \dfrac{d\varphi(t)}{dt} = A\phi(t) \\ \varphi(0) = y_0 \end{cases} \tag{2.46}$$

and ψ solution of the system

$$\begin{cases} \dfrac{d\psi(t)}{dt} = A\psi - Bf\left(B^* y(t)\right) \\ \psi(0) = 0 \end{cases} \tag{2.47}$$

Using (2.45) and (2.41) we get

$$\frac{1}{2}\|y(T)\|^2 \le \frac{1}{2}\|y(0)\|^2 \le \frac{1}{2e}\int_0^T \| B^*\varphi(t) \|_V^2 \, dt. \tag{2.48}$$

Since $y = \varphi + \psi$, we have

$$\| B^*\varphi(t) \|_V^2 \le 2\left(\| B^*\psi(t) \|_V^2 + \| B^* y(t) \|_V^2\right) \tag{2.49}$$

and

$$\int_0^T \| B^*\varphi(t) \|_V^2 \, dt \le 2\left\{\int_0^T \| B^*\psi(t) \|_V^2 \, dt + \int_0^T \| B^* y(t) \|_V^2 \, dt\right\} \tag{2.50}$$

According to (2.47), for $0 \le t \le T_0$, we have

$$\psi(t) = \int_0^t e^{(t-s)A} Bf(B^* y(s)) ds \tag{2.51}$$

As $S(t)$ is a contraction semigroup, by the Cauchy–Schwarz inequality, we have

$$\begin{aligned} \| B^*\psi(t) \|_V^2 &\le |||B^*|||^2 |||B|||^2 \left\{\int_0^T |||e^{(t-s)A}||| \; \| f(B^* y(s)) \| \, ds\right\}^2 \\ &\le |||B^*|||^2 |||B|||^2 \, T \int_0^T \| f(B^* y(s)) \|_V^2 \, ds \end{aligned} \tag{2.52}$$

Therefore there exists a constant $C_1 > 0$ such that

$$\int_0^T \| B^*\varphi(t) \|_V^2 \, dt \le C_1 \left\{\int_0^T \| f(B^* y(t)) \|_V^2 \, dt + \int_0^T \| B^* y(t) \|_V^2 \, dt\right\} \tag{2.53}$$

We consider the decomposition of $[0, T_0] = K_1 \cup K_2$ with

$$\begin{aligned} K_1 &= \{t \in [0, T] / \| B^* y(t) \|_V \ge 1\} \\ K_2 &= \{t \in [0, T] / \| B^* y(t) \|_V < 1\} \end{aligned} \tag{2.54}$$

From the hypotheses (2.39) and (2.40), we obtain

$$\int_{K_1} \| f(B^*y(t)) \|_V^2 \, dt + \int_{K_1} \| B^*y(t) \|_V^2 \, dt \leq \left(\frac{d^2}{c} + \frac{1}{c}\right) \int_0^{T_0} \langle f(B^*y(t)), B^*y(t)\rangle_V dt \tag{2.55}$$

and

$$\int_{K_2} \| f(B^*y(t)) \|_V^2 \, dt + \int_{K_2} \| B^*y(t) \|_V^2 \, dt \leq \int_0^T \xi(\langle f(B^*y(t)), B^*y(t)\rangle_V) dt \tag{2.56}$$

Applying Jensen's inequality, we get

$$\int_0^T \xi(\langle f(B^*y(t)), B^*y(t)\rangle_V) dt \leq T\xi(\frac{1}{T}\int_0^T \langle f(B^*y(t)), B^*y(t)\rangle_V) dt) \tag{2.57}$$

Using inequalities (2.45) and (2.53), we have

$$\tfrac{1}{2}\|y(T)\|^2 \leq C_0\{\int_0^T \langle f(B^*y(t)), B^*y(t)\rangle_V dt + \xi(\tfrac{1}{T}\int_0^T \langle f(B^*y(t)), B^*y(t)\rangle_V dt)\}. \tag{2.58}$$

If we note $h(s) = s + \xi(\frac{s}{T})$, we have

$$\frac{1}{2}\|y(T)\|^2 \leq C_0 h(\frac{1}{2}\|y_0\|^2 - \frac{1}{2}\|y(T)\|^2) \tag{2.59}$$

where $\widetilde{h}(t) = t + h(\dfrac{t}{T})$. The function $p = (C_0 h)^{-1}$ is increasing on $[0, +\infty)$ and with (2.59), we get

$$\frac{1}{2}\|y(T)\|^2 + p\left(\frac{1}{2}\|y(T)\|^2\right) \leq \frac{1}{2}\|y_0\|^2 \tag{2.60}$$

In addition, for $k \in \mathbb{N}$, we have

$$\frac{1}{2}\|y((k+1)T)\|^2 + p\left(\frac{1}{2}\|y((k+1)T)\|^2\right) \leq \frac{1}{2}\|y(kT)\|^2. \tag{2.61}$$

Since the estimate (2.60) is verified on intervals of the form $[kT, (k+1)T]$.

By applying Lemma 2.4 to the sequence $s_k = \dfrac{1}{2}\|y(kT)\|^2$, we get, for all $k \in \mathbb{N}$,

$$\frac{1}{2}\|y(kT)\|^2 \leq U(k) \tag{2.62}$$

For integer $k \neq 0$, for all $t > T$, and $0 \leq \tau < T$, we have $t = kT + \tau$, consequently we deduce, for all $t > T$,

$$\frac{1}{2}\|y(t)\|^2 \leq \frac{1}{2}\|y(kT)\|^2 \leq \frac{1}{2}U(k) \leq U\left(\frac{t-\tau}{T}\right) \leq U\left(\frac{t}{T}-1\right) \tag{2.63}$$

□

Remark 2.4 The stabilization problem can be seen as a special case of an output stabilization one. For this and for other results linking weak or strong stabilization of the output of the system (2.25) to that of (2.28), see, in particular, [17–19].

References

1. Pazy, A.: Semigroups of Linear Operators and Applications to Partial Differential Equations. Springer, New York (1983)
2. Prichard, A.J., Zabczyk, J.: Stability and stabilizability of infinite dimensional systems. Siam Rev. **23**, 25–51 (1981)
3. Engel, K.J., Nagel, R.: One Parameter Semigroups for Linear Evolution Equations. Springer, New York (2000)
4. Kato, T.: Perturbation Theory for Linear Operators, 2nd edn. Springer, Berline (1980)
5. Nagy, S.Z.: Stability theory for ordinary differential equations. J. Differ. Equ. **4**, 307–324 (1968)
6. Balakrishnan, A.V.: Strong stability and the steady state Riccati equation. Appl. Math. Optim. **7**, 335–345 (1981)
7. Triggiani, R.: On the stabilizability problem in Banach space. J. Math. Anal. Appl **52**, 383–403 (1975)
8. Hille, E., Phillips, R.: Functional Analysis and Semigroups. American Mathematical Society, Providence (1957)
9. Zabczyk, J.: Remarks on the algebraic Riccati equation in Hilbert space. Appl. Math. Optim. **3**, 251–258 (1976)
10. Datko, R.: Extending a theorem of A. M. Liapunov to Hilbert Space. J. Math. Anal. Appl. **32**, 610–616 (1970)
11. Datko, R.: A linear control problem in abstrat Hilbert space. J. Differ. Equ. **9**, 346–359 (1971)
12. Curtain, R.F., Zwart, H.: An Introduction to Infinite Dimentional Linear System. Springer, New York (1995)
13. Benchimol, C.: A note on weak stabilizability of contraction semi-groups. SIAM J. Control Optim. **16**, 373–379 (1978)
14. Banks, H., Kunisch, B.: The linear regulator problem for parabolic systems. SIAM J. Control Optim. **22**, 684–696 (1984)
15. Lasiecka, I., Tataru, D.: Uniform boundary stabilisation of semilinear wave equation with nonlinear boundary damping. J. Differ. Integr. Equ. **6**, 507–533 (1993)
16. Berrahmoune, L.: Asymptotic stabilization and decay estimate for distributed bilinear systems. Recerche di Matematica. fasc. **1**, 89–103 (2001)

17. Rabah, R., Ionescu, D.: Stabilization problem in Hilbert space. Int. J. Control **46**, 2035–2042 (1987)
18. Staffans, O.: Quadratic optimal control of stable systems through spectral factorization. Math. Control Signals Syst. **8**, 167–197 (1995)
19. Weiss, M., Weiss, G.: Optimal control of stable weakly regular linear systems. Math. Control Signals Syst. **10**, 287–330 (1997)

Chapter 3
Stabilization of Infinite Dimensional Semilinear Systems

This chapter concerns the study of the stability of certain classes of infinite dimensional systems commonly encountered in applications: these are infinite dimensional semi linear systems.

3.1 Well Posed Systems

In this section we present some results on the existence and the uniqueness of the solution of a class of non linear systems described by the equation

$$\begin{cases} \dfrac{dy(t)}{dt} &= Ay(t) + g(y(t), t) \\ \\ y(0) &= y_0 \in X \end{cases} \tag{3.1}$$

where $A : \mathcal{D}(A) \subset X \longrightarrow X$ is a linear operator that generates a C_0-semigroup $S(t)$, $t \geq 0$, in X, and $g : X \times \mathbb{R}^+ \longrightarrow X$ a given function.

First let recall some results on nonlinear semigroups.

Definition 3.1 A nonlinear semigroup $S(t)$ on X is a family of continuous operators $S(t) : X \longrightarrow X,\ t \geq 0$, that verify:
(i) $S(0) = Id_X$
(ii) $S(t+s) = S(t)S(s)$, for all $t, s \geq 0$
If in addition

$$||S(t)y - S(t)z|| \leq ||y - z||,\ \forall y, z \in X,$$

then $S(t)$ is said to be a contraction semigroup.

E. H. Zerrik and O. Castillo, *Stabilization of Infinite Dimensional Systems*,
Studies in Systems, Decision and Control 355,
https://doi.org/10.1007/978-3-030-68600-0_3

Also let recall that a set C is said to be positively invariant by $S(t)$, if $S(t)C \subset C$, for all $t \geq 0$, and it is said to be invariant if $S(t)C = C$, for all $t \geq 0$.

For $\phi \in X$, we define the following sets:

– The positive orbit associated with ϕ by $\mathcal{O}^+(\phi) = \bigcup_{t \geq 0} S(t)\phi$. In particular, the positive orbit associated with an equilibrium state is reduced itself.

– The set $\omega-$limit of ϕ by

$\omega(\phi) = \{\psi \in X$ such that there exists $t_n \longrightarrow +\infty;\ \ S(t_n)\phi \longrightarrow \psi$ when $n \longrightarrow +\infty\}$.

– The set weak $\omega-$limit of ϕ by

$\omega_w(\phi) = \{\psi \in X$ such that $\exists\ t_n \longrightarrow +\infty;\ \ S(t_n)\phi \rightharpoonup \psi$ when $n \longrightarrow +\infty\}$.

The last two sets represent the states that can be approached to infinity by elements of $\mathcal{O}^+(\phi)$.

In general, we have $\omega(\phi) \subset \omega_w(\phi)$, with equality in the case of finite dimension. In the case of bounded orbits, we have $\omega_w(\phi) \neq \emptyset$, which is not the case for the set $\omega(\phi)$ which is positively invariant.

The following result underlines this result

Theorem 3.1 ([1, 2]) *1. If $\mathcal{O}^+(\phi)$ is pre-compact then $\omega(\phi) \neq \emptyset$.*
2. If for all $t \geq 0$, $S(t)$ is sequentially weakly continuous, then $\mathcal{O}^+(\phi)$ bounded implies $\omega_w \neq \emptyset$ and invariant by $S(t)$.

The previous theorem can be used for the study of nonlinear semigroups of parabolic type and contraction semigroups of hyperbolic type with pre-compact orbits (see [3]). An important application concerns the Lasalle invariance principle which gives information on the structure of the sets $\omega-$ limit for dynamical systems having a Lyapunov function.

Theorem 3.2 ([4, 5] (Lasalle principal)) *Let V be a proper Lyapunov function. Then every Pre-compact trajectories of (3.1) tend strongly towards Λ, the largest invariant set by the dynamic of the system (3.1) and contained in the set $M_r =: \{y \in X$ such that $V(y(t)) = r,\ \forall t \geq 0\}$, for $r \geq 0$.*

To apply this principle in the study of the stability of a steady state y_1, it suffices to show that the set Λ is reduced to y_1.

For the stabilization of such systems, let recall the definitions

Definition 3.2 Let $T > 0$, a function $y \in \mathcal{C}([0, T], X)$ is said
1. weak solution to (3.1) on $[0, T]$ if the function $g(y(.), .) \in L^1(0, T; X)$ and for all $\varphi \in \mathcal{D}(A^*)$, the function $t \longrightarrow \langle y(t), \varphi\rangle$ is absolutely continuous on $[0, T]$ and verify

$$\frac{d}{dt}\langle y(t), \varphi\rangle = \langle y(t), A^*\varphi\rangle + \langle g(y(t), t), \varphi\rangle \quad \text{a.e on } [0, T]$$

2. strong solution of system (3.1) on $[0, T]$ if for all $y_0 \in \mathcal{D}(A)$, the solution $y(.)$ is continuous on $[0, T[$, of class C^1 on $]0, T[$ and verify (3.1).

The system (3.1) is considered well posed if it has a unique global weak solution.

The following result gives conditions for the existence of the weak solution to the system (3.1), see [6].

Theorem 3.3 ([6]) *1.If $g(y(.), .) \in L^1(0, T; X)$, then a solution of the integral equation*

$$y(t) = S(t)y_0 + \int_0^t S(t-s)g(y(s), s)ds \quad a.e. \ on \ [0, T] \tag{3.2}$$

is a weak solution of (3.1).

2. If $g : X \times [0, T] \to X$ is continuous on $[0, T]$, uniformly k-Lipshitz on X then the system (3.1) has a unique weak solution $y \in \mathcal{C}(0, T; X)$ and the application $y_0 \to y$ is Lipshitz from X in $\mathcal{C}(0, T; X)$.

Proof Let $y_0 \in X$, we define the application $F : \mathcal{C}(0, T; X) \to \mathcal{C}(0, T; X)$ by

$$F(y)(t) = S(t)y_0 + \int_0^t S(t-s)g(y(s), s)ds \tag{3.3}$$

The function g is uniformly Lipshitz, there exists $M > 0$ such that

$$\| F(y)(t) - F(z)(t) \| \leq Mk \| y - z \|_\infty$$

or again

$$\| F(y)^m(t) - F(z)^m(t) \| \leq \frac{(MkT)^m}{m!} \| y - z \|_\infty$$

For m quite large we have $\frac{(MkT)^m}{m!} < 1$. Therefore F has a fixed point $y \in \mathcal{C}(0, T; X)$, solution of the system (3.1).

Now let z be another weak solution of (3.1) associated to the initial condition z_0, we have

$$\begin{aligned} \| y(t) - z(t) \| &\leq \| S(t)y_0 - S(t)z_0 \| \\ &\quad + \int_0^t \| S(t)(g(y(s), s) - g(z(s), s)) \| ds \\ &\leq M \| y_0 - z_0 \| + Mk \int_0^t \| y(s) - z(s) \| ds \end{aligned}$$

By applying the Gronwall lemma we get

$$\| y(t) - z(t) \| \leq Me^{MkT} \| y_0 - z_0 \|$$

from where $\| y - z \| \leq Me^{MkT} \| y_0 - z_0 \|$. Therefore the solution is unique and continuous. □

Theorem 3.4 ([6]) *Let $g \in L^1([0,T];X)$, and on $[0,T]$, we consider the function*

$$R(t) = \int_0^t S(t-s)g(s)ds$$

If one of the following two conditions is satisfied:
1. $R(t)$ is differentiable on $[0,T]$ and $\dfrac{dR(t)}{dt} \in L^1(0,T;X)$
2. For all $t \in [0,T]$, $R(t) \in \mathcal{D}(A)$,
then any solution $y(t)$ of (3.2) is a strong solution of (3.1), for all $y_0 \in \mathcal{D}(A)$.

Theorem 3.5 ([6]) *If $g : X \times [0,+\infty[\to X$ is continuous for all $t > 0$ and uniformly continuous over any bounded interval of $\mathbb{R}^+$ and if, moreover, g is locally Lipschitz on X, then the system (3.1) has a unique weak solution y on $[0, t_{\max}[$ with $t_{\max} \le +\infty$ and $\lim\limits_{t \to t_{\max}} \|y(t)\| = +\infty$.*

Proof Let us first show that for $t_0 \ge 0$ and $y_0 \in X$, the system (3.1) has a weak solution on $[t_0, t_1]$ such that $t_1 - t_0 \le \delta(t_0, \|y_0\|)$ where

$$\delta(t_0, \|y_0\|) = \min\left\{1\ ,\ \frac{\|y_0\|}{K(t_0)\ \ k(K(t_0), t_0+1)\ +\ N(t_0)}\right\} \tag{3.4}$$

with

$$\begin{aligned} K(t_0) &= 2\|y_0\|M(t_0) \\ M(t_0) &= \max\{|||S(t)|||\ \text{on}\ [0, t_0+1]\} \\ N(t_0) &= \max\{\|g(t,0)\| : 0 \le t \le t_0+1\} \end{aligned}$$

and $k(c,t)$ is the Lipschitz constant of the function g.

We consider the function (3.3) on $\mathcal{C}([t_0,t_1];X)$ with $t_1 = t_0 + \delta(t_0, \|y_0\|)$. We have $F(B(0,K(t_0))) = B(0,K(t_0))$ is the ball centered in 0 and radius $K(t_0)$. Indeed

$$\begin{aligned} \|(Fy)(t)\| &\le M(t_0)\|y_0\| + \int_{t_0}^t |||S(t-s)|||(\|g(s,y(s)) - g(s,0)\| + \|g(s,0)\|)ds \\ &\le M(t_0)\|y_0\| + M(t_0)K(t_0)L(K(t_0), t_0+1))(t-t_0) + M(t_0)N(t_0)(t-t_0) \\ &\le M(t_0)\{\|y_0\| + K(t_0)L(K(t_0), t_0+1)(t-t_0) + N(t_0)(t-t_0)\} \\ &\le 2M(t_0)\|y_0\| = K(t_0) \end{aligned}$$

F is uniformly Lipschitz on $B(0, k(t_0))$ and of Lipschitz constant $L = L(K(t_0), t_0 + 1)$ and similarly to the previous theorem, we show that F admits a fixed point y in this ball, solution of (3.1) on the interval $[t_0, t_1]$.

Therefore if y is a weak solution on the interval $[0,\tau]$, it admits an extension on the interval $[0, \tau+\delta]$ with $\delta > 0$, defined on $[\tau, \tau+\delta]$, and solution of the following equation

$$y(t) = S(t-\tau)y(\tau) + \int_\tau^t S(t-s)g(s,y(s))ds\ \ \tau \le t \le \tau + \delta$$

Moreover, δ depends only on $\|y(\tau)\|$, $k(\tau)$ and $N(\tau)$.

Let $[0, t_{\max}[$ be the maximum interval of existence of the weak solution y of the system (3.1).

If $t_{\max} < \infty$, then $\lim\limits_{t \to t_{\max}} \|y(t)\| = \infty$, otherwise there exists an increasing sequence (t_m) which converges to $t_{\max}$ such that $\|y(t_m)\| \leq C$, for all m.

The solution y defined on $[0, t_m]$ admits an extension on $[0, t_m + \delta]$ with $\delta > 0$ which does not depend on t_m, thus y admits an extension for $t > t_{\max}$ which contradicts the definition of $t_{\max}$.

The uniqueness of the global solution of system (3.1), is a consequence of the Theorem 3.3. □

Remark 3.1 If y is bounded, the weak solution $y(t)$ is global ($t_{\max} = +\infty$).

The result that follows gives conditions on the existence of a global solution [6].

Theorem 3.6 ([6]) *1. If A generates a compact C_0-semigroup $S(t)$, $t \geq 0$, on X and if $g : [0, \infty[\times X \to X$ is such that the image of any bounded of $[0, \infty[\times X$ is a bounded of X, then for all $y_0 \in X$, the system (3.1) has at most a weak solution over a maximum interval $[0, t_{\max}[$.*

2. If, moreover, $t_{\max} < \infty$, then $\lim\limits_{t \to t_{\max}} \| y(t) \| = \infty$.

Proof First, we note that a weak solution y of (3.1) defined over a closed interval $[0, t_1]$ can be extended to a larger interval $[0, t_1 + \delta]$, $\delta > 0$.

The function $z(t) = y(t + t_1)$ is a weak solution of the system

$$\begin{cases} \dfrac{dz(t)}{dt} - Az(t) = g(t + t_1, z(t)) \\ \\ z(0) = y(t_1) \end{cases} \tag{3.5}$$

defined on an interval of length δ [6].

Let $[0, t_{\max}[$ be the maximum interval of existence of the weak solution y to (3.1), We will show that if $t_{\max} < \infty$ then $\| y(t) \| \to \infty$ when $t \to t_{\max}$.

For that, we will show that $t_{\max} < \infty$ implies $\overline{\lim\limits_{t \to t_{\max}}} y(t) = \infty$. Indeed if $t_{\max} < \infty$ and $\overline{\lim\limits_{t \to t_{\max}}} y(t) < \infty$, we can assume that $|||S(t)||| \leq M$ and $\| y(t) \| \leq K$ for $0 \leq t < t_{\max}$ where M and K are constants.

By assumption on g, there exists a constant N such that for $0 \leq t < t_{\max}$, $\| g(t, y(t)) \| \leq N$.

Now if $0 < \rho < t < \tau < t_{\max}$, then

$$\begin{aligned}
\| y(\tau) - y(t) \| \leq& \| S(\tau)y_0 - S(t)y_0 \| \\
&+ \| (\int_0^{t-p} + \int_{t-p}^{t})(S(\tau - s) - S(t - s))g(s, y(s))ds \| \\
&+ \| \int_t^{\tau} S(\tau - s)g(s, y(s))ds \| \\
\leq& \| S(\tau)y_0 - S(t)y_0 \| + N \int_0^{t-p} \| (S(\tau - s) - S(t - s) \| ds \\
&+ 2MN\rho + (\tau - t)MN.
\end{aligned} \tag{3.6}$$

Since $0 < \rho < t$ is arbitrary and $S(t)$ is continuous for the operators uniform topology, the right term of (3.6) tends to zero when t and τ tend to $t_{\max}$. So $\lim_{t \to t_{\max}} y(t) = y(t_{\max})$ exists, and from the first part of the proof, $y(t)$ can be extended beyond $t_{\max}$, which contradicts the fact that $t_{\max}$ is maximum. So the hypothesis $t_{\max} < \infty$ implies $\overline{\lim_{t \to t_{\max}}} y(t) = \infty$.

To finish the proof, we will show that $\lim_{t \to t_{\max}} y(t) = \infty$. Otherwise, there exists a sequence (t_n) and a constant K such that $\| y(t_n) \| \leq K$ for all n.

Let $M > 0$ be such that $\| T(t) \| \leq M$ for $0 \leq t \leq t_{\max}$ and

$$N = \sup\{\| g(t, x) \| : 0 \leq t \leq t_{\max}, \| x \| \leq M(K + 1)\}$$

Since $t \to \| y(t) \|$ is continuous and $\overline{\lim_{t \to t_{\max}}} \| y(t) \| = \infty$, there exists a sequence (h_n) such that $h_n \to 0$ when $n \to \infty$ and $\| y(t) \| \leq M(K + 1)$ for $t_n \leq t \leq t_n + h_n$ and $\| y(t_n + h_n) \| = M(K + 1)$.

But we have:

$$\begin{aligned}
M(K + 1) = \| y(t_n + h_n) \| \leq& \| S(h_n)y(t_n) \| \\
&+ N \int_{t_n}^{t_n + h_n} |||S(t_n + h_n - s)|||ds \\
\leq& MK + h_n NM
\end{aligned}$$

which is absurd when $h_n \to 0$ and so $\lim_{t \to t_{\max}} \| y(t) \| = 0$, which completes the proof.

□

The corollary which follows gives another result of existence of a global solution of the system (3.1).

Corollary 3.1 ([6]) *Suppose that A generates a compact C_0-semigroup $S(t)$ and that the function g is continuous in t and the image by g of any bounded of $X \times \mathbb{R}^+$ is a bounded of X. If, in addition, one of the following conditions holds*
1. There exists a function $k : X \times \mathbb{R}^+ \longrightarrow \mathbb{R}^+$, such that $||y(t)|| \leq k(t)$, $\forall t \in [0, t_1[$.
2. There exist two locally integrable functions k_1 and k_2: $X \times \mathbb{R}^+ \longrightarrow \mathbb{R}^+$, such that,

for all $y \in X$, *we have*

$$||g(y,t)|| \leq k_1(t)||y|| + k_2(t), \forall t \in [0, t_1[$$

then for all $y_0 \in X$, *the system (3.1) admits a global solution* $y(.)$.

Proof 1. This point follows from the previous theorem.
2. The system (3.1) admits a weak solution $y(t)$ defined on $[0, t_{\max}]$.
We know that $|||S(t)||| \leq Me^{\mu t}$, where $M \geq 0$ and $\mu \in \mathbb{R}$.
The function

$$\psi(t) = M \parallel y_0 \parallel + \int_0^t Me^{-\mu s} k_2(s) ds$$

is continuous on $\mathbb{R}^+$ and we have

$$\begin{aligned} \parallel y(t) \parallel e^{-\mu t} &\leq e^{-\mu t} \parallel S(t)y_0 \parallel + e^{-\mu t} \int_0^t \parallel S(t-s)g(y(s),s) \parallel ds \\ &\leq \psi(t) + \int_0^t Mk_1(s) \parallel y(s) \parallel e^{-\mu s} ds \end{aligned}$$

Applying the Gronwall lemma, we obtain

$$\parallel y(t) \parallel e^{-\mu t} \leq \psi(t) + \int_0^t k_1(s)\psi(s) \ \exp(M \int_s^t k_1(r)dr)ds$$

Consequently $\parallel y(t) \parallel$ is bounded by a function continuous, and the conclusion follows from 1. □

We have another result about the existence of a global solution [6], given below.

Theorem 3.7 *If* g *is monotonous and does not depend on* t, *then the system (3.1) admits a unique and global weak solution.*

The following result gives sufficient conditions for the weak solution of the system (3.1) be a strong one.

Theorem 3.8 ([6]) *If* $g : X \times [0, T] \to X$ *is of class* $\mathcal{C}^1$ *for every* $t \in [0, T]$, *then for all* $y_0 \in \mathcal{D}(A)$, *the weak solution of system (3.1) is a strong one.*

Proof First if g is of class $\mathcal{C}^1$ for all $t \in [0, T]$, then g is continuous with respect to t, uniformly continues on $[0, T]$, and Lipschitz with respect to the second variable. The Theorem 3.3 implies that the system (3.1) admits a unique weak solution.

We show that the weak solution is of class $\mathcal{C}^1$ on $[0, T]$. Let the functions

$$f(s) = \frac{\partial}{\partial y} g(s, y(s)) \text{ and}$$

$$h(t) = S(t)g(t, y(t)) - AS(t)y_0 + \int_0^t S(t-s)\frac{\partial}{\partial s} g(s, y(s))ds$$

By hypothesis and as $s \mapsto f(s)$, is continuous on $[0, T]$, then $h \in C([0, T]; X)$ and the function $f(s)y$ is continuous on $[0, T]$, uniformly lipschitz in y, ends the proof. □

In the previous result, we assume that $g : X \times [0, T] \to X$ is of class $\mathcal{C}^1$; which is a too strong assumption. In the result that follows, a weaker condition on g ensures the existence of a strong solution.

Theorem 3.9 ([6]) *If X is a reflexive Banach space and $g : X \times [0, T] \to X$ is Lipschitz with respect to the two variables,*

$$\|g(x_1, t_1) - g(x_2, t_2)\| \le k(|t_1 - t_2| + \|x_1 - x_2\|)$$

for all $t_1, t_2 \in [0, T]$ and for all $x_1, x_2 \in X$.

If for $y_0 \in \mathcal{D}(A)$, $y(.)$ is a weak solution of (3.1), then $y(.)$ is a strong solution.

The following result gives a characterization of the weak solution of the system (3.1) with respect to the weak topology of space X.

Theorem 3.10 ([6]) *If g is sequentially weakly continuous, ($y_n \rightharpoonup y \Longrightarrow g(y_n) \rightharpoonup g(y)$.) and if the system (3.1) has a weak solution $y(.)$ on $[0, T]$ and if there exist $M_0 > 0$ and $M_1 > 0$ such that*

$$||y_0|| \le M_0 \Longrightarrow ||y(t)|| \le M_1 \quad \textit{for all } t \in [0, T]$$

then for all $t \in [0, T]$, the application $y_0 \longrightarrow y(t)$ is sequentially weakly continuous.

Proof Let $y_n(t) = y(t; y_{0n})$ and $y(t) = y(t; y_0)$. As (y_{0n}) is bounded then $(y_n(t))$ is also bounded for each n and $t \in [0, T]$.

On the other hand, the image of any bounded by g remains bounded, thus there exists $C > 0$ such that $\|g(y_n(t))\| \le C$.

Let $t_n \to t$ in $[0, T]$, for $z \in X$, we set

$$a_r = \sup_{\|\phi\| \le 1, 0 \le s \le t} |\langle [S(t - s) - S(t_r - s)]\phi, z\rangle|$$

then $a_r \to 0$ when $r \to \infty$. Otherwise there exist two sequences (ϕ_μ) and (s_μ) such that $\phi_\mu \rightharpoonup \phi$, $s_\mu \to s$ in $[0, T]$, $t_\mu \to t$, and a number $\varepsilon > 0$ with

$$|\langle [S(t - s_\mu) - S(t_\mu - s_\mu)]\phi_\mu, z\rangle| \ge \varepsilon$$

But the application $(t, \phi) \mapsto S(t)\phi$ is weakly sequentially continuous on $\mathbb{R}^+ \times X$ (see [4]) then $S(t - s_\mu)\phi_\mu \rightharpoonup S(t - s)\phi$ and $S(t_\mu - s_\mu)\phi_\mu \rightharpoonup S(t - s)\phi$ and therefore $a_r \to 0$ when $r \to \infty$.

By the formula of the variation of the constant, there exist two positive constants C_1 and C_2 such that

$$
\begin{aligned}
|\langle y_n(t_r) - y_n(t), z\rangle| &\le |\langle [S(t_r) - S(t)]y_{0n}, z\rangle| \\
&\quad + \int_0^t |\langle [S(t_r - \tau) - S(t - \tau)]f(y_n(\tau)), z\rangle| d\tau \\
&\quad + \int_t^{t_r} |\langle S(t_r - \tau)g(y_n(\tau)), z\rangle| d\tau \\
&\le C_1 + C_2|t_r - t|
\end{aligned}
$$

Then $\langle y_n(t_r) - y_n(t), z\rangle \to 0$ uniformly when $r \to \infty$. Hence $y_n(t)$ is equicontinuous in $\mathcal{C}([0, T]; X_{\sigma(X,X')})$.

Since $(y_n(t))$ is uniformly bounded with respect to n then for all $t \in [0, T]$, $y_n(t)$ belongs to a bounded set of X provided with the metric deduced by the weak topology of X. Applying the Ascoli-Arzela theorem, there exist $\tilde{y} \in \mathcal{C}([0, T]; X_{\sigma(X,X')})$ and a sequence $(y_n(t))$ such that $y_n(t) \rightharpoonup \tilde{y}(t)$ uniformly on $[0, T]$ when $n \to \infty$. Since

$$
y_n(t) = S(t)y_{0n} + \int_0^t S(t - s)g(y_n(s))ds
$$

then for all $z \in X$, we have

$$
\langle y_n(t), z\rangle = \langle S(t)y_{0n}, z\rangle + \int_0^t \langle S(t - s)g(y_n(s)), z\rangle ds
$$

The function g is sequentially weakly continuous, by the Lebesgue dominated convergence theorem and when $n \to \infty$, for all $z \in X$, we get

$$
\langle \tilde{y}(t), z\rangle = \langle S(t)y_0, z\rangle + \int_0^t \langle S(t - s)g(\tilde{y}(s)), z\rangle ds
$$

We deduce that

$$
\tilde{y}(t) = S(t)y_0 + \int_0^t S(t - s)g(\tilde{y}(s))ds
$$

By the uniqueness of the solution of (3.1) we have $y(t) = \tilde{y}(t)$ on $[0, T]$.

To conclude, we suppose that $y_{0n} \rightharpoonup y_0$ and that $(y_n(t))$ does not converge to $y(t)$ in $\mathcal{C}([0, T]; X_{\sigma(X,X')})$, we can assume that there exists in $\mathcal{C}([0, T]; X_{\sigma(X,X')})$ a neighborhood $\mathcal{O}$ of $y(t)$ such that $(y_n(t)) \notin \mathcal{O}$. Consequently there exists a subsequence of $(y_n(t))$ which converges to $y(t)$, which is a contradiction. □

3.2 Stabilization of Infinite Dimensional Bilinear Systems

Let X be a real and separable Hilbert space and we consider a system described by the equation:

$$\begin{cases} \frac{dy(t)}{dt} &= Ay(t) + v(t)(By(t) + b) \\ y(0) &= y_0 \in X \end{cases} \tag{3.7}$$

where A generates a contraction semigroup $(S(t))_{t\geq 0}$, $B : X \to X$ a linear operator, $b \in X$ and $v(.)$ is a real valued control.

Definition 3.3 The system (3.7) is said to be globally strongly stabilizable if there exists a control $v(.) : [0, \infty[\to \mathbb{R}$ such that the system (3.7) admits a unique weak, global solution and $\lim_{t\to+\infty} \|y(t)\| = 0$, for all $y_0 \in X$.

For all $m \in \mathbb{N}$, let the operator

$$R_m(A) = \int_0^{\infty} e^{mt} S(t)dt$$

In case $b = 0$, we have the following result ([7]).

Proposition 3.1 *If B is a self adjoint and dissipative operator (or positive defined) then the control*

$$v_r(t) = \frac{r\langle By(t), y(t)\rangle}{1 + \epsilon\langle By(t), y(t)\rangle} \ \textit{with}\ \epsilon = signe\langle By(t), y(t)\rangle \ \textit{and}\ r > 0, \tag{3.8}$$

strongly stabilizes the system (3.7) if and only if

$$\{y \in X, \langle BS(t)y + b, S(t)y\rangle = 0, t \geq 0\} \cap G_u = \{0\}$$

where G_u is defined in Theorem 2.6

Proof We have $|v_r(t)| \leq r$ and according to the Theorem 3.10, the system (3.7) has a weak, unique and global solution. The operator

$$\widetilde{A} = -A + \frac{r\langle B., .\rangle}{1 + \epsilon\langle B., .\rangle} B$$

is maximum monotonous with $\mathcal{D}(\widetilde{A}) = \mathcal{D}(A)$. Therefore

$$\overline{\mathcal{D}(\widetilde{A})} = X$$

In addition, for all $\lambda > 0$ the operator $\lambda(Id + \widetilde{A})^{-1}$ is compact [8].

The operator $-\widetilde{A}$ generates a nonlinear C_0-semi group $\widetilde{S}(t)$, and the weak solution of the system (3.7) is given by

$$y(t) = \widetilde{S}(t)y_0 = S(t)y_0 + \int_0^t S(t-s)v_r(s)By(s)ds$$

moreover $\lim_{t\to+\infty} y(t) \in \omega(y_0) \subset \{y \in X | \ \ \|y\| = \sigma\}$ and $\sigma \leq \|y_0\|$.

It remains to show that $\omega(y_0) = \{0\}$. For this, let $y_0 \in \mathcal{D}(A)$, then $\omega(y_0) \subset \mathcal{D}(A)$, the weak solution $y(t)$ coincides with the strong solution of the system (3.7) and $y(t) \in \mathcal{D}(A)$ [7].

Let $y \in \omega(y_0)$. Since $\omega(y_0)$ is invariant by $\widetilde{S}(t)$, let the function

$$f(t) = \frac{1}{2}\|\widetilde{S}(t)y\|^2 = \frac{1}{2}\|z\|^2$$

Deriving the function f, we obtain

$$\frac{df}{dt} = \langle \widetilde{S}(t)y, A\widetilde{S}(t)y\rangle - \frac{r\langle B\widetilde{S}(t)y, \widetilde{S}(t)y\rangle^2}{1+\epsilon\langle B\widetilde{S}(t)y, \widetilde{S}(t)y\rangle} = 0 \tag{3.9}$$

Therefore

$$\langle \widetilde{S}(t)y, A\widetilde{S}(t)y\rangle = 0 \text{ and } r\langle B\widetilde{S}(t)y, \widetilde{S}(t)y\rangle = 0 \text{ for all } t \geq 0. \tag{3.10}$$

From the equation (3.9), we get $\widetilde{S}(t)y = S(t)z$ which implies that $\omega(y_0)$ is invariant by $S(t)$.

The equation (3.10) implies that for all $y \in \mathcal{D}(A)$, we have

$$\frac{d\|\widetilde{S}(t)y\|^2}{dt} = \langle \widetilde{S}(t)z, A\widetilde{S}(t)y\rangle = 0 \text{ and } \|\widetilde{S}(t)y\|^2 = \|y\|^2 \tag{3.11}$$

By the Theorem 2.6 there exist $y_1 \in X_u$ and $y_2 \in X_s$ such that $y = y_1 + y_2$. As X_u and X_s are invariant by $S(t)$ and $X = X_u \oplus X_s$, we have

$$\|y\|^2 = \|y_1\|^2 + \|y_2\|^2 \text{ and } \|S(t)y\|^2 = \|S(t)y_1\|^2 + \|S(t)y_2\|^2 \tag{3.12}$$

By (3.11) and (3.12) we obtain

$$\|S(t)y_2\|^2 = \|y_2\|^2 \tag{3.13}$$

For m integer large enough, $R_m(A)$ is compact and $y_2 \in X_s$, the Theorem 2.6 imply $\lim_{m\to+\infty} R_m(A)S(t)y_2 = 0$.

In the other hand as $S(t)$ is a contraction semigroup and $R_m(A)S(t) = S(t)R_m(A)$, then $mR_m(A)S(t)y_2$ converges uniformly to $S(t)y_2$ when $m \to +\infty$, which gives

$$\lim_{m\to+\infty}\lim_{t\to+\infty} mR_m(A)S(t)y_2 = \lim_{t\to+\infty}\lim_{m\to+\infty} mR_m(A)S(t)y_2$$

Therefore $\lim_{m\to+\infty} S(t)y_2 = 0$ and with (3.13) we obtain $y_2 = 0$, which implies $y = y_2 \in \{y \in X, \langle BS(t)y + b, S(t)y\rangle = 0, \ t \geq 0\} \cap G_u = \{0\}$.

Thus $\omega(y_0) = \{0\}$ for all $y_0 \in \mathcal{D}(A)$.

Let $y_0 \in X$, $\mathcal{D}(A)$ is dense in X, there exists a sequence $y_{0m} \in \mathcal{D}(A)$ such that $\lim\limits_{m \to +\infty} \|y_{0m} - y\| = 0$.

As $S(t)$ is a contraction semigroup, by triangular inequality we deduce that $\omega(y_0) = \{0\}$. Conversely, let $y_0 \in \{y \in X, \langle BS(t)y + b, S(t)y\rangle = 0, t \geq 0\} \cap G_u$ and $y(t)$ the solution of system (3.7) associated to y_0 given by

$$y(t) = S(t)y_0 + \int_0^t S(t-s)v_r(s)By(s)ds \tag{3.14}$$

As $\langle BS(t)y_0, S(t)y_0\rangle = 0$, for all $t \geq 0$, then $S(t)y_0$ is also solution of the integral equation (3.14), and by the uniqueness of the solution we have $y(t) = S(t)y_0$.

We have $y_0 \in G_u$ and the system (3.7) is strongly stable then $\|y_0\| = \|S(t)y_0\| = \|y(t)\|$. When $t \to \infty$ we obtain $y = 0$. □

In the case $b \neq 0$, we have the result (see [7]).

Proposition 3.2 *The control given by*

$$v_r(t) = \frac{r\langle b, y(t)\rangle}{1 + |\langle b, y(t)\rangle|} \tag{3.15}$$

strongly stabilizes the system (3.7) if and only if

$$\{y \in X, \langle BS(t)y + b, S(t)y\rangle = 0, t \geq 0\} \cap G_u = \{0\}$$

Proof (1) Sufficient condition: The operator $\widetilde{A} = -A + \dfrac{r\langle b, .\rangle}{1 + |\langle b, .\rangle|} b$ is maximum monotonous and $(\lambda Id - \widetilde{A})^{-1}$ is compact on X (see [9]). The operator $-\widetilde{A}$ generates a nonlinear C_0-semigroup $\widetilde{S}(t)$, and the weak solution of the system (3.7) is given by

$$y(t) = \widetilde{S}(t)y_0 = S(t)y_0 + \int_0^t S(t-s)v_r(s)By(s)ds \tag{3.16}$$

and we have

$$\omega(y_0) \subset \{y \in X \mid \| y \| = \sigma\} \quad \text{and} \quad \sigma \leq \| y_0 \|,$$

therefore $\lim\limits_{t \to \infty} y(t) \in \omega(y_0)$.

We will show that $\omega(y_0) = \{0\}$. Let $y_0 \in \mathcal{D}(A)$, then $\omega(y_0) \subset \mathcal{D}(A)$ and the weak solution of system (3.7) is a strong solution and $y(t) \in \mathcal{D}(A)$ (see [7]).

Since $\omega(y_0)$ is invariant by $\widetilde{S}(t)$. Deriving on $\omega(y_0)$ the function

$$f(t) = \frac{1}{2} \| \widetilde{S}(t)y \|^2 = \frac{1}{2} \| y \|^2$$

we obtain

$$\dot{f}(t) = \langle \widetilde{S}(t)y, \widetilde{A}\widetilde{S}(t)y\rangle - \frac{r\langle b\widetilde{S}(t)y, \widetilde{S}(t)y\rangle^2}{1+\epsilon\langle b\widetilde{S}(t)y, \widetilde{S}(t)y\rangle} = 0$$

thus, for all $t \geq 0$,

$$\langle \widetilde{S}(t)y, \widetilde{A}\widetilde{S}(t)y\rangle = 0 \ \text{ and } r\langle b\widetilde{S}(t)y, \widetilde{S}(t)y\rangle = 0 \tag{3.17}$$

As $\omega(y_0)$ is invariant by $S(t)$, for all $y \in \mathcal{D}(A)$ we have

$$\begin{cases} \dfrac{d \parallel S(t)y \parallel^2}{dt} = 2\langle S(t)y, AS(t)y\rangle = 0 \\ \quad\quad\text{and} \\ \parallel S(t)y \parallel^2 = \parallel y \parallel^2 \end{cases} \tag{3.18}$$

There exist $y_1 \in X_u$ and $y_2 \in X_s$ such that $y = y_1 + y_2$, thus $\parallel S(t)y \parallel^2 = \parallel y_2 \parallel^2$.

For m integer large enough, $R_m(A)$ is compact and since $y_2 \in X_s$, then $R_m(A)S(t)y_2$ tends to 0 when $m \to \infty$.

But $R_m(A)S(t) = S(t)R_m(A)$ and $S(t)$ is a contraction semigroup, so $mR_m(A)S(t)y_2$ converges uniformly to $S(t)y_2$ when $m \to \infty$ and therefore $\lim_{m\to\infty} S(t)y_2 = 0$ and as $\parallel S(t)y_2 \parallel = \parallel y_2 \parallel$, we deduce $y_2 = 0$. This implies that

$$y = y_2 \in \{y \in X, \langle BS(t)y + b, S(t)y\rangle = 0, t \geq 0\} \cap G_u = \{0\}$$

thus $\omega(y_0) = \{0\}$ for all $y_0 \in \mathcal{D}(A)$ and by density, it is also true for all $y_0 \in X$.

(2) Necessary Condition:
Let $y_0 \in \{y \in X, \langle BS(t)y + b, S(t)y\rangle = 0,\ t \geq 0\} \cap G_u$ and $y(t)$ the solution of system (3.7) associated to y_0, given by

$$y(t) = S(t)y_0 + \int_0^t S(t-s)v_r(s)By(s)ds$$

As $\langle b, S(t)y_0\rangle = 0$ for all $t \geq 0$, therefore $\widetilde{S}(t)y_0$ is also a solution of (3.7), and by the uniqueness of the solution we have $y(t) = S(t)y_0$, and since $y_0 \in G_u$, the system (3.7) is strongly stabilizable, then we have

$$\parallel y_0 \parallel = \parallel S(t)y_0 \parallel = \parallel y(t) \parallel \to 0 \text{ when } t \to \infty$$

which completes the proof. □

We consider the system (3.7) and assume that A is dissipative, (i.e $\langle Ay, y\rangle \leq 0$, for all $y \in \mathcal{D}(A)$), that A is closed and $\mathcal{D}(A) = \mathcal{D}(A^*)$. Furthermore we assume that $B : X \longrightarrow X$ is anti adjoint ($\langle By, y\rangle = -\langle y, By\rangle$).

Let the control

$$v(t) = -\langle b, y(t)\rangle \tag{3.19}$$

We show that the system (3.7) admits a unique and global weak solution. We need the following lemma.

Lemma 3.1 *1. If $y_n(t)$ the solution of the system (3.7) associated with the initial condition y_{0n} and if $y_{0n} \to y_0$ when $n \to +\infty$ then $y_n(t) \to y(t)$ when $n \to +\infty$ on $[0, T]$, the solution of (3.7) associated with the initial condition y_0.*
2. The system (3.7) admits a unique and global weak solution. If $y_0 \in \mathcal{D}(A)$ then both weak and strong solutions of the system (3.7) coincide, and we have:

$$\frac{d}{dt}\|y(t)\|^2 = 2(\langle Ay(t), y(t)\rangle - \langle b, y(t)\rangle^2) \leq 0 \tag{3.20}$$

We deduce that $\|y(t)\ |^2 \leq \|y_0\|^2$, so the solution is bounded.

Proof 1. This point follows from the fact that $y \mapsto -\langle b, y\rangle(By + b)$ is locally Lipschitz and from the Gronwall lemma.
2. For $y_0 \in X$, there exists $y_{0n} \in \mathcal{D}(\mathcal{A})$ such that $y_{0n} \to y_0$ when $n \to +\infty$, and in addition we have $\|y_n(t)\| \leq \|y_{0n}\|$, for all $t \in [0, T]$ and $n \in \mathbb{N}$, according to point 1, we have $\|y(t)\| \leq \|y_0\|$, for all $t \in [0, T]$.
Therefore the solution of the system (3.7) is bounded for any initial condition $y_0 \in X$, consequently the system (3.7) admits a weak and global solution. □

The following result concerns sufficient conditions for the weak convergence of the solution of system (3.7). We need the following lemma.

Lemma 3.2 *We have the following two results:*
1. $\lim_{t\to+\infty} \langle b, y(t)\rangle = 0$.
2. For all $t \geq 0$, $S(t)\omega_w(y_0) \subseteq \omega_w(y_0)$.

Proof 1. For $y_0 \in \mathcal{D}(A)$, integrating (3.20) we have

$$\|y(t)\|^2 = \|y(0)\|^2 + \int_0^t \langle Ay(s), y(s)\rangle ds - \int_0^t \langle b, y(s)\rangle^2 ds \text{ for all } t \geq 0$$

which gives $\int_0^t \langle b, y(s)\rangle^2 ds \leq \|y_0\|^2$, for all $t \geq 0$.

This gives $\int_0^{+\infty} \langle b, y(s)\rangle^2 ds < +\infty$, and moreover $\frac{d}{dt}\langle b, y(t)\rangle$ is bounded.

Hence $\lim_{t\to+\infty} \langle b, y(t)\rangle = 0$.

Now for $y_0 \in X$ there exists a sequence $(y_{0n}) \in \mathcal{D}(A)$ such that $y_{0n} \to y_0$ when $n \to +\infty$. From the above, for any $n \geq 0$ we have

$$\int_0^{+\infty} \langle b, y_n(s)\rangle^2 ds \leq \|y_{0n}\|^2$$

Fatou lemma gives $\int_0^{\infty} \langle b, y(s)\rangle^2 ds \leq \|y_0\|^2$.

We also verify that $\frac{d}{dt}\langle b, y(t)\rangle$ is bounded and therefore $\lim_{t\to+\infty} \langle b, y(t)\rangle = 0$.

2. We show that $\omega_w(y_0) = \{0\}$.

Let $\psi \in \omega(y_0)$, and $t_n \longrightarrow +\infty$ such that $y(t_n) \rightharpoonup \psi$ when $n \longrightarrow +\infty$. Then we have

$$y(t + t_n) = S(t)y(t_n) - \int_0^t S(t-s)\langle b, y(s+t_n)\rangle (By(s+t_n) + b)ds$$

It follows

$$|\langle \phi, y(t+t_n)\rangle - \langle S(t)\psi, \phi\rangle| \leq |\langle \phi, y(t+t_n)\rangle - \langle S(t)y(t_n), \phi\rangle|$$
$$+ |\langle S(t)y(t_n), \phi\rangle - \langle S(t)\psi, \phi\rangle|$$

We also have

$$|\langle \phi, y(t+t_n)\rangle - \langle S(t)y(t_n), \phi\rangle| \leq a|\int_0^t \langle b, y(s+t_n)\rangle ds|$$

or when $n \to +\infty$, $\langle b, y(s+t_n)\rangle \to 0$ and $|\langle S(t)y(t_n), \phi\rangle - \langle S(t)\psi, \phi\rangle| \to 0$.

Therefore $|\langle \phi, y(t+t_n)\rangle - \langle S(t)y(t_n), \phi\rangle| \to 0$ when $n \to +\infty$ which completes the proof. □

The following theorem characterizes the weak stabilization of the system (3.7).

Theorem 3.11 *We assume that $b \in \mathcal{D}((A^*)^k)$, for all $k \in \mathbb{N}$. If, moreover, the subspace generated by $\{b, A^*b, \cdots, (A^*)^k b, \cdots\}$ is dense in X, then $y(t) \rightharpoonup 0$ when $t \to +\infty$.*

Proof We have $\omega_w(y_0) = \{0\}$. Indeed, let $\psi \in \omega_w(y_0)$, by definition we have

$$\frac{d^k}{dt^k}\langle S(t)\psi, b\rangle = \langle S(t)\psi, (A^*)^k b\rangle = 0, \text{ for all } t \geq 0 \text{ and } k \in \mathbb{N}.$$

In particular, for $t = 0$, we have $\langle \psi, (A^*)^k b\rangle = 0$ for all $k \in \mathbb{N}$ and therefore $\omega_w(y_0) = \{0\}$. □

Now we consider the autonomous bilinear system described by the equation

$$\begin{cases} \frac{dy(t)}{dt} = Ay(t) + u(t)By(t) \\ y(0) = y_0 \in X \end{cases} \tag{3.21}$$

where A is the generator of a contraction semigroup $(S(t))$, and we consider the control,

$$u(t) = -\langle Ky(t), y(t)\rangle \tag{3.22}$$

where $K : X \to X$ a continuous linear operator. The following result gives a sufficient condition ensuring the existence of a global solution of the system (3.21) excited by the control (3.22).

Proposition 3.3 *If the condition*

$$Re(\langle Ky, y\rangle\langle y, By\rangle) \geq 0 \tag{3.23}$$

is satisfied for all $y \in X$, *then the solution* $y(t)$ *of the system (3.21) is global. In addition* $y(t)$ *depends continuously on* y_0.

Proof As the application $y \longrightarrow - < Ky, y > By$ is monotonous, according to the Theorem 3.7, the system (3.21) admits a unique weak solution $y \in \mathcal{C}([0, t_{\max}[; X)$ defined on a maximum interval $[0, t_{\max}[$. In addition, the mapping $y_0 \longrightarrow y(t)$ is continuous in $\mathcal{C}([0, t_{\max}[; \ X)$.

To have $t_{\max} = +\infty$, it remains to show that $y(t)$ is bounded on $[0, t_{\max}[$. Let $T \in [0, t_{\max}]$, on $[0, T]$ we consider the function

$$F(t) = - < Dy(t), y(t) > By(t)$$

Let a sequence of functions $F_n \in \mathcal{C}^1([0, T]; X)$ such that $F_n \longrightarrow F$ in $C([0, T]; X)$. For $y_0 \in X$, there exists $y_{0n} \in \mathcal{D}(A)$ such that $y_{0n} \longrightarrow y_0$ and we consider the system

$$\begin{cases} \dfrac{dy_n(t)}{dt} = Ay_n(t) + F_n(t) \\ \\ y_n(0) = y_{0n} \end{cases} \tag{3.24}$$

Since $S(t)$ is a contraction semigroup, we have

$$\begin{aligned} ||y_n(t) - y(t)|| &\leq ||y_{0n} - y_0|| + \int_0^T ||F_n(t) - F(t)||dt \\ &\leq ||y_{0n} - y_0|| + T \sup_{t\in[0,T]} ||F_n(t) - F(t)||, \ \forall t \ \in [0, T] \end{aligned}$$

It follows that $y_n \longrightarrow y$ in $\mathcal{C}([0, T]; X)$.

As $y_n(t) \in \mathcal{D}(A)$, for any $t \in [0, T]$ we have

$$\frac{d}{dt}||y_n(t)||^2 = 2Re(< Ay_n(t), y_n(t) >) + 2Re(< F_n(t), y_n(t) >)$$

Since the operator A is dissipative, we get

$$||y_n(T)||^2 - ||y_n(0)||^2 \leq 2\int_0^T Re(< F_n(t), y_n(t) >)dt$$

Finally when $n \longrightarrow +\infty$, we obtain

$$||y(T)||^2 \leq ||y_0||^2 + 2\int_0^T Re(< F(t), y(t) >)dt$$

and using (3.22), we deduce that $||y(T)|| \leq ||y_0||$. So when T tends to $t_{\max}$; the solution $y(t)$ remains bounded, and therefore it is global. □

For weak stabilization, we have the following result [1].

Proposition 3.4 *Suppose that the operator B is compact and that the condition*

$$\{\langle BS(t)y, S(t)y\rangle = 0\} \Rightarrow y = 0 \tag{3.25}$$

is verified, then the control (3.22) weakly stabilizes the system (3.21).

Proof We set $F(y) = -\langle By, y\rangle By$, using the Proposition 3.3, the system (3.21) admits a weak and global solution.

By Theorem 2.4 (see [1]), $\omega_w(y_0) \neq \emptyset$, for all $y_0 \in X$ and $\psi \in \omega_w(y_0)$, we have $\langle S(t)\psi, F(S(t)\psi)\rangle = 0$ for all $t \geq 0$ and by (3.25) we obtain $\psi = 0$. □

For the strong stability of the system (3.21) we have the following result [8] .

Theorem 3.12 *If A is of compact resolvent and B self-adjoint and verify (3.25), if in addition one of the following two conditions is satisfied*

$$\langle By, y\rangle \geq 0 \tag{3.26}$$

or

$$\langle By, y\rangle \leq 0 \tag{3.27}$$

for all $y \in X$, then the control (3.22) strongly stabilizes the system (3.21).

The following result (see [10]) gives conditions for the strong stabilization of the system (3.21).

Theorem 3.13 *If $S(t)$ is a contraction semigroup and B is a self-adjoint operator verifying (3.26), if in addition, the condition*

$$\|y\|^2 \leq \int_0^T \langle BS(t)y, S(t)y\rangle dt \tag{3.28}$$

is satisfied for some $T > 0$, then the system (3.21) is strongly stabilizable and we have the estimation

$$\|y(t)\| = O(\frac{1}{\sqrt{t}}) \quad when \quad t \to +\infty$$

Proof We consider the system

$$\begin{cases} \dfrac{dy(t)}{dt} = Ay(t) - \langle By(t), y(t)\rangle By(t) \\ y(0) \quad = 0 \end{cases} \tag{3.29}$$

Since $S(t)$ is a contraction semigroup we have

$$\begin{aligned} \|y(t)\| &\le \|B^{\frac{1}{2}}\| \int_0^T |\langle By(s), y(s)\rangle| \|B^{\frac{1}{2}} y(s)\| ds \\ &\le \|B^{\frac{1}{2}}\| \int_0^T \|B^{\frac{1}{2}} y(s)\|^3 \, ds \end{aligned}$$

It gives

$$\|y(t)\|^2 \le T^{\frac{1}{2}} \|B^{\frac{1}{2}}\|^2 (\int_0^T \|B^{\frac{1}{2}} y(s)\|^4 ds)^{\frac{3}{2}}$$

and with (3.28), there exists $c > 0$ such that

$$\begin{aligned} \|y(0)\|^2 &\le 2(\|z(0)\|^2 + \|y(0) - z(0)\|^2) \\ &\le c[(\int_0^T \|B^{\frac{1}{2}} y(s)\|^4 ds)^{\frac{3}{2}} + (\int_0^T \|B^{\frac{1}{2}} y(s)\|^4 ds)^{\frac{1}{2}}] \end{aligned}$$

With the same approach as for (3.20), we obtain

$$\|y(T)\|^2 + 2 \int_0^T \langle By(s), y(s)\rangle^2 ds \le \|y(0)\|^2$$

and therefore we have

$$\begin{aligned} \|y(T)\|^2 \le c[(\int_0^T \|B^{\frac{1}{2}} y(s)\|^4 ds)^{\frac{3}{2}} &+ (\int_0^T \|B^{\frac{1}{2}} y(s)\|^4 ds)^{\frac{1}{2}} \\ &+ \int_0^T \|B^{\frac{1}{2}} y(s)\|^4 ds] \end{aligned} \tag{3.30}$$

Il follows that

$$\|y(T)\|^2 \le cf(\|y_0\|^2 - \|y(T)\|^2) \tag{3.31}$$

with $f(t) = t^{\frac{1}{2}} + t + t^{\frac{3}{2}}$. Thus

$$\frac{1}{c}\|y(T)\|^2 + f^{-1}(\|y(T)\|^2) \le \|y(0)\|^2 \tag{3.32}$$

and moreover, we have $f^{-1}(t) \sim t^2$ when $t \to 0$.

We set $h(t) = \|y(t)\|^2$ and we use the same techniques on the intervals of type $[kT, (k+1)T]$, $k \in \mathbb{N}^*$, we obtain

$$\frac{1}{c}h(k(T+1)) + f^{-1}(h(k(T+1))) \leq h(kT) \tag{3.33}$$

Now for $k \in \mathbb{N}$ we set $s_k = h(kT)$.

By Lemma 2.4 with (3.33) we obtain $s_k \leq S(k)s_0$, .

For $t > 0$, we carry out the Euclidean division of t by T we obtain $t = qT + r$ with $q \in \mathbb{R}$ and $0 \leq r < T$.

Since the functions $h(t)$ and $S(t)\xi$ are decreasing, then

$$h(t) \leq h(qT) \leq S(\frac{t-r}{T})h(0) \leq S(\frac{t}{T} - 1)S(0)$$

As $f_1(t + f_0(t)) = f_0(t)$, we have $\lim_{t \to +\infty} f_1(t) = 0$.

The function $f_3(t) = \displaystyle\int_t^{\xi} \frac{ds}{f_1(s)}$, $t > 0$ is decreasing with $f_3(\xi) = 0$ and $f_3(0^+) = +\infty$. We deduce that $f_3([0, +\infty[) = [0, +\infty[$ and that the solution of system (2.38) is $U(t) = f_3^{-1}(t)$, with $\lim_{t \to +\infty} U(t) = 0$.

Let $0 < \varepsilon < 1$, there exists $\eta > 0$ such that if $t < \eta$ we have: $|f_1(t) - ct^2| < \varepsilon ct^2$. In addition there exists $t_\varepsilon > 0$ such that for all $t > t_\varepsilon$ we have $0 < U(t) < \eta$. For $t \geq t_\varepsilon$ we have $-f_1(U(t)) \leq c(\varepsilon - 1)U(t)^2$. thus

$$\frac{dU(t)}{dt} + c\,(\varepsilon - 1)U(t)^2 \leq 0$$

Integrating over $[t_\varepsilon, t]$ we obtain the estimation $U(t) = O(\frac{1}{t})$ and so $h(t) = O(\frac{1}{t})$. □

In the following we present an extension of the space state decomposition method to the stabilization of bilinear system.

We consider the system (3.21) and we assume that X can be decomposed as in (2.15), with $dim(X_u) < +\infty$,

Theorem 3.14 ([11]) *If the semigroup $S_s(t)$ verify (2.32) and B is a compact operator and verify the condition*

$$[\langle BS(t)y, S(t)y\rangle = 0] \Rightarrow y = 0 \tag{3.34}$$

then the control

$$u(t) = r\langle By(t), y(t)\rangle \tag{3.35}$$

strongly stabilizes the system (3.21).

Proof 1. Existence of a solution.
The control (3.35) ensures the existence of a weak solution of the system (3.21) defined on a maximum interval $[0, t_{\max}[$, $y(t)$ given by

$$y(t) = S(t)y_0 + \int_0^t r\langle By(s), y(s)\rangle S(t-s)By(s)ds \tag{3.36}$$

Let the function $F(y) = r\langle By, y\rangle By$. Since $S(t)$ is a contraction semigroup, for all $y_0 \in \mathcal{D}(A)$ we have

$$\frac{d}{dt}\|y(t)\|^2 \le -2\langle F(y(t)), y(t)\rangle \le \|y_0\|^2$$

Furthermore considering (3.36) and by the fact that $S(t)$ is a contraction semigroup and Gronwall's lemma, the mapping $y_0 \to y(t)$ is continuous. Therefore, for all $y_0 \in X$, $\|y(t)\| \le \|y_0\|$, so the solution $y(t)$ of the system (3.21) is global.
2. The system (3.21) is strongly stabilizable. Indeed, as X is reflexive and $\|y(t)\| \le \|y_0\|$, then there exists a sequence $t_n \to +\infty$, such that $y(t_n) \rightharpoonup y \in X$ when $n \to +\infty$.

In addition for $T > 0$, there exists a constant $c(T, \|y_0\|) > 0$ such that

$$\int_0^T |\langle BS(t)y_0, S(t)y_0\rangle|dt \le c(T, \|y_0\|)\|y_0\| \left(\int_0^T |\langle F(y(t)), y(t)\rangle|dt\right)^2 \tag{3.37}$$

We take $y_0 = y(t_n)$ in (3.37), with dominated convergence Lebegue theorem we get $\langle S(t)By, S(t)y\ rangle = 0$, for all $t \ge 0$, and with (3.34), it comes $y = 0$. Hence $y(t) \rightharpoonup 0$ when $t \to +\infty$. Finally since $dim X_u < +\infty$, $y_u(t) \to 0$ when $t \to +\infty$.

It remains to study the behavior of $y_s(t)$. For that, with (2.32), there exists $\alpha > 0$ such that, for all $t \ge 0$,

$$\|S_s(t)y_s\| \le \alpha e^{-\delta t} \tag{3.38}$$

Thus, for all $0 \le \tilde{t} \le t$, we have

$$\|y_s(t)\| \le \alpha e^{-\delta(t-\tilde{t})}y_s(\tilde{t}) + \alpha \int_{\tilde{t}}^t e^{-\delta(t-\tau)}\|F(y(\tau))\|d\tau \tag{3.39}$$

Let $\epsilon > 0$ and $\tilde{t} > 0$, since F is sequentially continuous then $\|F(y(t))\| < \epsilon$ and by (3.39) we obtain

$$\begin{aligned}\|y_s(t)\| &\le \alpha e^{-\delta(t-\tilde{t})}y_s(\tilde{t}) + \alpha\epsilon \int_{\tilde{t}}^t e^{-\delta(t-\tau)}d\tau \\ &\le \alpha e^{-\delta(t-\tilde{t})}y_s(\tilde{t}) + \frac{\alpha\epsilon}{\delta}\end{aligned}$$

Thus $y_s(t) \to 0$ when $t \to +\infty$. As $y(t) = y_s(t) + y_u(t)$, the system (3.21) is strongly stabilizable. □

Remark 3.2 We obtain the same result of the above theorem with the control

$$u(t) = \begin{cases} \dfrac{r}{\|y(t)\|^2}\langle By(t), y(t)\rangle & \text{if } y(t) \neq 0 \\ 0 & \text{otherwise} \end{cases} \tag{3.40}$$

For exponential stabilization we have the following result [11].

Theorem 3.15 *If $S_u(t)$ is a C_0 -semigroup of isometries and $S_s(t)$ verify (2.32). If the condition*

$$[\langle B_u S_u(t) y_u, S_u(t) y_u\rangle = 0] \Rightarrow y_u = 0 \tag{3.41}$$

is verified, then the control

$$u(t) = \begin{cases} \dfrac{r}{\|y_u(t)\|^2}\langle By_u(t), y_u(t)\rangle & \textit{if } y_u(t) \neq 0 \\ 0 & \textit{otherwise} \end{cases} \tag{3.42}$$

exponentially stabilizes the system (3.21).

Proof We set

$$F(y_u(t)) = \begin{cases} \dfrac{r}{\|y_u(t)\|^2}\langle By_u(t), y_u(t)\rangle & \text{if } y_u(t) \neq 0 \\ 0 & \text{otherwise} \end{cases} \tag{3.43}$$

then the system

$$\begin{cases} \dfrac{dy_u(t)}{dt} = A_u y_u(t) + F(y_u(t)) B_u y_u(t) \\ y_u(0) = y_{0u} \end{cases} \tag{3.44}$$

admits a unique and global weak solution $y_u(t) = \widetilde{S}_u(t) y_{0u}$ with $\widetilde{S}_u(t)$ is a non linear semigroup on X_u.

For $m \in \mathbb{N}$, we integrate between m and $m+1$ the relation

$$\frac{d}{dt}\|y_u(t)\|^2 \leq -2|F(y_u(t))|^2 \|y_u(t)\|^2 \tag{3.45}$$

we obtain

$$\|y_u(m+1)\|^2 - \|y_u(m)\|^2 \leq -2\int_m^{m+1} |F(y_u(\tau))|^2 \|y_u(_t au)\|^2 d\tau$$

and since $\|y_u(t)\|$ is decreasing, we have

$$\|y_u(m+1)\|^2 - \|y_u(m)\|^2 \leq -2\int_m^{m+1} |F(y_u(\tau))|^2 \parallel y_u(\tau) \parallel^2 d\tau \qquad (3.46)$$

moreover

$$\|y_u(m+1)\|^2 - \|y_u(m)\|^2 \leq -2 \parallel y_u(m+1) \parallel^2 \int_m^{m+1} |F(y_u(\tau))|^2 d\tau.$$

We set $\eta = 2 \inf_{\|y_{0u}\|=1} \int_0^1 |F(\widetilde{S}_u(\tau)y_{0u})|^2 d\tau \geq 0.$

Since B_u is linear, $F(\alpha y_u) = F(y_u)$, for all $\alpha \in \mathbb{C}$, $y_u \in X_u$.

As the solution of the system (3.44) is unique, then for all $t \geq 0$,

$$\widetilde{S}_u(t)(\alpha y_u) = \alpha \widetilde{S}_u(t) y_u$$

Which implies that for all $y_{0u} \in X_u$ such that $y_{0u} \neq 0$ we have

$$\int_0^1 |F(\widetilde{S}_u(\tau)y_{0u})|^2 d\tau \geq \frac{1}{2}\eta$$

But $\|\widetilde{S}_u(t)y_{0u}\| \leq \|y_{0u}\|$ then $\int_m^{m+1} |F(\widetilde{S}_u(\tau)y_{0u})|^2 d\tau \geq \frac{1}{2}\eta$, for all $m \in \mathbb{N}$.

Thus by (3.46) it follows

$$\|y_u(m+1)\|^2 - \|y_u(m)\|^2 \leq -\eta \|y_u(m+1)\|^2$$

and therefore we have $\|y_u(m)\| \leq \dfrac{\|y_{0u}\|}{(1+\eta)^{\frac{m}{2}}}$.

As $\|y_u(t)\|$ is decreasing we get

$$\|y_u(t)\| \leq \|y_u(m)\| \leq e^{\frac{ln(1+\eta)}{2}} e^{-\frac{ln(1+\eta)}{2}t} \|y_{0u}\|$$

We have $\eta > 0$, otherwise $\eta = 0$ then there exists $y_u \in X_u$ with $\|y_u\| = 1$ and $\int_0^1 |F(\widetilde{S}_u(\tau)y_u)|^2 d\tau = 0$ since the mapping $y_{0u} \mapsto \int_0^1 |F(\widetilde{S}_u(\tau)y_{0u})|^2 d\tau$ is continuous and $\dim X_u < +\infty$.

Thus $F(\widetilde{S}_u(t)y_u) = 0$, for all $0 \leq t \leq 1$. therefore $\widetilde{S}_u(t)y_u = S_u(t)y_u$,for all $0 \leq t \leq 1$. Thus $< B_u S_u(t)y_u, S_u(t)y_u\rangle = 0$ for all $0 \leq t \leq 1$.

We deduce that (see [12]), for all $y_u \in X_u$,

$$\int_0^1 | < B_u S_u(t) y_u, S_u(t) y_u > | dt \geq \eta \| y_u \|^2$$

Consequently $y_u = 0$, contradiction. We conclude that $y_u(t) \to 0$ exponentially when $t \to +\infty$.

Let us show that $y_s(t) \to 0$ exponentially when $t \to +\infty$.

First we show that $y_s(t)$ is a weak, unique and global solution. The control (3.42) ensures the existence of a weak and unique solution of the system (3.21) over a maximum interval $[0, t_{\max}[$, given by

$$y_s(t) = S_s(t) y_{0s} + \int_0^t u(\tau) S_s(t-\tau) B_s y_s(\tau) d\tau \text{ for all } t \in [0, t_{\max}[\tag{3.47}$$

By (3.38) and (3.47), for all $t \in [0, t_{\max}[$, there exists β such that

$$\| y_s(t) \| \leq \beta \, e^{-\delta t} \| y_{0s} \| + \alpha \| B \| \int_0^t e^{-\delta(t-\tau)} |u(\tau)| \| y_s(\tau) \| d\tau$$

and thus

$$z(t) \leq \beta \| y_{0s} \| + r\beta \| B \|^2 \int_0^t z(\tau) d\tau$$

with $z(t) = \| y_s(t) \| \, e^{\delta t}$.

By Gronwall's lemma we get $z(t) \leq \beta \| y_{0s} \| e^{r\alpha \| B \|^2 t}$. This means that for all $t \leq t_{\max}$,

$$\| y_s(t) \| \leq \beta \| y_{0s} \| e^{(r\beta \| B \|^2 - \delta) t} \tag{3.48}$$

For r such that $r\beta \| B \|^2 < \delta$, $y_s(t)$ is bounded therefore $y_s(t)$ is a global solution of the system (3.21). In addition, the estimate (3.48) is verified for all $t \geq 0$, which allows to conclude. □

3.3 Stabilization of Infinite Dimensional Semilinear Systems

This section concerns existence and uniqueness of the solution of semi linear systems and stabilization results.

3.3.1 Existence and Uniqueness Results

Let us consider the system

$$\begin{cases} \dfrac{dy(t)}{dt} = Ay(t) + f(y(t), t) \\ y(0) = y_0 \in X \end{cases} \tag{3.49}$$

where $A : \mathcal{D}(A) \subset X \longrightarrow X$ is a linear operator that generates a C_0−semigroup $S(t), t \geq 0$ X, and $f : X \times \mathbb{R}^+ \longrightarrow X$ a given function.

Definition 3.4
1. A function $y : [0, t_1] \longrightarrow X$ is said a strong solution of (3.49) on $[0, t_1]$ if the function $y(.)$ is of class C^1 on $]0, t_1[$ and verify (3.49).

2. A function $y \in C([0, t_1]; X)$ is said a weak solution to (3.49) on $[0, t_1]$ if $f(y(.), .) \in L^1(0, t_1; X)$, for all $\varphi \in \mathcal{D}(A^*)$, the function $< y(t), \varphi >$ is absolutely continuous on $[0, t_1]$ and verify

$$\frac{d}{dt} < y(t), \varphi >=< y(t), A^*\varphi > + < f(y(t), t), \varphi > \quad p.p \text{ on } [0, t_1]$$

The following result gives a characterization of the weak solution to the system (3.49).

Theorem 3.16 ([6]) *A function $y : [0, t_1] \longrightarrow X$ is a weak solution of system (3.49) on $[0, t_1]$ if and only if $f(y(.), .) \in L^1(0, t_1; X)$ and y verify the relationship (called formula of the constant variation)*

$$y(t) = S(t)y_0 + \int_0^t S(t-s)f(y(s), s)ds \quad p.p \text{ on } [0, t_1]$$

The following result gives sufficient conditions for the existence and uniqueness of a weak solution for system (3.49) as well as its dependence on the initial state.

Theorem 3.17 ([6]) *Suppose that f is continuous on $[0, t_1]$, and uniformly Lipschitz on X. For each $y_0 \in X$, the system (3.49) has a weak solution $y \in C([0, t_1]; X)$.*

In addition, the correspondence $y_0 \longrightarrow y(.)$ is continuous from X to $C([0, t_1]; X)$.

Previous assumptions may be relaxed and lead to local existence results.

Theorem 3.18 ([1, 6]) *Suppose that f is continuous with respect to t on $I\!R^+$, uniformly continuous over all bounded intervals and locally Lipschitz with respect to y.*

Then for all $y_0 \in X$, the system (3.49) has a weak solution $y(.)$ on $[0, t_{max}[$, and if $y(t)$ is bounded with respect to t then this solution is global ($t_{max} = +\infty$).

If in addition, f does not depend on t and monotonous, then for $t_1 \in]0, t_{max}[$ the mapping $y_0 \longrightarrow y(.)$ is continuous from X into $C([0, t_1]; X)$.

The above result gives information about the continuity of the application $y_0 \longrightarrow y(.)$ for the topology defined by the norm of X. The following theorem gives similar results for the weak topology of X.

Theorem 3.19 ([1]) *Suppose that f is sequentially weakly continuous, $(y_n \rightharpoonup y \Longrightarrow f(y_n) \rightharpoonup f(y))$ and that (3.49) has a weak solution $y(.)$ on $[0, T]$.*

If there exist $M_0, M_1 > 0$; $||y_0|| \leq M_0 \Longrightarrow ||y(t)|| \leq M_1, \forall t \in [0, T]$, then for all $t \in [0, T]$; the mapping $y_0 \longrightarrow y(t)$ is sequentially weakly continuous.

The result that follows gives a sufficient conditions which allows regularity of the weak solution of the system (3.49).

Theorem 3.20 ([6]) *Suppose that f is continuously differentiable in t on $\mathbb{R}^+$, uniformly over each bounded intervals and locally Lipschitz in x, then for each $y_0 \in \mathcal{D}(A)$ the weak solution $y(.)$ of the system (3.49) is a strong one.*

For stabilization problems, we need the global existence results that we give in the the following result

Theorem 3.21 ([6]) *Suppose that A generates a compact C_0−semigroup and that f is continuous and that the image of a bounded set of $\mathbb{R}^+ \times X$ be a bounded one of X.*

Then for each $y_0 \in X$, the system (3.49) has a global solution $y(.)$ in one of the following cases:

1. There exists a function $k : \mathbb{R}^+ \times X \longrightarrow]0, +\infty[$, such that $||y(t)|| \leq k(t), \forall t \in [0, t_1[$.

2. There exist two functions locally integrable $k_1, k_2 : \mathbb{R}^+ \times X \longrightarrow]0, +\infty[$, such $||f(t, y)|| \leq k_1(t)||y|| + k_2(t)$, for all $t \in [0, t_1[, \;\; y \in X$.

Let us consider the nonlinear autonomous evolution system

$$\begin{cases} \frac{dy(t)}{dt} = Ay(t) + f(y(t)) \\ y(0) = y_0 \in \mathcal{D}(A) \end{cases} \tag{3.50}$$

Definition 3.5 $y_1 \in \mathcal{D}(A)$ is an equilibrium state if $Ay_1 + f(y_1) = 0$

To give an interpretation to the previous definition, we evoke the evolution of the temperature $y(t)$ of a solid, which intuitively becomes stationary when t tends to infinity or again $\dfrac{dy}{dt} = 0$. We are interested in the stability of a given equilibrium state y_1.

Definition 3.6 1. An equilibrium state y_1 is stable in the Lyapunov sense if

$$\forall \varepsilon > 0, \text{ there exists } \eta > 0 \text{ such that } ||y_0 - y_1|| < \eta \Longrightarrow ||y(t) - y_1|| < \varepsilon, \; \forall t > 0.$$

2. An equilibrium state y_1 is exponentially (resp. strongly, weakly) stable for the system (3.50) if it is stable in the Lyapunov sense and if there exists $R > 0$, such that $||y_0 - y_1|| < R$ implies that $||y(t) - y_1||$ tends to 0 exponentially (resp. strongly, weakly) when $t \longrightarrow \infty$.

If this convergence holds for any initial state, the stability is said to be global.

Remark 3.3 1. An equilibrium state is stable in the Lyapunov sense if a weak disturbance in the initial conditions leads to a slight disturbance of the later trajectory.

2. The second definition consists in requiring in addition that the system returns asymptotically to the equilibrium state.

The stability, as it was defined, is characterized in a different way according to the description given on the system. However, Lyapunov's method remains the general approach allowing the study of the stability of any system without having to determine analytically (spectral approach) or numerically the trajectories. The original idea of this method comes from mechanic where we involve the concept of total energy to analyze the stability of a system. The generalization of this idea consists in constructing a so-called Lyapunov scalar function verifying conditions related to the decreasing of the system energy.

Definition 3.7 ([4] *(Lyapunov function)*) A function $V : X \longrightarrow I\!R$ continuously differentiable is called Lyapunov function if:
1. $V(y) \geq 0, \ \forall y \in X$
2. $< Ay + f(y), V'(y) > \leq 0$, for all $y \in \mathcal{D}(A)$ where V' indicates the derivative of V.

If in addition $\lim\limits_{||y|| \longrightarrow \infty} V(y) = \infty$, then V is said proper.

Condition 2. imply that $V(y)$ is decreasing along any trajectory of system (3.50).

The problem of stabilizing infinite dimensional semi linear system was solved in the case of a contraction semigroup ([1]), and the used method is based on nonlinear semigroup tools.

3.3.2 Stabilization Results

In this paragraph, we recall some results concerning the question of stabilization of the following homogeneous semi linear systems:

$$\begin{cases} \dfrac{dy(t)}{dt} = Ay(t) + v(t)By(t) \\ y(0) = y_0 \in X \end{cases} \tag{3.51}$$

where $A : \mathcal{D}(A) \subset X \longrightarrow X$ is a Linear operator that generates a C_0- semigroup $S(t), t \geq 0$ in X, B is an a nonlinear operator from X to X, and $v : [0, +\infty[\longrightarrow \mathbb{C}$ is a real valued control.

Definition 3.8 The system (3.51) is said to be exponentially (resp. strongly, weakly) stabilizable if there exists a feedback control $v(t)$ such that
– For any initial state y_0, the system (3.51) has only one weak and global solution $y(t)$.
– $\{0\}$ is a stable equilibrium state for (3.51).

– For all $y_0 \in X$, $y(t) \longrightarrow 0$ exponentially (resp. strongly, weakly) when $t \longrightarrow +\infty$.

Note that the stabilization defined above is often called global stability and that we can consider any equilibrium state in place of 0.

The natural approach for the stabilization uses the derivative (formally) of $||y(t)||^2$ along the trajectory of system (3.51), which provides

$$\frac{d}{dt}||y(t)||^2 = 2 < Ay(t), y(t) > +2v(t) < y(t), By(t) >$$

It is therefore natural to consider the control $v(t) = - < y(t), By(t) >$, since this choice makes possible to obtain energy dissipation inequality

$$\frac{d}{dt}||y(t)||^2 \leq -2 < y(t), By(t) >^2$$

So, we are going to be interested in controls of type

$$v(t) = - < y(t), Ky(t) > \tag{3.52}$$

where $K : X \longrightarrow X$ is a continuous operator.

The following result gives a sufficient condition ensuring the existence of a global solution for system (3.51) excited by (3.52). This result was proved for $K = B$, a linear operator ([13]).

Proposition 3.5 *Suppose that $S(t)$ is a contraction semigroup. If*

$$Re(< y, Ky >< By, y >) \geq 0, \forall y \in X \tag{3.53}$$

then the solution $y(t)$ of the system (3.51) is defined for all $t \geq 0$. Furthermore $y(t)$ depends continuously on y_0.

Proof We know from the previous existence and uniqueness results that the system (3.51) has only one weak solution $y \in \mathcal{C}([0, t_{\max}[; X)$ defined on a maximum interval $[0, t_{\max}[$, and that the correspondence $y_0 \longrightarrow y(t)$ is continuous in $\mathcal{C}([0, t_{\max}[; X)$, since the application $y \longrightarrow - < Ky, y > By$ is monotone.

We show that $y(t)$ is bounded on $[0, t_{\max}[$ to conclude that $t_{\max} = +\infty$.

Let $T \in [0, t_{\max}]$ and consider the function $F(t) = - < Ky(t), y(t) > By(t)$ for $t \in [0, T]$, and let $F_n \in \mathcal{C}^1([0, T]; X)$ such that $F_n \longrightarrow F$ in $\mathcal{C}([0, T]; X)$.

Let $y_{0n} \longrightarrow y_0$ with $y_{0n} \in \mathcal{D}(A)$ and consider the following system

$$\begin{cases} \dfrac{dy_n(t)}{dt} = Ay_n(t) + F_n(t) \\ y_n(0) = y_{0n} \end{cases} \tag{3.54}$$

The semigroup $S(t)$ is a contraction, thus

$$\begin{aligned}||y_n(t) - y(t)|| &\leq ||y_{0n} - y_0|| + \int_0^T ||F_n(t) - F(t)||dt \\ &\leq ||y_{0n} - y_0|| + T \sup_{t\in[0,T]} ||F_n(t) - F(t)||\end{aligned}$$

It follows that $y_n \longrightarrow y$ in $\mathcal{C}([0, T]; X)$.

As $y_n \in \mathcal{D}(A)$, for all $t \in [0, T]$ we have

$$\frac{d}{dt}||y_n(t)||^2 = 2Re(< Ay_n(t), y_n(t) >) + 2Re(< F_n(t), y_n(t) >)$$

As the operator A is dissipative, gives

$$||y_n(T)||^2 - ||y_n(0)||^2 \leq 2\int_0^T Re(< F_n(t), y_n(t) >)dt$$

When $n \longrightarrow +\infty$, we obtain $||y(T)||^2 \leq ||y_0||^2 + 2\int_0^T Re(< F(t), y(t) >)dt$, and using (3.53), we deduce that $||y(T)|| \leq ||y_0||$, and when T tends to $t_{\max}$; the solution $y(t)$ remains bounded, therefore it is global. □

Ball and Slemrod [4] showed that it is possible to stabilize a large class of semi linear systems using the following feedback control

$$v(t) = - < By(t), y(t) > \tag{3.55}$$

Theorem 3.22 *Suppose that $S(t)$ is a contraction semigroup and that $B : X_w \longrightarrow X$ is sequentially continuous ($y_n \rightharpoonup y$ implies $By_n \longrightarrow By$), and that*

$$< BS(t)\psi, S(t)\psi > = 0,\ \forall t \geq 0 \Longrightarrow \psi = 0 \tag{3.56}$$

then the control (3.55) weakly stabilizes system(3.51)

Proof The existence and uniqueness of a global solution of the system (3.51) are assured by the previous proposition taking $K = B$.

We consider the function $F(t) = - < y(t), By(t) > By(t)$. where $y \in X$. With the same techniques used in the proof of previous proposition we have

$$||y(t)||^2 \leq ||y_0||^2 + 2\int_0^t Re < F(t), y(t) > dt \leq ||y_0||^2 \tag{3.57}$$

As the operator A is dissipative, we get $||y(t)|| \leq ||y_0||$.

The semigroup $S(t)$ is sequentially continuous, and the second point of Theorem 3.1, imply that $\omega_w(y_0)$ is nonempty.

Let $\psi \in \omega_w(y_0)$, and a seauence $(t_n) \longrightarrow \infty$ such that $S(t_n)y_0 \rightharpoonup \psi$ when $n \longrightarrow \infty$.

Using (3.57), we have $\displaystyle\int_0^{+\infty} Re(< F(t), y(t) >)dt < \infty$, and by Cauchy criteria, we deduce

$$\lim_{n \longrightarrow \infty} \int_{t_n}^{t_n+t} < F(S(s)z_0), S(s)z_0 > ds = 0, \ t \in \mathbb{R}^+$$

By semigroup property we obtain

$$\lim_{n\rightarrow\infty} \int_0^t < F(S(s)S(t_n)y_0), S(s)S(t_n)y_0 > ds = 0$$

Moreover, if $y_n \rightharpoonup y$, then (y_n) is bounded, and as B is sequentially continuous from X_w to X, we have $F(y_n) \longrightarrow F(y)$.

Using (3.57), we deduce that the mapping $y_0 \longrightarrow y(t)$ is sequentially weakly continuous. consequently, for Therefore, for all $s \in [0, t[$, we have

$$\begin{aligned}
&\lim_{n\rightarrow\infty} \int_0^t < F(S(s)S(t_n)y_0), S(s)S(t_n)y_0 > ds \\
&= \int_0^t < F(S(s)\psi), S(s)\psi > ds
\end{aligned}$$

The dominated convergence theorem allows

$$\int_0^t < F(S(s)\psi), S(s)\psi > ds = 0$$

Since F is continuous, we have $< F(S(t)\psi), S(t)\psi >= 0$, for all $t \in \mathbb{R}^+$, more again $< BS(t)\psi, S(t)\psi >= 0$, for all $t \geq 0$ and (3.56) allows to conclude. □

The following result gives a sufficient condition for strong stabilization of system (3.51) and allows an estimation of the stabilisation error (see [11]).

Theorem 3.23 *If B is locally lipschitz and (3.28) is verified, then the control (3.55) strongly stabilizes the system (3.51) and, for t large enough, we have the estimation*

$$\|y(t)\|^2 = O(\frac{1}{t}) \tag{3.58}$$

Proof According to Proposition 3.3 system (3.51)we has a weak solution $y(t)$ defined on a maximal interval $[0, t_{\max}[$, and if this solution is bounded then it is global.

With the same techniques using to prove (3.57) we have, for all $t \in [0, t_{max}[$,

$$\|y(t)\|^2 \leq \|y_0\|^2 + 2\int_0^t |\langle By(s), y(s)\rangle|^2 ds \tag{3.59}$$

therefore $\|y(t)\| \leq \|y_0\|$, and we conclude that the solution $y(t)$ is global.

We set $z(t) = y(t) - S(t)y_0$ and with the constant variation formula we obtain

$$z(t) = -\int_0^t \langle By(s), y(s)\rangle S(t-s)By(s)ds$$

then there exists $C_1 > 0$ that depends on B, T and $\|y_0\|$ such that

$$\|z(t)\|^2 \leq C_1 \int_0^T |\langle By(s), y(s)\rangle|^2 ds, \text{ for all } t \leq T. \tag{3.60}$$

Using the triangular inequality gives:

$$\begin{aligned}|\langle BS(t)y_0, S(t)y_0\rangle| \leq\ & |\langle BS(t)y_0 - By(t), y(t) - z(t)\rangle| \\ & + |\langle By(t), y(t)\rangle| + |\langle By(t) - Bz(t), z(t)\rangle| \\ & + |\langle Bz(t), z(t)\rangle|\end{aligned}$$

As B is locally Lipschitz and $\|y(t)\| \leq \|y_0\|$, there exists $\beta > 0$ such that

$$\begin{aligned}|\langle BS(t)y_0, S(t)y_0\rangle| &\leq \beta\|S(t)y_0 - y(t)\|\|y(t) - z(t)\| \\ &\quad + |\langle By(t), y(t)\rangle| + \beta\|y(t) - z(t)\|\|z(t)\| + \beta\|z(t)\|^2 \\ &\leq 2\beta\|z(t)\|\|y_0\| + |\langle By(t), y(t)\rangle| + \beta\|z(t)\|^2\end{aligned}$$

and by (3.60), for all $t \in [0, T]$, we have the inequality

$$|\langle BS(t)y_0, S(t)y_0\rangle| \leq 2\beta C_1 C_2\|y_0\| + |\langle By(t), y(t)\rangle| + \beta C_1^2\alpha^2 \tag{3.61}$$

with $C_2 = \int_0^T |\langle By(s), y(s)\rangle|^2 ds$ depends on B, T and $\|y_0\|$.

Integrating (3.61) between 0 and T and using (3.28) and (3.60) we have

$$\|y(T)\|^2 \leq C_2(\sqrt{C_1} + C_1) \tag{3.62}$$

By (3.59) and (3.60) we obtain $2C_2 \leq \|y_0\|^2 - \|y(T)\|^2$, which gives the estimation

$$\|y(T)\|^2 \leq \frac{1}{2}C_3\ f_0(\|y_0\|^2 - \|y(T)\|^2) \tag{3.63}$$

with $f_0(t) = t + \sqrt{t}$, for all $t \geq 0$, and if $f_1(.)$ indicates the inverse of $\frac{1}{2}C_3 f_0(.)$, then (3.63) becomes

$$f_1(\|y(T)\|^2) + \|y(T)\|^2 \leq \|y_0\|^2$$

Instead of the interval $[0, T]$ we consider the intervals $[kT, (k+1)T]$ with $k \in \mathbb{N}^*$ to have the inequality

$$f_1(\|y((k+1)T)\|^2) + \|y((k+1)T)\|^2 \leq \|y(kT)\|^2, \text{ for all } k \in \mathbb{N}^* \tag{3.64}$$

Following the same manner to prove Theorem 3.13, we deduce the desired estimate. □

Remark 3.4 The above result is an improvement of the one of Theorem 3.13.

Now we consider the system

$$\begin{cases} \dfrac{dy(t)}{dt} = Ay(t) + v(t)f(y(t)) \\ \\ y(0) = y_0 \in X \end{cases} \tag{3.65}$$

where $A : \mathcal{D}(A) \subset X \longrightarrow X$ is a linear operator that generates a compact C_0-semigroup $S(t), t \geq 0$ in X and $f : X \to X$.

Let note $\mathcal{B}(0, a)$ the ball of center 0 and of radius $a > 0$, we have the result [8].

Theorem 3.24 *If f is a function locally Lipschitz and $r(.) : X \to \mathbb{R}^+$ is a function of class C^p, $p \geq 1$ verifying*

$$\forall a > 0 \ , \ \exists r_a > 0 \ , \quad y \in \mathcal{B}(0, a) \quad : \ r(y) \geq r_a$$

If $(S(t))_{t\geq 0}$ is compact and if

$$\{y \in H | \ \langle f(S(t)y), S(t)y\rangle = 0\} = \{0\}$$

then the control

$$v(t) = -r(y) < f(y), y >$$

strongly stabilises the system (3.65).

Proof If $T(t)$ is a non linear semigroup of generator $A + v(t)f(.)$, the system (3.65) has a unique weak and global solution $y(t) = T(t)y_0$.

Let the function $f(y) = \frac{1}{2} \parallel y \parallel^2$, as A est dissipative, then $f(y)$ is a Lyapunov function

$$f(T(t)y_0) - f(T(s)y_0) \geq 0 \quad \text{for all} \quad (t, s), 0 \leq t \leq s$$

Indeed, if $y_0 \in \mathcal{D}(A)$, the solution $y(t) = T(t)y_0$ of system (3.65) is absolutely continuous and $y \in \mathcal{D}(A)$.

Deriving $f(t)$ along the trajectories $y(t)$ and since $\langle Ay, y\rangle \leq 0$ for all $y \in \mathcal{D}(A)$, we obtain

$$\frac{df(T(t)y_0)}{dt} = \langle AT(t)y_0 + v(t)f(T(t)y_0), T(t)y_0\rangle$$

$$\leq -r(T(t)y_0)\langle f(T(t)y_0), T(t)y_0\rangle^2$$

Therefore for all $0 \leq t \leq s$, we have

$$f(T(s)y_0) - f(T(t)y_0) \leq -\int_t^s r(T(\tau)y_0)\langle T(\tau)y_0, T(\tau)y_0\rangle d\tau$$

If $y_0 \notin \mathcal{D}(A)$, the the above inequality is verified as the density of $\mathcal{D}(A)$ in X. It follows that $(T(t)y_0)_{t\geq 0}$ is bounded and the system is stable in the Lyapunov sense, moreover we have

$$\int_0^{\infty} r(T(\tau)y_0)\langle f(T(\tau)y_0), T(\tau)y_0\rangle^2 d\tau < \infty \tag{3.66}$$

□

For the proof we need the following Lemma [8].

Lemma 3.3 *Under the hypothesis of Theorem 3.24, for all $y_0 \in X$ we have*
1. The set $\omega_w(y_0)$ corresponding to the orbit $(T(t)y_0)_{t\geq 0}$ is nonempty.
2. For all $y^* \in \omega(y_0)$ and every sequence (t_n) of $\mathbb{R}^+$, with $\lim_{n\to+\infty} t_n = +\infty$, we have

$$\lim_{n\to+\infty} T(t_n)y_0 = y^* \Rightarrow \lim_{n\to+\infty} T(t_n + t)y_0 = T(t)y^*, \text{ for all } t \geq 0.$$

In particular $\omega(y_0)$ is positively invariant by $T(t)$.

Proof • We begin with the proof of the lemma.
1. We know that the system (3.65) has a bounded solution, then the set of positive orbit $O^+(y_0)$ of system (3.65) is pre-compact [14] and the Theorem 3.1 imply that $\omega(y_0)$ is a nonempty and invariant by $T(t)$ in X.
2. Let $y^* \in \Omega(y_0)$ and a sequence $(t_n)_{n\geq 0}$ of $\mathbb{R}^+$, such that $\lim_{n\to+\infty} t_n = +\infty$, and $\lim_{n\to+\infty} T(t_n)(y_0) = y^*$.
By the the constant variation formula we have

$$T(t_n + t)y_0 = T(t)T(t_n)y_0 + \int_0^t T(t-s)v_{t_n+s}f(T(t_n+s)y_0)ds$$

$$= T(t)T(t_n)y_0$$

$$-\int_{t_n}^{t_n+t} T(t+t_n-s)r(T(s)y_0)\langle f(T(s)y_0), T(s)y_0\rangle f(T(s)y_0)ds$$

Applying Schwarz inequality gives

$$\| T(t_n+t)y_0 - T(t)T(t_n)y_0 \|^2 \leq \left(\int_{t_n}^{t_n+t} r(T(s)y_0) \| f(T(s)y_0) \|^2 \, ds \right)$$

$$\times \left(\int_{t_n}^{t_n+t} r(T(s)y_0)\langle f(T(s)y_0), T(s)y_0 \rangle^2 ds \right)$$

therefore

$$\| T(t_n+t)y_0 - T(t)T(t_n)y_0 \|^2$$

$$\leq ct \int_{t_n}^{t_n+t} r(T(s)(y_0))\langle f(T(s)(y_0)), T(s)(y_0) \rangle^2 ds \tag{3.67}$$

where $c = \sup_{y \in f(0, \|y_0\|)} r(y) \| f(y) \|^2$.

But as $\| S(t)y_0 \| \leq \| y_0 \|$, thus $c < +\infty$. By formula (3.66) and (3.67) we obtain

$$\lim_{n \to +\infty} \| T(t_n+t)y_0 - T(t)T(t_n)y_0 \|^2 = 0$$

Finally $\lim_{n \to +\infty} T(t_n)y_0 = y^*$, and $(T(t))_{t \geq 0}$ is a C_0-semigroup uniformly bounded, and the triangular inequality gives:

$$\lim_{n \to +\infty} T(t_n+t)y_0 = T(t)y^* \quad , \text{for all} t > 0$$

• Now we finish the proof of Theorem 3.24.

Let $y_0 \in X$, according to Theorem 3.1, we know that $\omega(y_0)$ is nonempty. It remains to prove that $\omega(y_0) = \{0\}$.

Let $y_0 \in X$, $y^* \in \Omega(y_0)$, we consider $y^*(t) = T(t)y^*$ the solution of system (3.65) that corresponds to the initial condition y^*, and the sequence $(t_n)_{n \geq 0}$ of $\mathbb{R}^+$, such that $\lim_{n \to +\infty} (t_n) = +\infty$, we have $\lim_{n \to +\infty} S(t_n)(y_0) = y^*$.

By Theorem 3.1 we know that $\lim_{n \to +\infty} S(t_n+s)y_0 = S(t)y^*$, for all $s \geq 0$.

By the continuity of $y \to r(y)\langle f(y), y \rangle^2$, we obtain

$$\lim_{n \to +\infty} r(S(t_n+t)(y_0))\langle f(S(t_n+t)(y_0)), S(t_n+t)(y_0) \rangle^2$$

$$= r(S(t)y^*)\langle f(S(t)y^*), S(t)y^* \rangle^2$$

According to dominated convergence Lebesgue theorem, we deduce that

$$\int_0^t r(T(s)y^*)\langle f(T(s)y^*), T(s)y^*\rangle^2 ds$$

$$= \lim_{n\to+\infty} \int_0^t r(T(t_n+s)y_0)\langle f(S(t_n+s)y_0), T(t_n+s)y_0\rangle^2 ds$$

$$= \lim_{n\to+\infty} \int_{t_n}^{t_n+t} r(T(s)(y_0))\langle f(T(s)y_0), T(s)(y_0)\rangle^2 ds \;=\; 0$$

As $r(T(s)y^*) < \langle f(T(s)y^*), T(s)y^*\rangle$ is a function continuous and positive then $\langle f(T(s)y^*), T(s)y^*\rangle = 0$ for all $s > 0$. But

$$\{y \in X \text{such that } \langle f(T(s)y), T(s)y\rangle = 0,\ s > 0\} = \{0\}$$

thus $y^* = 0$, and therefore $\omega(y_0) = \{0\}$. □

References

1. Ball, J.M., Slemrod, M.: Feedback stabilization of distributed semilinear control systems. Appl. Math. Optim. **5**, 169–179 (1979)
2. Dafermos, C.M.: Uniform processes and semicontinuous Liapunov functionals. J. Differ. Equ. **11**, 401–415 (1972)
3. Konishi, C.M.: Sur la compacité des semi-groupes non linéaires dans les espaces de Hilbert. Proc. Japan Acad. **48**, 278–280 (1972)
4. Ball, J.M.: On the asymptotic behavior of generalized processes with application to nonlinear evolutions equations. J. Differ. Equ. **27**, 224–265 (1978)
5. Lasalle, J.P.: Stability theory for ordinary differential equations. J. Differ. Equ. **4**, 307–324 (1968)
6. Pazy, A.: Semigroups of Linear Operators and Applications to Partial Differential Equations. Springer Verlag, New York (1983)
7. Bounit, H.: Comments in the feedback stabilization of bilinear control systems. Appl. Math. Lett. **16**, 847 851 (2003)
8. Bounit, H., Hammouri, H.: Feedback stabilisation for class of distributed semilinear systems. Nonlinear Anal. Theory Methods Appl. **37**, 953–969 (1999)
9. Bounit, H., Hammouri, H.: Bounded feedback stabilisation and globale separation principle of distributed systems. IEEE. Trans. Automat. Contr. **42** (1997)
10. Berrahmoune, L.: Asymptotic stabilization and decay estimate for distributed bilinear systems, Recerche di Matematica. fasc. 1, 89–103 (2001)
11. Ouzahra, M.: Exponential stabilization of distributed semilinear systems by optimal control. J. Math. Anal. Appl. **380**, 117–123 (2011)
12. Quinn, J.P.: Stabilization of bilinear systems by quadratic feedback control. J. Math. Anal. Appl. **75**, 66–80 (1980)
13. Berrahmoune, L., El boukfaoui, Y., Erraoui, M.: Remarks on the feedback stabilization of system affine in control. Eur. J. Control **7**, 17–28 (2001)
14. Pazy, A.: A class of semilinear equations of evolution Isreal. J. Math. **20**, 23–36 (1975)

Chapter 4
Regional Stabilization of Infinite Dimensional Linear Systems

This chapter, we concerns the concept of regional stability for infinite dimensional linear systems evolving on a spacial domain $\Omega \subset \mathbb{R}^n$: it is about studying the asymptotic behavior of the system not on its whole evolution domain but on an internal or boundary subregion $\omega \subset \Omega$.

In general, any system is considered to be stable or unstable. However, there exist systems that are unstable on their whole geometrical domains Ω, but do not behave the same way on any subregion ω of Ω. In addition, we can need to stabilize a system or improve its stability degree only on a subregion of its evolution domain, which will require a lower cost. Furthermore, when we have to stabilize the system on the whole domain, we will need to know its state on Ω, while the regional stabilization requests the state only on the subregion of interest ω of Ω .

4.1 Internal Regional Stability

In this section, we give the definitions and characterizations concerning the notion of regional stability. Next, we characterize a control of minimum cost that assures the stabilization on a target region, ω ,with a bounded state on the residual part $\Omega \setminus \omega$. The obtained results are illustrated by some examples.

4.1.1 Definitions

Let Ω be a regular domain of $\mathbb{R}^n$, $n \geq 1$, and let $Q = \Omega \times]0, \infty[$.

E. H. Zerrik and O. Castillo, *Stabilization of Infinite Dimensional Systems*,
Studies in Systems, Decision and Control 355,
https://doi.org/10.1007/978-3-030-68600-0_4

We consider a system described by the following state equation

$$\begin{cases} \dfrac{dy(t)}{dt} = Ay(t) & Q \\ y(.,0) = y_0 \in X & \Omega \end{cases} \tag{4.1}$$

where A an operator of domaine $\mathcal{D}(A)$ generates a C_0-semigroup $S(t), t \geq 0$ on a functional state space $X = L^2(\Omega)$, endowed with the inner product $<,>$, and the corresponding norm $||.||$.

Now lets consider an open internal region $\omega \subset \Omega$ of positive Lebesgue measure, and let χ_ω be the operator restriction to ω. We note the operator $i_\omega = \chi_\omega^* \chi_\omega$.

Definition 4.1 The system (4.1) is said (regionally)

1. weakly stable (r.w.s.) on ω, if for any initial state $y_0 \in X < \chi_\omega S(t)y_0, z >_{L^2(\omega)} \longrightarrow 0$ when $t \longrightarrow \infty$, $\forall z \in X$.
2. strongly stable (r.s.s) on ω, if for any initial state $y_0 \in X$, we have $||\chi_\omega S(t)y_0||_{L^2(\omega)} \longrightarrow 0$ when $t \longrightarrow \infty$.
3. exponentially stable (r.e.s.) on ω, if $\forall\ y_0 \in X$, $||\chi_\omega S(t)y_0||_{L^2(\omega)}$ exponentially tends to 0 when $t \longrightarrow \infty$.

Remark 4.1 1. This concept is general because for $\omega = \Omega$ we obtain the usual definitions of stability.
2. The above definitions show that we are interested by the asymptotic behavior of the system (4.1) only on ω.
3. Note that r.e.s. $\Longrightarrow$ r.s.s. $\Longrightarrow$ r.f.s.
The converse is not true as shown by the following examples:

4.1.2 Examples

Example 4.1 1. Let us consider the system defined on $\Omega =]0, 2[$ by

$$\begin{cases} \dfrac{dy(t)}{dt} = (x-1)y \text{ on } Q = \Omega \times]0, \infty[\\ y(.,0) = y_0(x) \quad \text{ on } \Omega \end{cases} \tag{4.2}$$

It is clear that this system is unstable on Ω, but exponentially stable on any subregion $\omega =]0, a[\subset]0, 1[$. Indeed we have

$$||\chi_\omega S(t)y_0||_{L^2(\omega)} \leq e^{t(a-1)}||y_0||,\ \forall\ y_0 \in L^2(\Omega).$$

2. On $\Omega =]0, \infty[$ Let us consider the system described by the equation

$$\begin{cases} \dfrac{\partial y(t)}{\partial t} + \dfrac{\partial y(t)}{\partial x} = 0 \text{ on } Q \\ \\ y(.,0) = y_0 \qquad \text{on } \Omega \end{cases} \tag{4.3}$$

The operator $A = -\dfrac{\partial}{\partial x}$ of domain $\mathcal{D}(A) = \{y \in H^1(\Omega);\ y(0) = 0\}$ generates a C_0-semigroup $S(t), t \geq 0$ given by

$$(S(t)y_0)(x) = \begin{cases} y_0(x-t), \text{ if } x \geq t \\ 0, \qquad\quad\ \text{if } x < t \end{cases}$$

For $a > 0$ and $\omega =]a, +\infty[$, $(a > 0)$ we have $||\chi_\omega S(t)y_0|| = ||\chi_\omega y_0||$, $\forall t \geq 0$, thus the system (4.3) is not strongly stable on ω. But it is weak stable on ω, indeed when $t \to \infty$

$$| < \chi_\omega S(t)y, z > |^2 \leq \int_0^{+\infty} y(x)^2 dx \int_{\sup(t,a)}^{+\infty} z(x)^2 dx \to 0$$

3. Let us consider the system defined on $\Omega =]0, 1[$ by

$$\begin{cases} \dfrac{dy(t)}{dt} = Ay(t) \qquad \text{on } Q \\ \\ y(x,0) = y_0 \in L^2(\Omega) \end{cases} \tag{4.4}$$

with

$$Ay = \lambda_1 < y, 1 > 1 + \sum_{(n,m)\in\Lambda} \lambda_{n,m} < y, h_{(n,m)} > h_{(n,m)}, \quad y \in L^2(\Omega), \tag{4.5}$$

where $\lambda_1 = -1$, $\Lambda = \{(n, m) \in I\!N^2; m < 2^n\}$ and $(\lambda_{n,m})_{(n,m)\in\Lambda}$ are given by $\lambda_{1,0} = 1$, and $\lambda_{n,m} = -\frac{1}{n^2+m^2}$ if $(n, m) \neq (1, 0)$, $h_{(n,m)}$ are the Haar functions defined for $(n, m) \in \Lambda$ and $x \in \Omega$ by

$$h_{(n,m)}(x) = \begin{cases} 2^{\frac{n}{2}}, \quad \text{if } x \in [\dfrac{m}{2^n}, \dfrac{m+\frac{1}{2}}{2^n}[\\ \\ -2^{\frac{n}{2}}, \text{ if } x \in [\dfrac{m+\frac{1}{2}}{2^n}, \dfrac{m+1}{2^n}[\\ 0, \quad\ \text{otherwise} \end{cases} \tag{4.6}$$

The family $\{1\} \cup \{h_{(n,m)}\}_{(n,m)\in\Lambda}$ forms an orthonormal basis of $L^2(\Omega)$.

The system (4.4) is strongly stable on $\omega =]\frac{1}{2}, 1[$. Indeed $||S(t)y_0||^2 = e^{-t}\langle y, \mathbb{1}\rangle \mathbb{1}$ $+ \sum_{(n,m)\in\Lambda} e^{\lambda_{n,m}t}\langle y, h_{(n,m)}\rangle h_{(n,m)} \to 0, \quad y \in L^2(\Omega)$.

Now for all $y_0 \in L^2(\Omega)$ we have $||S(t)z_0|| \leq 1$ and $||S(t)h_{(n,m)}|| \to 1$, when $n, m \to \infty$ so $||S(t)|| = 1$. This shows that the system (4.4) is not exponentially stable on ω.

4. Let us consider the system defined on $\Omega =]0, +\infty[$ by

$$\begin{cases} \dfrac{\partial y(t)}{\partial t} - \dfrac{\partial y(t)}{\partial x} = 0 \quad \text{on } Q \\ \\ y(.,0) = y_0 \in H^1(\Omega) \end{cases} \tag{4.7}$$

The operator $A = \dfrac{\partial}{\partial x}$ of domain $\mathcal{D}(A) = \{y \in H^1(\Omega);\ y(0) = 0\}$ generates a C_0−semigroup $S(t), t \geq 0$ on $H^1(\Omega)$ given by $(S(t)y_0)(x) = y_0(x+t)$.

For $\omega =]a, b[$, and $y \in L^2(\Omega)$, $||\chi_\omega S(t)y||^2 = \displaystyle\int_{a+t}^{b+t} y(x)^2 dx \to 0$, when $t \to \infty$.

On the other hand we have $||\chi_\omega S(t)|| = 1,\ \forall t \geq 0$, therefore the system (4.7) is not exponentially stable on ω.

5. The following example is about an unstable system on the whole domain Ω but stable on certain regions ω of the evolution domain Ω.

Let us consider a system of a spacial domain $\Omega =]0, 2[$ and described by the equation

$$\begin{cases} \dfrac{\partial y(t)}{\partial t} = (x-1)y \quad \text{on } Q \\ \\ y(.,0) = y_0 \in L^2(\Omega) \end{cases} \tag{4.8}$$

It is clear that the system (4.8) is not stable on Ω, while for a region $\omega =]0, a[\subset]0.1[$, we have

$$||\chi_\omega S(t)y_0||_{L^2(\omega)} \leq e^{t(a-1)}||y_0||, \quad for\ all\ y_0 \in L^2(\Omega)$$

So the system (4.8) is stable on ω.

4.1.3 Characterization of Regional Stability

In this paragraph we give results that characterize the regional stability.

We consider the sets

$$\sigma_\omega^1(A) = \{\lambda \in \sigma(A) \mid Re(\lambda) \geq 0, Ker(A - \lambda I) \not\subset Ker(i_\omega)\}$$

and

$$\sigma_\omega^2(A) = \{\lambda \in \sigma(A) \mid Re(\lambda) < 0, Ker(A - \lambda I) \not\subset Ker(i_\omega)\}$$

Proposition 4.1 *1. If the system (4.1) is weakly stable on ω then $\sigma_\omega^1(A) = \emptyset$.*
2. Let us suppose that the state space X has an orthonormal basis $(\varphi_n)_n$ of eigenfunctions of A. If $\sigma_\omega^1(A) = \emptyset$ and if there exists $\alpha > 0$ such that, for all $\lambda \in \sigma_\omega^2(A)$, $Re(\lambda) \leq -\alpha$, then the system (4.1) is exponentially stable on ω.

Proof 1. Suppose there exist $\lambda \in \sigma(A)$ and $\varphi \in N(A - \lambda I)$ such that $Re(\lambda) \geq 0$ and $\chi_\omega \varphi \neq 0$.
For $y_0 = \varphi$, the solution of (4.1) is $S(t)\varphi = e^{t\lambda}\varphi$.
So $< \chi_\omega S(t)\varphi, \varphi >_{L^2(\omega)} \geq ||\chi_\omega \varphi||_{L^2(\omega)} \neq 0$ and (4.1) is not strongly stable on ω.
2. Without loss of generality, we can assume that the eigenvalues of A are simple. In this case, for all $y_0 \in X$, we have

$$\begin{aligned}\chi_\omega S(t) y_0 &= \sum_{n=1}^{+\infty} e^{\lambda_n t} < y_0, \varphi_n > \chi_\omega \varphi_n \\ &= \chi_\omega \sum_{\lambda_n \in \sigma_\omega^2(A)} e^{\lambda_n t} < y_0, \varphi_n > \varphi_n\end{aligned}$$

Now if there exists $\alpha > 0$, such that $Re(\lambda) \leq -\alpha$, for all $\lambda \in \sigma_\omega^2(A)$, then

$$||\chi_\omega S(t) y_0||_{L^2(\omega)} \leq |||\chi_\omega||| e^{-\alpha t} ||y_0||$$

and therefore the system (4.1) is exponentially stable on ω. □

Example 4.2 Lets consider the system (4.4) with $\omega =]\frac{1}{2}, 1[$. We have

$$\sigma_\omega^1(A) = \emptyset \text{ and } \sigma_\omega^2(A) = \{-1\}$$

hence system (4.4) is exponentially stability on ω.

Remark 4.2 For $\omega = \Omega$, by the first point the stability is ensured if the dynamic operator of the system has no eigenvalues with positive real parts.

In the global case ($\omega = \Omega$), the exponential stability is equivalent to the upper bound

$$\int_0^\infty ||S(t)y||^2 dt < +\infty \text{ for all } y \in X. \tag{4.9}$$

Now, is the exponential stability over a region $\omega \subset \Omega$ equivalent to the regional upper bound

$$\int_0^\infty ||\chi_\omega S(t)y||^2 dt < \quad \text{for all } y \in L^2(\Omega)$$

We always have the result: exponential stability on ω implies

$$\int_0^{\infty} ||\chi_\omega S(t)y||^2 dt < +\infty \text{ for all } y \in L^2(\Omega).$$

The converse is not always true as shown in the following example.

Example 4.3 Let's reconsider system (4.7), as we've seen the semigroup $S(t), t \geq 0$ is given by $(S(t)y_0)(x) = y_0(x+t)$. And for $\omega =]a, b[,\ a < b$, we don't have exponential stability on ω. However, for all $y \in L^2(\Omega)$, we have

$$\int_0^{+\infty} ||\chi_\omega S(t)y||^2 dt = \int_a^b \int_x^{+\infty} y(t)^2 dt dx < \infty.$$

Before characterizing regional stability, we introduce the following result.

Proposition 4.2 *The regional upper bound*

$$\int_0^{\infty} ||\chi_\omega S(t)y||^2 dt < +\infty, \quad \textit{for all } y \in L^2(\omega)$$

is true if and only if the Lyapunov equation

$$< Ay, Py > + < Py, Ay > + < i_\omega y, y > = 0 \tag{4.10}$$

admits a solution P, a self adjoint and positive operator.

Proof Suppose that $\int_0^{\infty} ||\chi_\omega S(t)y||^2_{L^2(\omega)} dt < \infty$, for all $y \in X$, then the operator

$$P(y) = \int_0^{\infty} (\chi_\omega S(t))^* (\chi_\omega S(t)) y \, dt \tag{4.11}$$

is well defined and solution of (4.10).

Conversely if (4.10) admits a solution, then

$$\int_0^{\infty} ||\chi_\omega S(t)y||^2_{L^2(\omega)} dt < \infty \; for \; ally \in X.$$

Proposition 4.3 *Suppose that A generates a semigroup $S(t), t \geq 0$ verifying*

$$||\chi_\omega S(t+s)y||_{L^2(\omega)} \leq |||\chi_\omega S(t)|||.||\chi_\omega S(s)y||_{L^2(\omega)} \; \textit{for all } \; t, s \geq 0, \;\; y \in X. \tag{4.12}$$

then the system (4.1) is exponentially stable on ω, if and only if

$$\int_0^{\infty} ||\chi_\omega S(t)y||^2_{L^2(\omega)} dt < \infty \; \textit{for all } y \in X.$$

Proof If the system (4.1) is exponentially stable on ω, then

$$\int_0^\infty ||\chi_\omega S(t)y||^2_{L^2(\omega)} dt < \infty$$

for all $y \in X$. The operator

$$P(y) = \int_0^\infty (\chi_\omega S(t))^*(\chi_\omega S(t))y\, dt$$

is well defined and verify (4.10).

Conversely, if (4.10) has a solution then

$$\int_0^\infty ||\chi_\omega S(t)y||^2_{L^2(\omega)} dt < \infty \text{ for all } y \in X.$$

therefore for $n \geq 1$, we can define the operator

$$S_n : X \longrightarrow L^2(0, \infty, L^2(\omega))$$

$$y \longrightarrow 1_{[0,n]}(t)\chi_\omega S(t)y$$

It follows that $||S_n y||^2 \leq k_n ||y||^2$, where $k_n = \int_0^n |||\chi_\omega S(t)|||^2 dt$ and according to Banach Steinhaus' theorem, there exists $\gamma > 0$ such that $|||S_n||| \leq \gamma,\ \ n \geq 1$.

Let $\alpha > 0$, $M_1 > 0$ be such that $|||S(t)||| \leq M_1 e^{\alpha t},\ \ t \geq 0$. For $t_0 > 0$, the family $(|||\chi_\omega S(t)|||)_{t \leq t_0}$ is bounded and for $t > t_0$, we have

$$\frac{1 - e^{-2\alpha t}}{2\alpha} ||\chi_\omega S(t)y||^2_{L^2(\omega)} = \int_0^t e^{-2\alpha(t-s)} ||\chi_\omega S(t)y||^2_{L^2(\omega)} ds$$

$$\leq \int_0^t e^{-2\alpha(t-s)} ||\chi_\omega S(s)y||^2_{L^2(\omega)} |||\chi_\omega S(t-s)|||^2 ds$$

$$\leq M_1^2 \gamma^2 ||y||^2$$

We deduce that there exists $k > 0$ such that $|||\chi_\omega S(t)||| \leq k,\ \ t \geq 0$.

On the other hand, by (4.12) we get

$$t||\chi_\omega S(t)y||^2_{L^2(\omega)} \leq \int_0^t ||\chi_\omega S(s)y||^2_{L^2(\omega)} |||\chi_\omega S(t-s)|||^2 ds.$$

then $t||\chi_\omega S(t)y||^2_{L^2(\omega)} \leq k^2\gamma^2 ||y||^2$.

Therefore there exists t_0 such that $\ln |||\chi_\omega S(t)||| < 0$, for all $t \geq t_0$. We deduce that $\sigma_0 =: \inf_{t>0} \frac{\ln |||\chi_\omega S(t)|||}{t} < 0$.

We show that

$$\sigma_0 = \lim_{t\to+\infty} \frac{\ln |||\chi_\omega S(t)|||}{t} \tag{4.13}$$

Let $t_0 > 0$, and set $M_2 = \sup_{t\in[0,t_0]} |||S(t)|||$. For $t \geq t_0$, there exists $n \in \mathbb{N}$ such that $nt_0 \leq t < (n+1)t_0$. Thus

$$\frac{\ln |||\chi_\omega S(t)|||}{t} \leq \frac{\ln |||\chi_\omega S(nt_0)|||}{t} + \frac{\ln |||S(t-nt_0)|||}{t}$$

With the property (4.12) we have

$$\frac{\ln |||\chi_\omega S(t)|||}{t} \leq \frac{nt_0}{t}\frac{\ln |||\chi_\omega S(t_0)|||}{t_0} + \frac{\ln M_2}{t}$$

It follows that

$$\lim_{t\to\infty}\sup \frac{\ln |||\chi_\omega S(t)|||}{t} \leq \frac{\ln |||\chi_\omega S(t_0)|||}{t_0}$$

t_0 being arbitrary, we get

$$\lim_{t\to\infty}\sup \frac{\ln |||\chi_\omega S(t)|||}{t} \leq \inf_{t>0} \frac{\ln |||\chi_\omega S(t)|||}{t} \leq \lim_{t\to+\infty}\inf \frac{\ln |||\chi_\omega S(t)|||}{t}$$

which shows (4.13) and therefore

$$\forall \sigma \in]0, -\sigma_0[, \exists\, M_3 \;/\; |||\chi_\omega S(t)||| \leq M_3 e^{-\sigma t},\ t \geq 0,$$

Consequently the system is regionally exponentially stable on ω. □

Remark 4.3 The condition (4.12) is not necessary for regional stability because any system exponentially stable on Ω is also on ω without condition (4.12). This condition somehow translates the continuity of $\chi_\omega S(t)$ with respect to $\chi_\omega y$. Especially if $\omega = \Omega$, then (4.12) is verified and the exponential stability on Ω is equivalent to the condition (4.9) (see [1]).

Example 4.4 Let us consider the system defined on $\Omega =]0, +\infty[$ by

$$\begin{cases} \dfrac{\partial y(t)}{\partial t} + \dfrac{\partial y(t)}{\partial x} = 0 & \text{on } \Omega\times]0,\infty[\\ y(.,0) = y_0 = y_0 \in H^1(\Omega) \end{cases} \tag{4.14}$$

The operator $A = -\dfrac{\partial}{\partial x}$ of domain $\mathcal{D}(A) = \{y \in H^1(\Omega);\ y(0) = 0\}$ generates a C_0-semigroup $S(t), t \geq 0$ given by

$$(S(t)y_0)(x) = \begin{cases} y_0(x-t) & \text{if } x \geq t \\ 0 & \text{if } x < t \end{cases}$$

On the region $\omega =]0, a[$ $(a > 0)$, the solution of (4.14) verify

$$||\chi_\omega S(t)y||^2_{L^2(\omega)} \leq ||\chi_\omega y||_{L^2(\omega)}$$

which gives (4.12) by applying it to $y = S(t)y_0$. On the other hand the operator $Py = (a-x)i_\omega y$ is solution of (4.10). Indeed, for $y \in \mathcal{D}(A)$, we have

$$\begin{aligned} &< -\frac{\partial y}{\partial x}, (a-x)i_\omega y > + < (a-x)i_\omega y, -\frac{\partial y}{\partial x} > \\ &\quad = -2Re\left(\int_0^a \frac{\partial y(x)}{\partial x}(a-x)\overline{y(x)}dx\right) \\ &\quad = -Re([(a-x)|y(x)|^2]_0^a + \int_0^a |y(x)|^2 dx) \\ &\quad = -||\chi_\omega y||^2_{L^2(\omega)} \end{aligned}$$

which shows that the system (4.14) is exponentially stable on ω.

Remark 4.4 1. Strong stability on ω does not cause convergence of the integral $\int_0^\infty ||\chi_\omega S(t)y||^2 dt$, as we can see by considering the system (4.7) and the region $\omega =]a, \infty[$, $a \geq 0$.

2. If the system (4.7) is strongly stable on ω, this leads to the following question. Is

$$\int_0^\infty ||\chi_\omega S(t)y||^2_{L^2(\omega)} dt < +\infty \quad \text{for all } y \in L^2(\Omega)$$

implies the strong stability on ω? This is the subject of next result.

Proposition 4.4 *If the system (4.1), with the output $z(t) = \chi_\omega y(t)$, is regionally observable on ω and if (4.10) admits a positive solution, then the system (4.1) is strongly stable on ω.*

Proof If the operator $P(y) = \int_0^\infty (\chi_\omega S(t))^*(\chi_\omega S(t))y\, dt$ verify (4.10) then

$$\int_0^\infty ||\chi_\omega S(t)y||^2_{L^2(\omega)} dt < \infty \quad \text{for all } y \in X$$

and the operator P is well defined.

We have $< PS(t)y, S(t)y >= \int_t^\infty ||\chi_\omega S(\tau)y||^2_{L^2(\omega)}\, d\tau \to 0$ when $t \to \infty$.
The system (4.1) is assumed to be observable on ω, we have

$$< Py, y > \geq \alpha ||\chi_\omega y||^2_{L^2(\omega)}$$

which allows to conclude. □

Proposition 4.5 *If the Eq. (4.10) admits a positive solution and $\chi_\omega S(t)$ is uniformly bounded, then the system (4.1) is strongly stable on ω.*

Proof Let $y \in \mathcal{D}(A)$, according to the previous lemma we have $< PS(t)y, S(t)y > \to 0$ when $t \to \infty$. Which imply that $PS(t)y \to 0$ when $t \to \infty$, and with (4.10), we have

$$||\chi_\omega S(t)y||^2_{L^2(\omega)} = - < AS(t)zy, PS(t)y > - < PS(t)y, AS(t)y >$$

Thus $\lim_{t\to\infty} ||\chi_\omega S(t)y||_{L^2(\omega)} = 0$.

Now let $y \in X$, and $(y_n) \subset \mathcal{D}(A)$ such that $y_n \to y$ when $n \to \infty$. So there exists $M > 0$ such that

$$\begin{aligned}||\chi_\omega S(t)y|| &\leq ||\chi_\omega S(t)(y - y_n)|| + ||\chi_\omega S(t)y_n|| \\ &\leq M||(y - y_n)|| + ||\chi_\omega S(t)y_n||.\end{aligned}$$

therefore $\limsup_{t\to+\infty} ||\chi_\omega S(t)y|| \leq M||y - y_n||$, for all $n \geq 1$. This implies that $||\chi_\omega S(t)y|| \to 0$ when $t \to \infty$. □

The following result characterizes the regional strong stability by a dissipativity condition. Let $R \in \mathcal{L}(X)$, be a positive and self adjoint operator; there exists $c > 0$ such that

$$\langle Ry, y\rangle \geq c||\chi_\omega y||_{L^2(\omega)}, \; y \in \mathcal{D}(A) \tag{4.15}$$

and consider the Lyapunov equation

$$\langle Ay, Py\rangle + \langle Py, Ay\rangle + \langle Ry, y\rangle = 0, \; y \in \mathcal{D}(A) \tag{4.16}$$

So we have the following result.

Proposition 4.6 *Suppose there exists a positive and self-adjoint operator $P \in \mathcal{L}(X)$ verifying (4.10). If, moreover,*

$$Re(< \chi_\omega Ay, y >_{L^2(\omega)}) \leq 0, \quad y \in \mathcal{D}(A) \tag{4.17}$$

then (4.1) is strongly stable on ω.

Proof Let P be a solution of (4.16) and the function $V(y) = < Py, y >$. $y \in L^2(\Omega)$.

For $y_0 \in \mathcal{D}(A)$, we have

$$\int_0^t ||\chi_\omega S(s)y_0||^2_{L^2(\omega)}ds = V(y_0) - V(y(t)).$$

and according to (4.15), we get

$$\int_0^t ||\chi_\omega S(s)y_0||^2_{L^2(\omega)}ds \leq \frac{V(y_0)}{c}.$$

The inequality (4.17) gives $\frac{d}{dt} < i_\omega S(t)y_0, S(t)y_0 >_{L^2(\omega)} \leq 0$. Thus

$$t||\chi_\omega S(t)y_0||^2_{L^2(\omega)} \leq \int_0^t ||\chi_\omega S(s)y_0||^2_{L^2(\omega)}ds \leq \frac{V(y_0)}{c}$$

Since V and i_ω are continuous, then $||\chi_\omega S(t)y_0||^2_{L^2(\omega)} \leq \frac{V(y_0)}{c.t}$, for all $t > 0$ and $y_0 \in X$, hence (4.1) is strongly stable on ω. □

Remark 4.5 1. In the case where $\omega = \Omega$, we have the above result without the condition (4.17).
2. The above results give necessary and sufficient conditions for regional stability of the system (4.1) on ω, without taking into account the behavior of the state of the system in the residual part $\Omega\backslash\omega$. The following proposition gives precisions on the behavior of the state on the residual part.

Proposition 4.7 *1. If the system (4.1) is exponentially stable on ω and if there exists $d > 0$ such that*

$$||\chi_\omega y||^2_{L^2(\omega)} \geq d\ Re(< Ay, y >),\ \ y \in \mathcal{D}(A) \tag{4.18}$$

then the state remains bounded on Ω.
2. If the system (4.1) is weakly stable on ω and if there exists $e > 0$ such that

$$Re(< \chi_\omega Ay, y >_{L^2(\omega)}) \geq e\ Re(< Ay, y >),\ \ y \in \mathcal{D}(A) \tag{4.19}$$

then the state remains bounded on Ω.

Proof 1. Let $y_0 \in \mathcal{D}(A)$, we have

$$Re\ (< AS(t)y_0, S(t)y_0 >) = \frac{1}{2}\frac{d||S(t)y_0||^2}{dt}$$

According to (4.18), we have

$$\int_0^t ||\chi_\omega S(s)y_0||^2_{L^2(\omega)}ds \geq \frac{d}{2}||S(t)y_0||^2 - \frac{d}{2}||y_0||^2$$

and as the system (4.1) is exponentially stable on ω, then

$$\int_0^{+\infty} ||\chi_\omega S(t)y_0||^2_{L^2(\omega)}dt < \infty$$

thus

$$||S(t)y_0|| \leq \frac{2}{d} < Py_0, y_0 > +||y_0||^2 \text{ for all } t \geq 0,$$

where $P \in \mathcal{L}(X)$ is solution of (4.10).

The density of $\mathcal{D}(A)$ in X.

2. Is deduced using the same approach as for the first point. □

Remark 4.6 1. The condition (4.18) can be replaced by:

$$||\chi_\omega y||^2_{L^2(\omega)} \geq d\ Re\ (< \chi_{\Omega\backslash\omega}Ay, y >),\ \ y \in \mathcal{D}(A)$$

and the condition (4.19) can be replaced by:

$$Re(< \chi_\omega Ay, y >_{L^2(\omega)}) \geq e\ Re(< \chi_{\Omega\backslash\omega}Ay, y >),\ \ y \in \mathcal{D}(A) \text{ with } e > 0.$$

2. In Lyapunov's equation (4.10), the operator i_ω can be replaced by a positive and self adjoint operator $R \in \mathcal{L}(X)$ verifying

$$< Ry, y > \geq c\ ||\chi_\omega y||^2_{L^2(\omega)},\ \ y \in X \text{ with } c > 0.$$

3. In the global case ($\omega = \Omega$), the stability exponential ensures the uniqueness of the solution of the Lyapunov equation

$$< Ay, Py > + < Py, Ay > + < y, y > = 0,\ \ y \in \mathcal{D}(A)$$

In the regional case this result is not true, as we can see it in the case of the system (4.14) where (4.10) admits two distinct solutions $Py = (a - x)i_\omega y$ and that given by (4.11).

We can also consider the example where $Ay = \frac{-1}{2}\chi_\omega y$, ω is a region strictly included in Ω, in which case we have both solutions $P = Id$ and $P = i_\omega$.

4.2 Internal Regional Stabilization

In this section we will characterize a control which stabilizes an infinite dimensional linear system on a given region ω of the system evolution domain Ω.

4.2.1 Definitions

Lets consider the system

$$\begin{cases} \dfrac{dy(t)}{dt} = Ay(t) + Bv(t) \quad \text{on } Q \\ \\ y(.,0) = y_0 \in X \end{cases} \tag{4.20}$$

A generates a C_0-semigroup on X and $B \in \mathcal{L}(V, X)$, V being the controls set assumed to be a Hilbert space.

Definition 4.2 The system (4.20) is said regionally weakly (respectively strongly, exponentially) stabilizable on $\omega \subset \Omega$, if there exists an operator $K \in \mathcal{L}(X, V)$ such that the system

$$\begin{cases} \dfrac{dy(t)}{dt} = (A + BK)y(t) \; Q \\ \\ y(.,0) = y_0 \in X \end{cases} \tag{4.21}$$

be regionally weakly (respectively strongly, exponentially) stable on ω.

Remark 4.7 1. Regional stabilization can be seen as an output stabilization problem with a partial observation: $z = \chi_\omega y$.
2. If the system (4.20) is stabilized on $\omega \subset \Omega$, then it is stabilized on each $\omega' \subset \omega$ by applying the same control.
3. The cost of regional stabilization of a system is less than that of stabilization over the whole domain. Indeed, if we consider the functional cost

$$q(u) = \int_0^{+\infty} ||u(t)||_V^2 dt \tag{4.22}$$

where $u \in V_{ad}(\omega) = \{u \in L^2(0, +\infty; V)$ is a control that stabilizes (4.20) exponentially on ω and $q(u) < \infty\}$, then we have $V_{ad}(\Omega) \subset V_{ad}(\omega)$ hence

$$\min_{V_{ad}(\omega)} q(u) \leq \min_{V_{ad}(\Omega)} q(u)$$

We may need to improve the degree of stability of a system on a privileged region of the system geometrical domain, as shown in the following example:

Example 4.5 We consider the system defined on $\Omega =]0, +\infty[$ by

$$\begin{cases} \dfrac{dy(t)}{dt} = Ay(t) + Bv(t) \quad \text{on } Q \\ \\ y(.,0) = y_0 \in H^1(\Omega) \end{cases} \tag{4.23}$$

where $A = -\frac{\partial}{\partial x}$, of domain $\mathcal{D}(A) = \{y \in H^1(\Omega);\ y(0) = 0\}$ and the operator $B \in \mathcal{L}(V, H^1(\Omega))$.

The control $v(.) = 0$ exponentially stabilizes the system (4.23) on $\omega =]0, a[$, $a > 0$ with a bounded state on Ω ($||y(t)|| = ||y_0||$), weakly stabilises the system but not exponentially on Ω.

On the other hand if we consider the cost function (4.22), then we have

$$\min_{V_{ad}(\omega)} q(v) = 0 < \min_{V_{ad}(\Omega)} q(v)$$

4.2.2 *Characterizations*

In this paragraph we characterize regional stabilization by an approach based on the decomposition of the system state space X.

We consider the system (4.20) and assume that the space $X = X_u \oplus X_s$ and that (4.20) can be decomposed into two subsystems:

$$\begin{cases} \dfrac{dy_u(t)}{dt} = A_u y_u(t) + \Pi B v(t) \\ y_{0u} = \Pi y_0, \quad y_u = \Pi y \end{cases} \tag{4.24}$$

and

$$\begin{cases} \dfrac{dy_s(t)}{dt} = A_s y_s(t) + (Id - \Pi) B v(t) \\ y_{0s} \quad = (Id - \Pi) y_0 \ , \quad y_s = (Id - \Pi) y \end{cases} \tag{4.25}$$

where the operators A_s and A_u are restrictions of A to X_s and X_u respectively and Π is the projection operator defined by (2.16) and we have

$$\sigma(A_s) = \sigma_s(A) \ , \quad \sigma(A_u) = \sigma_u(A)$$

and A_u is a bounded operator.

If $S_u(t)$ and $S_s(t)$ are the restrictions of $S(t)$ to X_u and X_s, which are respectively the C_0- semigroups generated by A_u and A_s, the solutions of (4.24) and (4.25) are given by

$$y_u(t) = S_u(t) y_{0u} + \int_0^t S_u(t-\tau) \Pi B v(\tau) d\tau \tag{4.26}$$

and

$$y_s(t) = S_s(t) y_{0s} + \int_0^t S_s(t-\tau)(Id - \Pi) B v(\tau) d\tau \tag{4.27}$$

If the operator A_s verify the spectral growth assumption

$$\lim_{t \to +\infty} \frac{\ln |||S_s(t)|||}{t} = \sup Re(\sigma(A_s)) \tag{4.28}$$

then stabilizing the system (4.20) turns up to stabilizing (4.24) (see [2]).

The following result extends this approach to the regional case.

Proposition 4.8 *Suppose that A_s verifies the inequality*

$$\lim_{t \to +\infty} \frac{\ln |||\chi_\omega S_s(t)|||}{t} \leq \sup Re(\sigma(A_s)) \tag{4.29}$$

1. *If there exists an operator $K_u \in \mathcal{L}(X_u, V)$, such that the control $v(t) = K_u i_\omega y_u(t)$, exponentially stabilizes the system (4.24) (respectively strongly) on ω, then the system (4.20) is exponentially (respectively strongly) stabilized on ω, by using the operator $K = (K_u i_\omega, 0)$.*
2. *If the system (4.24) is exponentially (respectively strongly) stabilized on Ω by a control $v(t) = K_u y_u(t)$, with $K_u \in \mathcal{L}(X, V)$, then the system (4.20) is regionally exponentially (respectively strongly) stabilized on ω by using the operator $K = (K_u, 0)$.*

Proof 1. Exponential stabilization.

According to the spectral decomposition, we have $\sup Re(\sigma(A_s)) \leq -\delta$. As A_s verify (4.29) then there exist $a > 0$ and $\mu > 0$ such that

$$|||\chi_\omega S_s(t)||| \leq a\, e^{-\mu t},\ t \geq 0 \tag{4.30}$$

The solution of system (4.24) is therefore

$$y_u(t) = e^{F_u t} y_{0u}, \quad \text{with} \quad F_u = A_u + \Pi B K_u i_\omega \in \mathcal{L}(X_u)$$

and there exist $0 < \alpha < \mu$, and $b > 0$ such that

$$||\chi_\omega y_u(t)||_{L^2(\omega)} \leq b\, e^{-\alpha t} ||y_{0u}||$$

So for $v(t) = K_u i_\omega y_u(t)$, we have

$$||v(t)||_V \leq b\, e^{-\alpha t}\ |||K_u|||.||y_{0u}||$$

Using (4.27) and (4.30), there exists $c > 0$ such that

$$\begin{aligned} ||\chi_\omega y_s(t)||_{L^2(\omega)} &\leq a\, e^{-\mu t} ||y_{0s}|| + c\, ||y_{0u}|| \int_0^t e^{-\mu(t-\tau)} e^{-\alpha\tau} d\tau \\ &\leq a\, e^{-\mu t} ||y_{0s}|| + c\, ||y_{0u}|| \frac{e^{-\mu t} - e^{-\alpha t}}{\alpha - \mu} \end{aligned}$$

Consequently the state of the system (4.20) excited by $v(t) = K y(t)$ satisfy

$$||\chi_\omega y(t)||_{L^2(\omega)} \leq \left(a\, e^{-\mu t} + c\, \frac{e^{-\mu t} - e^{-\alpha t}}{\alpha - \mu} + b\, e^{-\alpha t}\right) ||y_0||$$

This shows that the system (4.20) is exponentially stabilizable on ω.

2. Strong stabilization.

We have $\lim_{t \to +\infty} \|\chi_\omega y_u(t)\|_{L^2(\omega)} = 0$. Considering the control $v(t) = K_u i_\omega y_u(t)$, we have

$$\forall \varepsilon > 0, \exists\, t_0 > 0,\ t > t_0 \Longrightarrow \|v(t)\|_V < \frac{\varepsilon \mu}{2}$$

By (4.27), (4.30); there exists $d > 0$ such that

$$\|\chi_\omega y_s(t)\|_{L^2(\omega)} \leq d e^{-\mu t} \|y_{0s}\| + d \int_0^t e^{-\mu(t-\tau)} \|v(\tau)\|_V d\tau$$

We will show that $\lim_{t \to +\infty} \int_0^t e^{-\mu(t-\tau)} \|v(\tau)\|_V d\tau = 0$. For $t \geq t_0$, we have

$$\begin{aligned}\int_0^t e^{-\delta(t-\tau)} \|v(\tau)\|_V d\tau &= \int_0^{t_0} e^{-\mu(t-\tau)} \|v(\tau)\|_V d\tau + \int_{t_0}^t e^{-\mu(t-\tau)} \|v(\tau)\|_V d\tau \\ &\leq \gamma e^{-\mu(t-t_0)} + \frac{\varepsilon}{2}\end{aligned}$$

where $\gamma = \frac{1}{\sqrt{2\mu}} \int_0^{t_0} \|v(t)\|_V dt$, so $\lim_{t \to +\infty} ||\chi_\omega y_s(t)||_{L^2(\omega)} = 0$.

Consequently the solution of the system (4.20) associated with the control $v(t) = K y(t)$ satisfy

$$||\chi_\omega y(t)||_{L^2(\omega)} \leq ||\chi_\omega y_u(t)||_{L^2(\omega)} + ||\chi_\omega y_s(t)||_{L^2(\omega)}$$

The system (4.20) is therefore strongly stabilizable on ω.

2. The second point is demonstrated by the same approach. □

Remark 4.8 The inequality (4.29) may be strict, this is the case for example, if we consider the operator A defined by (4.5) with $\lambda_{1,0} = -1$ and $\lambda_{n,m} = \lambda_1 = -2$, for $(n, m) \neq (1, 0)$.

The following result is a consequence of the previous proposition and the equivalence between the controllability and exponential stabilization of the system (4.24) (see [2]).

Corollary 4.1 *If the condition (4.29) is verified and that*

1. *the space X_u is of finite dimension,*
2. *the system (4.24) is controllable on X_u, then the system (4.20) is exponentially stable on ω.*

The following example shows that it is possible that a system be regionally stabilized without being on the whole evolution domain.

Example 4.6 Lets consider the system (4.4) excited by $g(.)v(t)$ with

$$g =< 1, h_{(1,0)} > h_{(1,0)}+ < 1, h_{(2,0)} > h_{(2,0)}$$

The system (4.4) can be decomposed into (4.24) and (4.25) with

$$A_u y_u =< y_u, h_{(1,0)} > h_{(1,0)}+ < y_u, h_{(2,0)} > h_{(2,0)}, \quad y \in L^2(\Omega) \tag{4.31}$$

and

$$A_s y_s = - < y_s, 1 > 1 - \sum_{(n,m)\in\Lambda-\{(1,0),(2,0)\}} < y_s, h_{(n,m)} > h_{(n,m)}$$

By a rank argument, the unstable part (4.31) which is of finite dimension is not controllable, and therefore the system (4.4) cannot be stabilized on Ω. But for $\omega =]\frac{1}{2}, 1[$, the operator A_s verify (4.29) and $\chi_\omega y_u(t) = 0$. So the system (4.4) may be stabilized on ω by the null control.

4.2.3 *Regional Stabilizing Control*

In this paragraph, we propose to look for a control ensuring the regional stabilization. The used approach is based on using the Riccati equation.

Let us consider the system (4.20) with the same hypotheses on A and B. For an operator $K \in \mathcal{L}(X, V)$, $S^K(t)$, $t \geq 0$, indicates the semigroup generated by $A_K = A + BK$.

Let $R \in \mathcal{L}(X)$ be a positive, self adjoint operator and verify

$$\exists c > 0 \mid < Ry, y >\geq c \, ||\chi_\omega y||_{L^2(\omega)} \quad , \quad y \in X \tag{4.32}$$

For $y \in \mathcal{D}(A)$, we consider the following Riccati equation

$$< Ay, Py > + < Py, Ay > - < B^* Py, B^* Py > + < Ry, y >= 0 \tag{4.33}$$

and suppose that there exists an operator $P \in \mathcal{L}(X)$, positive and self adjoint, solution of the Eq. (4.33). Let note the operator $K = -B^* P$, The following results come from the Proposition 4.7.

Proposition 4.9 *(1) If the condition*

$$Re(< \chi_\omega (A + BK)y, y >_{L^2(\omega)}) \leq 0, \quad y \in \mathcal{D}(A) \tag{4.34}$$

is verified, then (4.20) is strongly stabilizable on ω.

If, in addition, we have

$$Re(< \chi_{\omega}(A + BK)y, y >_{L^2(\omega)}) \geq e\ Re(< (A + BK)y, y >)$$

then the state of the system (4.21) remains bounded on Ω.

(2) If $S^K(t)$ *satisfies (4.12), then the system (4.20) is exponentially stabilizable on* ω.

If, moreover, there exists $d > 0$ *such that*

$$< Ry, y > \geq d\ Re(< (A + BK)y, y >) \qquad (4.35)$$

then the state of the system (4.21) remains bounded on Ω.

In the case of conservative systems, consider the system (4.20) with

$$< Ay, y > + < y, Ay >= 0, \quad y \in \mathcal{D}(A) \qquad (4.36)$$

If the controls space is $V = X$ and there exists $c > 0$ such that for all $y \in X$,

$$||B^* y|| \geq c\ ||\chi_\omega y||_{L^2(\omega)}$$

Considering $R = BB^*$ and $P = Id$, we have the following result.

Corollary 4.2 *1. If the condition (4.34) is verified, then the control*

$$v(t) = -B^* y(t)$$

strongly stabilizes the system (4.20) on ω.

2. *If* $S^K(t)$ *verify (4.12), then control* $v(t)$ *exponentially stabilizes the system (4.20) on* ω.

The following result concerns the strong stabilization.

Proposition 4.10 *If (4.33) admits a positive solution* P, *then the control*

$$v(t) = -B^* P y(t)$$

strongly stabilizes the system (4.20) on ω *if and only if* $|||\chi_{\omega} S^K(t)|||$ *is bounded, with* $K = -B^* P$.

Proof Since (4.33) has a solution $P \geq 0$, then

$$\begin{aligned} &\langle A_K S^K(t)y, PS^K(t)y\rangle + \langle PS^K(t)y, A_K S^K(t)y\rangle \\ &\qquad + \langle B^* PS^K(t)y, B^* PS^K(t)y\rangle + \langle RS^K(t)y, S^K(t)y\rangle = 0 \end{aligned}$$

Thus

$$\frac{d}{dt}\langle PS^K(t)y, S^K(t)y\rangle + ||B^*PS^K(t)y||^2 + \langle RS^K(t)y, S^K(t)y\rangle = 0$$

Which gives

$$\begin{aligned}\int_0^t ||B^*PS^K(\tau)y||^2 d\tau + \int_0^t \langle RS^K(\tau)y, S^K(\tau)y\rangle d\tau \\ = \langle Py, y\rangle - \langle S^K(t)y, S^K(t)y\rangle \\ \leq \langle Py, y\rangle\end{aligned}$$

So $\int_0^\infty \langle RS^K(t)y, S^K(t)y\rangle dt < \infty$ and therefore the operator

$$G(y) = \int_0^\infty \left(R^{\frac{1}{2}} S^K(t)\right)^* \left(R^{\frac{1}{2}} S^K(t)\right) y dt$$

is well defined and we have $\langle Gy, y\rangle = \int_0^\infty \langle RS^K(t)y, S^K(t)y\rangle dt$. So

$$\begin{aligned}\langle GS^K(t)y, S^K(t)y\rangle\rangle &= \int_0^\infty \langle RS^K(\tau)y, S^K(\tau)y\rangle d\tau \\ &= \int_t^\infty \langle RS^K(\tau)y, S^K(\tau)y\rangle d\tau\end{aligned}$$

We deduce that $\lim_{t\to\infty} \langle GS^K(t)y, S^K(t)y\rangle = 0$, therefore $\lim_{t\to\infty} GS^K(t)y = 0$.

Since G is solution of Lyapunov's equation

$$\langle A_K y, Gy\rangle + \langle Gy, A_K y\rangle + \langle Ry, y\rangle = 0$$

it follows that for all $y \in D(A)$, we have

$$\langle A_K S^K(t)y, GS^K(t)y\rangle + \langle GS^K(t)y, A_K S^K(t)y\rangle + \langle RS^K(t)y, S^K(t)y\rangle = 0$$

When $t \to 0$, $GS^K(t)y \to 0$, it follows $\langle RS^K(t)y, S^K(t)y\rangle \to 0$, which is equivalent to $RS^K(t)y \to 0$ (since R is positive and self-adjoint).

Consequently $||\chi_\omega S^K(t)y||^2 \to 0$, for all $y \in D(A)$, and the conclusion follows with the same considerations as in the Proposition 4.5. □

Example 4.7 We consider the system defined on $\Omega =]0, 1[$ by

$$\begin{cases} \dfrac{\partial y(t)}{\partial t} = -i\dfrac{\partial^2 y(t)}{\partial x^2} + v(t) i_\omega y(t) & \text{on} \quad Q \\ y(., 0) = y_0 & \text{on} \quad \Omega \end{cases} \tag{4.37}$$

The operator $A = -i\dfrac{\partial^2}{\partial x^2}$ with domain $\mathcal{D}(A) = H^2(\Omega) \cap H_0^1(\Omega)$, $B = i_\omega$ and $i^2 = -1$.

The operator A is self adjoint and $Re(< Ay, y >) = 0$, for all $y \in \mathcal{D}(A)$ and so (4.36) is verified.

On $\omega =]0, a[\subset \Omega$, for $y \in \mathcal{D}(A)$, we have

$$Re(< \chi_\omega(A + BK)y, y >_{L^2(\omega)}) = -\int_\omega |y(x)|^2 dx \leq 0$$

thus (4.34) is verified.

We therefore conclude that the control $v(t) = -i_\omega y(t)$ strongly stabilizes (4.37) on ω.

4.2.4 Low Cost Regional Stabilization Problem

We will look for the control ensuring stabilization of system (4.20) on a region ω and minimizing a functional cost.

Let $R \in \mathcal{L}(X)$ be a positive and self adjoint operator. The problem of regional stabilization at low cost consists in finding a control $u \in L^2(0, +\infty; V)$ which stabilizes (4.20) on ω minimizing the functional

$$q(v) = \int_0^{+\infty} < Ry(t), y(t) > dt + \int_0^{+\infty} ||v(t)||_V^2 dt \tag{4.38}$$

Suppose that for all initial state y_0, there exists a control v such that (4.38) is finite. It is known [3] that if the Riccati equation (4.33) has a positive definite solution P, then the control

$$v^*(t) = -B^* P y^*(t) \tag{4.39}$$

minimizes (4.38), which results exponential stabilization of the system (4.20) on Ω.

Proposition 4.11 *If there exists $e > 0$ such that*

$$< Ry, y > \geq e \; ||\chi_\omega y||^2_{L^2(\omega)} \; for \; all \; y \in X \tag{4.40}$$

and if the Eq. (4.33) has a positive solution P, then, noting $K = -B^* P$, we have

1. *If the condition (4.34) is verified, the control (4.39) strongly stabilizes the system (4.48) on ω.*
2. *If $S^K(t)$ verify (4.12), then the control (4.39) exponentially stabilizes the system (4.20) on ω.*

The proof follows from the result of the previous proposition.

Remark 4.9 In the previous result if we suppose, moreover, that the operator R satisfies

$$\langle Ry, y\rangle \geq e\ Re(\langle Ay, y\rangle) \ , \ y \in \mathcal{D}(A) \tag{4.41}$$

and if BB^*P is positive, this is the case for example when P commutes with BB^*, then (4.41) implies (4.35); in this case the state of the system (4.20) remains bounded on the residual part $\Omega\backslash\omega$.

Example 4.8 On $\Omega =]0, +\infty[$, we consider a system described by the equation

$$\begin{cases} \dfrac{\partial y(t)}{\partial t} = -\dfrac{\partial y(t)}{\partial x} + v(x,t) \text{ on } Q \\ y(,0) = y_0 \in L^2(\Omega) \end{cases} \tag{4.42}$$

The operator $A = -\dfrac{\partial}{\partial x}$ with domain $\mathcal{D}(A) = \{y \in H^1(\Omega);\ y(0) = 0\}$. On $\omega =]0, 1[$ we consider the functional (4.38) with

$$R = ((1-x)^2 + 1)i_\omega$$

that verify (4.40). The operator $P = (1-x)i_\omega$ is solution of (4.33) and the operator $K = -(1-x)i_\omega$ satisfy (4.34).

The control

$$v^*(x,t) = -(1-x)i_\omega y(t)$$

minimizes (4.38) and strongly stabilizes the system (4.42) on ω with a cost

$$q(v^*) = < Py_0, y_0 > = \int_0^1 (1-x)|y_0(x)|^2 dx$$

Example 4.9 On $\Omega =]0, \infty[$, let a system described by the equation

$$\begin{cases} \dfrac{\partial y(x,t)}{\partial t} = -\dfrac{\partial y(x,t)}{\partial x} + \chi_\omega a(x)v(t) \text{ on } Q \\ y(.,0) = y_0 \ \in \ L^2(\Omega) \end{cases} \tag{4.43}$$

The operator $A = \dfrac{\partial}{\partial x}$ with domain $\mathcal{D}(A) = \{y \in H^1(\Omega) \mid y(x) = 0\}$ and $v \in V = L^2(\Omega)$. Since $y(\infty) = 0$, for all $y \in \mathcal{D}(A)$, we have

$$Re(\langle Ay, y\rangle) = -\int_0^\infty \frac{dy(x)}{dx} y(x)dx = 0$$

If $\omega =]a, b[\subset \Omega$ and $a(.) \geq \alpha > 0$, then the control $v^*(t) = -i_\omega y(t)$ strongly stabilizes the system (4.43) on ω and minimizes the functional

$$q(v) = \int_0^{+\infty} ||\chi_\omega y(t)||^2_{L^2(\omega)} dt + \int_0^{+\infty} ||v(t)||^2_V dt$$

and $q(u^*) = ||y_0||^2$.

Remark 4.10 We can also replace (4.36) by the condition

$$\langle i_\omega Ay, y\rangle + \langle y, i_\omega Ay\rangle = 0, \quad \forall y \in D(A), \tag{4.44}$$

and if $||B^* i_\omega y|| \geq \alpha ||\chi_\omega y||$, the control $v^*(t) = -B^* i_\omega y^*(t)$ stabilizes (4.20) on ω. However the relationship (4.36) can be verified without condition (4.44), as shown in the previous example since for any $\omega =]a, b[\subset \Omega$, there exists $y \in D(A)$ such that $\displaystyle\int_a^b \frac{dy(x)}{dx} y(x)dx \neq 0$.

Example 4.10 We consider the system defined on $\Omega =]0, 1[$ by the Schrodinger equation

$$\begin{cases} i\dfrac{\partial y(t)}{\partial t} = \dfrac{\partial^2 y(t)}{\partial x^2} + u(t) i_\omega y(t) \quad Q \\ \\ y(., 0) = y_0 \in L^2(\Omega) \end{cases} \tag{4.45}$$

The operator $A = -i\dfrac{\partial^2}{\partial x^2}$ with domain $\mathcal{D}(A) = H^2(\Omega) \cap H_0^1(\Omega)$, is self adjoint and therefore verify (4.36).

The eigenvalues and the eigenvectors of A are given by

$$\lambda_j = -i\pi^2 j^2 \quad \text{and} \quad \varphi_j(x) = \sqrt{2}\sin(j\pi x), \;\; j \geq 1.$$

Let $\omega \subset \Omega$, the control $v^*(t) = -i_\omega y(t)$ strongly stabilizes (4.45) on ω and minimizes the functional

$$q(v) = \int_0^{+\infty} ||\chi_\omega y(t)||^2_{L^2(\omega)} dt + \int_0^{+\infty} ||v(t)||^2_V dt$$

and $q(u^*) = ||\chi_\omega y_0||^2_{L^2(\omega)}$.

The stabilized system on ω becomes

$$\frac{\partial y}{\partial t} = -i\frac{\partial^2 y}{\partial x^2} + i_\omega y(t) = \begin{cases} -i\dfrac{\partial^2 y}{\partial x^2} + y(t) \;\; \text{on } \omega \\ \\ -i\dfrac{\partial^2 y}{\partial x^2} \qquad\qquad \text{on } \Omega \setminus \omega \end{cases}$$

Let $y = y_1 + iy_2$, where y_1 and y_2 are the real and imaginary parts of the state y, we obtain

$$\begin{cases} \dfrac{\partial y_1(t)}{\partial t} = \Delta y_2(t) - i_\omega y_2(t) \quad \text{on } Q \\ \\ \dfrac{\partial y_2(t)}{\partial t} = -\Delta y_1(t) + i_\omega y_1(t) \text{ on} \end{cases}$$

To summarize the previous approach, numerical resolution of an optimal stabilization problem is obtained through the following steps

1.	Resolution of the Eq. (4.33)	$\longrightarrow$	P
2.	Computation of the control given by (4.39)	$\longrightarrow$	v^*
3.	Resolution of the Eq. (4.20)	$\longrightarrow$	$y^*(t)$

4.3 Numerical Examples

In this section we will present numerical examples illustrating the established results, in the case of one-dimensional and two-dimensional systems.

4.3.1 One Dimensional Case

On $\Omega =]0, 1[$, we consider the system described by the equation

$$\begin{cases} \dfrac{\partial y(t)}{\partial t} = (1-x)y(t) + v(x,t) \text{ on } Q \\ \\ y(.,0) = y_0 \in L^2(\Omega) \end{cases} \tag{4.46}$$

where $v(t) \in V = L^2(\Omega),\ t \geq 0$ and $y_0 = x|0.5 - x|$.

The uncontrolled system ($v(.) = 0$) is unstable on Ω (see Fig. 4.1).

We are looking for the control that stabilizes the system (4.46) on $\omega =]0, 0.5[$ minimizing the functional (4.38) in which we take $R = i_\omega$.

The operator $P = ((1-x) + \sqrt{1+(1-x)^2})i_\omega$ is solution of (4.33) and the control $v^*(t) = -Py(t)$ minimizes (4.38). The associated solution of (4.46) (see Fig. 4.2) is given by

$$y(x,t) = \begin{cases} e^{-t\sqrt{1+(1-x)^2}}\ y_0(x) & \text{if } x \in \omega \\ e^{t(1-x)}\ y_0(x) & \text{otherwise} \end{cases}$$

The system (4.46) is exponentially stabilized on the region ω.

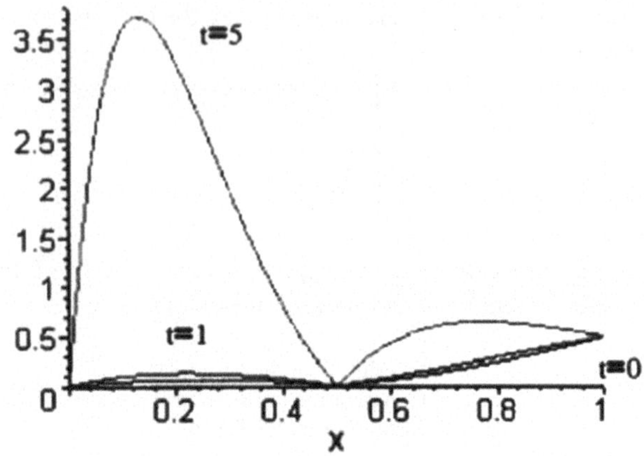

Fig. 4.1 The evolution of the autonomous system on Ω

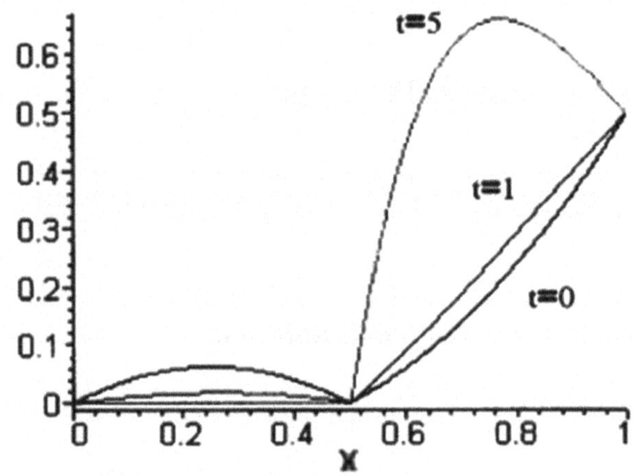

Fig. 4.2 The evolution of the system stabilized on ω

4.3.2 Two Dimensional Case

On $\Omega =]0, 2[\times]0, 2[$, we consider the system

$$\begin{cases} \dfrac{\partial y(t)}{\partial t} = f(x, z)y(t) + v(t; x, z) \text{ on } Q \\ y(., 0) = y_0 \in L^2(\Omega) \end{cases} \tag{4.47}$$

with

$$f(x, z) = \begin{cases} (x-1)^2 & \text{if } (x, z) \in]0, 1[\times]0, 2[\\ 0 & \text{otherwise} \end{cases}$$

and we assume that $v \in V = L^2(\Omega)$. It is clear that the uncontrolled system ($v(.) = 0$) is instable on Ω.

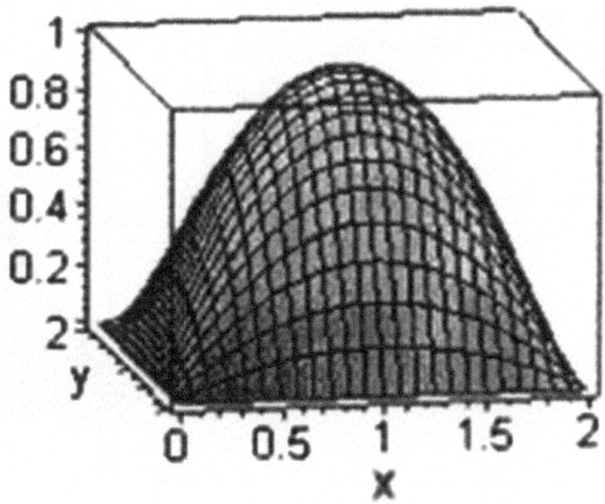

Fig. 4.3 The state of the system at $t = 0$

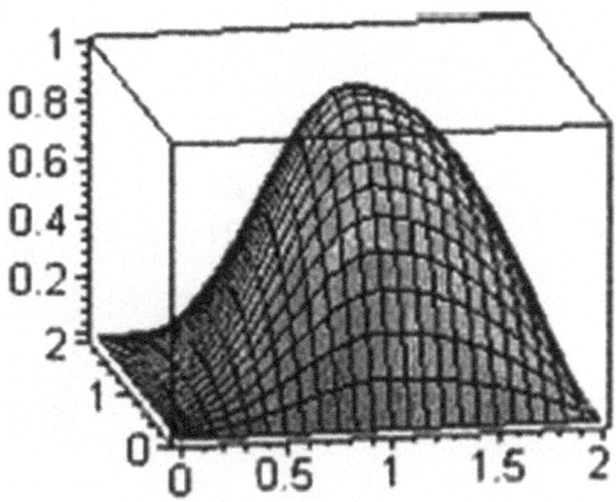

Fig. 4.4 Stabilized state on ω and bounded state on $\Omega\backslash\omega$ at $t = 1$

We look for a control that minimizes (4.38) in which we take $Ry = 3f(.)^2 y$, verifying (4.40) that stabilizes the system (4.47) on the region $\omega =]0, 1[\times]0, 2[$.

The operator $Py = 3f(.)y$ is solution of the Eq. (4.33) and the solution of system (4.47) associated to the control $v^*(t) = -Py(t)$, is given by

$$y(t; x, z) = \begin{cases} e^{-2t(x-1)^2} y_0(x, z) & \text{if } (x, z) \in]0, 1[\times]0, 2[\\ \\ y_0(x, y) & \text{otherwise} \end{cases}$$

Consequently the system (4.47) is exponentially stabilized on ω and the state remains bounded on $\Omega\backslash\omega$ (see Figs. 4.3, 4.4, 4.5 and 4.6).

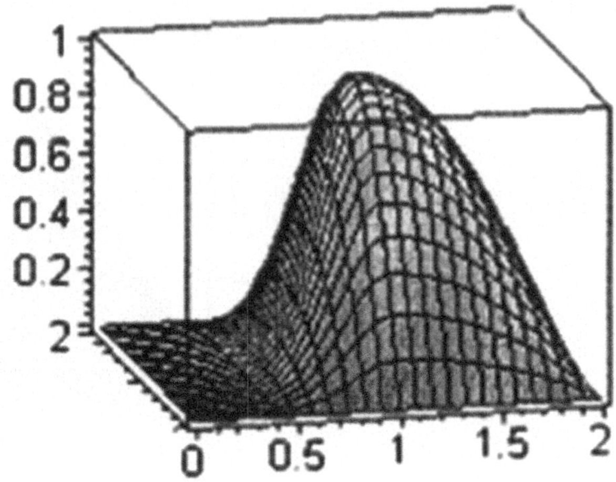

Fig. 4.5 Stabilized state on ω and bounded state on $\Omega\backslash\omega$ at $t = 5$

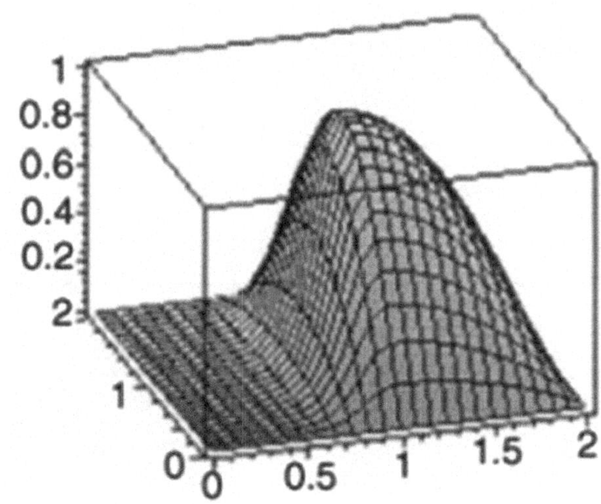

Fig. 4.6 Stabilized state on ω and bounded state on $\Omega\backslash\omega$ at $t = 10$

4.4 Regional Boundary Stabilization

In this section results of the previous section are expanded to the case where the target region Γ is located on the boundary $\partial\Omega$ of Ω ($\Gamma \subset \mathbb{R}^{n-1}$). It is then a question of stabilizing an unbounded output. To make sense of this output, we're going to use the notion of admissible region. Let Ω be an open bounded domain of $\mathbb{R}^n$ of boundary $\partial\Omega$. We still consider the system (4.1), with the same assumptions on the operator A and B.

$$\begin{cases} \dfrac{dy(t)}{dt} = Ay(t) \qquad \text{on } Q \\ \\ y(.,0) = y_0 \in X \end{cases} \tag{4.48}$$

Consider a region $\Gamma \subset \bar{\Omega}$ a surface to stabilize and let the restriction operator

$$\begin{aligned} \chi_{\Gamma} : X &\longrightarrow L^2(\Gamma) \\ y &\longrightarrow \chi_{\Gamma} y = y_{|\Gamma} \end{aligned}$$

Let χ_{Γ}^* be the adjoint operator of χ_{Γ} and $i_{\Gamma} = \chi_{\Gamma}^* \chi_{\Gamma}$.

We suppose that $L^2(\Gamma)$ is provided with its usual scalar product $(y, z) = \int_{\Gamma} \bar{y} z \, d\sigma$, the associated norm is given by $||z||_{L^2(\Gamma)} = \left(\int_{\Gamma} z^2 d\sigma \right)^{\frac{1}{2}}$ ($d\sigma$ indicates the superficial measure defined on $\partial\Omega$ and induced by the Lebesgue measure). We are then interested in the asymptotic behavior of the output

$$z(t) = \chi_{\Gamma} y(t) \tag{4.49}$$

We will see that it is possible to make sense of $z(t)$ for any trajectory $y(t)$ of the system (4.1).

Definition 4.3 The region Γ is said admissible for $S(t)$ if

1. The operator $\chi_{\Gamma} : (\mathcal{D}, ||.||_A) \longrightarrow L^2(\Gamma)$ is continuous, $\mathcal{D}(A)$ is provided with the graph norm $||.||_A$).
2. There exists $\alpha > 0$ such that

$$||z(.)||_{L^2(0,\infty;L^2(\Gamma))} = ||\chi_{\Gamma} S(.) y_0||_{L^2(0,\infty;L^2(\Gamma))} \leq \alpha ||y_0|| \text{ for all } y_0 \in \mathcal{D}(A). \tag{4.50}$$

Remark 4.11 The admissibility of Γ is in fact the admissibility of the operator χ_{Γ} in the sense of Weiss [4]. It allows to extend the application $y_0 \longrightarrow \chi_{\Gamma} S(.) y_0$ in a continuous application from X to $L^2(0, \infty, X)$, so (4.49) is well defined for any $y_0 \in X$.

Definition 4.4 If Γ is admissible, the system (4.48) with the output (4.49), is said weakly (respectively strongly, exponentially) stable on Γ, if $z(t)$ tends to 0 weakly (respectively strongly, exponentially) when $t \to \infty$.

Example 4.11 On $\Omega =]0, 2[\times]0, 2[$, we consider the system described par the following equation

$$\begin{cases} \dfrac{dy(t)}{dt} = \left(\dfrac{1}{2} - z \right) y(t) & \text{on } \Omega \times]0, \infty[\\ y(., 0) = y_0(x, z) \in H^1(\Omega) \end{cases} \tag{4.51}$$

The operator $Ay = (\frac{1}{2} - z) y$, with domain

$$\mathcal{D}(A) = \{ y \in H^1(\Omega) \, / y = 0 \text{ on } \Gamma_0 = \{0\} \times [0, 1] \}$$

$\mathcal{D}(A)$ is a closed space in $H^1(\Omega)$, provided with its usual inner product, it is therefore a Hilbert space.

Let $\Gamma = \{0\} \times [0, 2]$ and $\omega_a =]0, 2[\times]a, 2[$, with $a > \dfrac{1}{2}$.

For $y_0 \in \mathcal{D}(A)$, the solution of (4.51) verify

$$||\chi_\Gamma S(t)y_0||^2_{L^2(\Gamma)} = \int_1^2 e^{2(\frac{1}{2}-z)t}|y_0(0, z)|^2 dz \leq e^{-t} \int_1^2 |y_0(0, z)|^2 dz$$

We deduce, by density and continuity of the trace operator on $H^1(\Omega)$, that the system (4.51) is strongly stable on Γ.

Let ω be a region such that $\Gamma \subset \partial\omega$, (there exists $b > 0$ with $0 < c < \dfrac{1}{2}$) and let $\omega_0 =]0, b[\times]c, \dfrac{1}{2}[\subset \omega$. We have $||\chi_\omega y(t)|| \geq ||\chi_{\omega_0} y_0||_{L^2(\omega_0)}$, which shows that system (4.51) is not strongly stable on ω.

The following result establish the link between the regional stability of the system (4.1) on Γ and the properties of the spectrum point $\sigma(A)$ of A.

For that we suppose that A has a decreasing sequence of real eigenvalues$(\lambda_n)_{n\geq 1} \rightarrow -\infty$, this is the case for instance if A is self adjoint and has compact resolvent set. In this case A has a finite number of positive eigenvalues, each associated with a finite dimension proper subspace (see [2]).

Proposition 4.12 *1. If the system (4.48) is weakly stable on* Γ *then*

$$(\forall \lambda \in \sigma(A))\,; \quad Re\,(\lambda) \geq 0 \implies \forall \varphi \in Ker(A - \lambda\, I),\ \chi_\Gamma \varphi = 0 \tag{4.52}$$

2. *If the space* X *has an orthonormal basis* $(\varphi_n)_{n\geq 1}$ *of eigenfunctions of* A *associated to eigenvalues* $(\lambda_n)_{n\geq 1}$ *such that*

$$\exists\, \delta > 0\ \forall \lambda \in \sigma(A),\ (\exists \varphi \in Ker(A - \lambda\, I),\ |\ \chi_\Gamma \varphi \neq 0) \implies Re\,(\lambda) \leq -\delta \tag{4.53}$$

then the system (4.1) is exponentially stable on Γ.

Proof 1. Suppose there exist $\lambda \in \sigma(A)$ and $\varphi \in Ker(A - \lambda I)$ such that $Re(\lambda) \geq 0$ and $\chi_\Gamma \varphi \neq 0$, then

$$\langle \chi_\Gamma S(t)\varphi, \varphi\rangle_{L^2(\Gamma)} \geq ||\chi_\Gamma \varphi||_{L^2(\Gamma)} \neq 0\,, \quad \text{for all } t \geq 0$$

So the system (4.57) is not weakly stable on Γ.

2. We can assume that the eigenvalues of A are simple.
For $y_0 \in X$ and $t \geq 0$, we have

$$S(t)y_0 = \sum_{\lambda_n \in Sp(A)} e^{\lambda_n t}\langle y_0, \varphi_n\rangle \varphi_n$$

By continuity of $\chi_\Gamma : (D(A), ||.||_A) \longrightarrow L^2(\Gamma)$, there exists $h > 0$ such that

$$||\chi_\Gamma S(t) y_0||_{L^2(\Gamma)} \leq h \sum_{\substack{\lambda_n \in \sigma(A) \\ Ker(A-\lambda_n I) \not\subset Ker(\chi_\Gamma)}} \exp\{Re(\lambda_n)t\} |\langle y_0, \varphi_n \rangle| . ||\varphi_n||_A$$

So

$$||\chi_\Gamma S(t) y_0||_{L^2(\Gamma)} \leq h \sum_{\substack{\lambda_n \in \sigma(A) \\ Ker(A-\lambda_n I) \not\subset Ker(\chi_\Gamma)}} \left[\exp\{\frac{Re(\lambda_n)t}{2}\}\right]^2 ||y_0|| . ||\varphi_n||_A$$

We deduce that for all $t > 2$, we have

$$\begin{aligned} ||\chi_\Gamma S(t) y_0||_{L^2(\Gamma)} &\leq h \exp\left\{-\frac{\delta}{2} t\right\} ||y_0|| \times \sum_{\substack{\lambda_n \in \sigma(A) \\ Ker(A-\lambda_n I) \not\subset Ker(\chi_\Gamma)}} ||\varphi_n||_A e^{Re(\lambda_n)} \\ &\leq h \exp\left\{-\frac{\delta}{2} t\right\} ||y_0|| \times \sum_{\substack{\lambda_n \in \sigma(A) \\ ker(A-\lambda_n I) \not\subset Ker(\chi_\Gamma)}} (1 + |\lambda_n|^2)^{\frac{1}{2}} e^{Re(\lambda_n)} \end{aligned}$$

But according to (4.53), we have

$$\sum_{\substack{\lambda_n \in \sigma(A) \\ Ker(A-\lambda_n I) \not\subset Ker(\chi_\Gamma)}} (1 + |\lambda_n|^2)^{\frac{1}{2}} e^{Re(\lambda_n)} < \infty$$

and so (4.48) is exponentially stable on Γ. □

Example 4.12 On $\Omega =]0, 1[\times]0, 1[$, we consider the system described by the equation

$$\begin{cases} \dfrac{dy(t)}{dt} = Ay(t) & Q \\ y(0) = y_0 \in H^2(\Omega) \end{cases} \tag{4.54}$$

where

$$Ay = \Delta y + 4\pi^2 \langle y, \cos(x) \rangle \cos(\pi x) - < y, 1\!\!1 > 1\!\!1$$

with Δ is the Laplacian operator. The operator A of domain $\mathcal{D}(A) = \{y \in H^2(\Omega) \mid \frac{\partial y}{\partial \nu} = 0\}$, generates a C_0-semigroup $S(t)$.

The eigenfunctions and the eigenvalues of A are given by

$$\varphi_{k,\ell}(x, z) = \begin{cases} 2\cos(k\pi x)\cos(\ell\pi z) & \text{if } (k, l) \in \mathbb{N}^2 \text{ and } (k, \ell) \neq (0, 0) \\ 1 & \text{otherwise} \end{cases}$$

and

$$\lambda_{(k,\ell)} = \begin{cases} -1 & \text{if } (k,\ell) = (0,0) \\ 0 & \text{if } (k,\ell) = (1,0) \\ -(k^2+\ell^2)\pi^2 & \text{if } (k,\ell) \notin \{(0,0),(1,0)\} \end{cases}$$

We will show that the region $\Gamma = \{\frac{1}{2}\} \times [0, 1]$ is admissible for $S(t)$.

The operator $\chi_\Gamma : (\mathcal{D}(A), ||.||_A) \longrightarrow (L^2(\Gamma), ||.||_{L^2(\Gamma)})$ is continuous, since the operator $\chi_\Gamma : H^1(\Omega) \longrightarrow L^2(\Gamma)$ is continuous, there exists $\gamma > 0$ such that

$$||\chi_\Gamma y||_{L^2(\Gamma)} \leq \gamma ||y||_{H^1(\Omega)} \ , \quad \text{for } y \in \mathcal{D}(A) \subset H^1(\Omega) \tag{4.55}$$

According to Green's formula, we have

$$||\nabla y||^2_{(L^2(\Omega))^2} = -\int_\Omega \langle Ay, y\rangle dx - |\langle y, 1\!\!1\rangle|^2 + 4\pi^2 |\langle y, \varphi_{(0,1)}\rangle|^2 \ \forall y \in D(A)$$

and using Schwartz's inequality, we get

$$\begin{aligned} ||\nabla y||^2_{(L^2(\Omega))^2} &\leq ||Ay||.||y|| + ||y||^2 + 4\pi^2 ||y||^2 \\ &\leq (2 + 4\pi^2) ||y||^2_A \end{aligned}$$

So for $y \in \mathcal{D}(A)$, we have

$$\begin{aligned} ||\chi_\Gamma y||_{L^2(\Omega)} &\leq \gamma ||y||_{H^1(\Omega)} \leq \gamma \left(||y||^2_{L^2(\Omega)} + ||\nabla y||^2_{(L^2(\Omega))^2} \right)^{\frac{1}{2}} \\ &\leq \gamma \left(||y||^2_{L^2(\Omega)} + (2+4\pi^2)||y||^2_A \right)^{\frac{1}{2}} \leq \gamma (3 + 4\pi^2) ||y||_A \end{aligned}$$

To show (4.50), we consider

$$\begin{aligned} S(t) y_0 = e^{-t} \langle y_0, 1\!\!1 \rangle 1\!\!1 + \langle y_0, \varphi_{(1,0)} \rangle \varphi_{(1,0)} \\ + \sum_{(k,l) \notin \{(0,0),(1,0)\}} e^{\lambda_{(k,l)} t} \langle y_0, \varphi_{(k,l)} \rangle \varphi_{(k,l)} \end{aligned} \tag{4.56}$$

As the series (4.56) is uniformly convergent in $H^1(\Omega)$, we can write

$$\begin{aligned} \chi_\Gamma S(t) y_0 = e^{-t} \langle y_0, 1\!\!1 \rangle \chi_\Gamma 1\!\!1 + \langle y_0, \varphi_{(1,0)} \rangle \chi_\Gamma \varphi_{(1,0)} \\ + \sum_{(k,l) \notin \{(0,0),(1,0)\}} e^{\lambda_{(k,l)} t} \langle y_0, \varphi_{(k,l)} \rangle \chi_\Gamma \varphi_{(k,l)} \end{aligned}$$

and since $\chi_\Gamma \varphi_{(1,0)} = 0$, it follows

$$\chi_\Gamma S(t)y_0 = e^{-t}\langle y_0, \mathbb{1}\rangle \chi_\Gamma \mathbb{1} + \sum_{(k,l)\notin\{(0,0),(1,0)\}} e^{\lambda_{(k,l)}t}\langle y_0, \varphi_{(k,l)}\rangle \chi_\Gamma \varphi_{(k,l)}$$

According to (4.55), we have

$$\begin{aligned}
||\chi_\Gamma S(t)y_0|| &\le e^{-t}|\langle y_0, \mathbb{1}\rangle|.||\chi_\Gamma \mathbb{1}||_{L^2(\Gamma)} \\
&+ \sum_{(k,l)\notin\{(0,0),(1,0)\}} e^{\lambda_{(k,l)}t}|\langle y_0, \varphi_{(k,l)}\rangle|.||\chi_\Gamma \varphi_{(k,l)}||_{L^2(\Gamma)} \\
&\le \gamma e^{-t}|\langle y_0, \mathbb{1}\rangle|.||\mathbb{1}||_{H^1(\Omega)} \\
&+ \gamma \sum_{(k,l)\notin\{(0,0),(1,0)\}} e^{\lambda_{(k,l)}t}|\langle z_0, \varphi_{(k,l)}\rangle|.||\varphi_{(k,l)}||_{H^1(\Omega)}
\end{aligned}$$

It follows

$$\begin{aligned}
||\chi_\Gamma S(t)y_0|| &\le \gamma e^{-t}|\langle y_0, \mathbb{1}\rangle|||\mathbb{1}||_{H^1(\Omega)} \\
&+\gamma \sum_{(k,l)\notin\{(0,0),(1,0)\}} e^{-(k^2+l^2)\pi^2 t}|\langle y_0, \varphi_{(k,l)}\rangle| \left[1+4\pi^2(k^2+l^2)\right]^{\frac{1}{2}} \\
&\le \gamma \left[e^{-t} + \sum_{(k,l)\notin\{(0,0),(1,0)\}} e^{-(k^2+l^2)\pi^2 t} \left[1+4\pi^2(k^2+l^2)\right]^{\frac{1}{2}} \right] ||y_0||
\end{aligned}$$

Consequently $\displaystyle\int_0^{+\infty} ||\chi_\Gamma S(t)y_0||^2 dt < \infty$ therefore Γ is admissible for $S(t)$.

On the other hand, the only positive or zero eigenvalue is $\lambda_{(1,0)} = 0$, and is such that $\chi_\Gamma \varphi_{(1,0)} = 0$, and $\lambda_{(k,l)} \le -1$ for $(k, l) \ne (1, 0)$.

The previous proposition applies and we conclude that the system (4.54) is exponentially stable on Γ.

Remark 4.12 The system (4.54) is not stable on any neighborhood region $\omega \subset \Omega$ of Γ.

4.5 Optimal Control for Regional Boundary Stabilization

This section extends the previous work on regional stabilization on a region internal to the evolution domain to a boundary region.

Consider the system

$$\begin{cases} \dfrac{dy(t)}{dt} = Ay(t) + Bv(t) \quad Q \\ y(., 0) = y_0 \in \mathcal{D}(A) \end{cases} \tag{4.57}$$

where A is a linear operator with domain $\mathcal{D}(A)$ and generates a C_0 -semi group $S(t)$, $t \geq 0$ in the state space X, assumed to be a Hilbert space, provided with the inner product $\langle .,. \rangle$ and the associated norm $\|\cdot\|$. The control $v(.)$ belongs to an Hilbert space V and $B \in \mathcal{L}(V, X)$.

The problem of optimal stabilization consists in finding a control which ensures the regional stabilization of (4.57) on an admissible region Γ, and solution of minimization problem

$$\begin{cases} \min q(u) = \int_0^{+\infty} \langle Ry(t), y(t)\rangle dt + \int_0^{+\infty} \|v(t)\|_V^2 dt \\ v \in V_{ad} = \{v \in L^2(0, +\infty; V);\ q(v) < \infty\} \end{cases} \tag{4.58}$$

where $R : X \longrightarrow X$ is a bounded linear operator verifying

$$\langle Ry, y\rangle \geq c\, \|\chi_\Gamma y\|_{L^2(\Gamma)} \text{ for all } y \in \mathcal{D}(A) \text{ and } c > 0. \tag{4.59}$$

Remark 4.13 The cost of stabilizing a system on a boundary region Γ is less than the one on an internal region ω, such that $\Gamma \subset \partial\omega$. Indeed, if we consider the minimization of (4.22) on the sets of admissible controls $V_{ad}(\Gamma)$ and $V_{ad}(\omega)$, we have $0 \in V_{ad}(\Gamma)$, but for any internal region ω neighborhood of Γ we have $0 \notin V_{ad}(\omega)$. This imply the inequality

$$\min_{V_{ad}(\Gamma)} q(v) < \min_{V_{ad}(\omega)} q(v)$$

which shows that the cost is lower on Γ. This remark apply to the case of system (4.51) with a boundary region $\Gamma = \{0\} \times [0, 2]$.

The problem (4.58) will be solved by two approaches: the first one is a direct and the second one is based on internal stabilization techniques.

4.5.1 Direct Approach

The considered approach in this paragraph is based on the Lyapunov techniques using the Riccati equation.

Theorem 4.1 *Let $K = -B^*P$, where P is a positive solution of the following Riccati equation*

$$\langle Ay, Py\rangle + \langle Py, Ay\rangle - \langle B^*Py, B^*Py\rangle + \langle Ry, y\rangle = 0 \tag{4.60}$$

and suppose that for $y \in \mathcal{D}(A)$,

$$Re\left(\langle \chi_\Gamma (A + BK)y, \chi_\Gamma y\rangle_{L^2(\Gamma)}\right) \leq 0 \tag{4.61}$$

then

1. *The control*

$$v^*(t) = -B^* P y^*(t) \tag{4.62}$$

is solution of the problem (4.58) and strongly stabilizes the system (4.57) on Γ with the estimation

$$\|\chi_\Gamma y(t)\| = O\left(\frac{1}{\sqrt{t}}\right) \quad \textit{when} \quad t \to +\infty$$

2. *If there exists $\beta > 0$, such that*

$$||\chi_\Gamma y||^2_{L^2(\Gamma)} \geq \beta Re\left(\langle (A + BK)y, y\rangle\right) \tag{4.63}$$

then the state of the system (4.57) is stabilized on Γ and remains bounded on Ω.

Proof 1. For $y_0 \in \mathcal{D}(A)$, we have

$$\begin{aligned}\frac{d}{dt}\left(\langle PS^K(t)y_0, S^K(t)y_0\rangle\right) &= \langle PAS^K(t)y_0, S^K(t)y_0\rangle \\ &+\langle PS^K(t)y_0, AS^K(t)y_0\rangle \\ &= -\langle RS^K(t)y_0, S^K(t)y_0\rangle\end{aligned}$$

Integrating by parts, for any $t \geq 0$ we get,

$$\langle Py_0, y_0\rangle - \langle PS^K(t)y_0, S^K(t)y_0\rangle = \int_0^t \langle RS^K(s)y_0, S^K(s)y_0\rangle ds$$

This last relation is verified for all y_0 of X, since the operators P, R and $S^K(t)$ are bounded. As $P \geq$ is positive, we deduce that $q(v) < \infty$, in other words $V_{ad} \neq \emptyset$ for all initial state y_0, therefore (4.62) is the unique control solution of (4.58), and according to (4.59) we deduced that

$$\int_0^t ||\chi_\Gamma S^K(s)y_0||^2_{L^2(\Gamma)} ds \leq \langle Py_0, y_0\rangle,$$

On the other hand, (4.61) allows to write

$$\frac{d}{dt}||\chi_\Gamma S^K(t)y_0||^2_{L^2(\Gamma)} \leq 0$$

It follows that

$$t||\chi_\Gamma S^K(t)y_0||^2_{L^2(\Gamma)} \leq \int_0^t ||\chi_\Gamma S^K(s)y_0||^2_{L^2(\Gamma)} ds$$

therefore

$$||\chi_\Gamma S^K(t)y_0||^2_{L^2(\Gamma)} \leq \frac{\langle P y_0, y_0\rangle}{t} > 0 \tag{4.64}$$

Let $y_0 \in X$ and $(y_{0n})_{n\geq 1} \subset D(A)$ such that $y_{0n} \longmapsto y_0$ in X. The Equation (4.60) implies the condition (4.50), thus for all $n \geq 1$

$$\int_0^{+\infty} ||\chi_\Gamma S^K(t)(y_{0n} - y_0)||^2_{L^2(\Gamma)} dt \leq \alpha ||y_{0n} - y_0||^2$$

when $n \to \infty$, gives

$$\int_0^{+\infty} ||\chi_\Gamma S^K(t)(y_{0n} - y_0)||^2_{L^2(\Gamma)} dt \longrightarrow 0 \; a.e. \; t \geq 0$$

Therefore there exists a subsequence $(y_{0\ell(n)})_{n\geq 1}$ of $(y_{0n})_{n\geq 1}$ such that $\chi_\Gamma S^K(t)y_{0\ell(n)} \longrightarrow \chi_\Gamma S^K(t)y_0, \quad a.e. \; t \geq 0$, (see [5]).

From (4.64), we get

$$||\chi_\Gamma S^K(t)y_{0\ell(n)}||^2_{L^2(\Gamma)} \leq \frac{\langle P y_{0\ell(n)}, y_{0\ell(n)}\rangle}{t} \tag{4.65}$$

it follows

$$||\chi_\Gamma S^K(t)y_0||^2_{L^2(\Gamma)} \leq \frac{\langle P y_0, y_0\rangle}{t}, \quad a.e. \; , \; t > 0 \tag{4.66}$$

Therefore $||\, \chi_\Gamma S^K(t)y_0|| = \mathrm{O}\left(\frac{1}{\sqrt{t}}\right)$, for all $y_0 \in X$.

2. From (4.63) and using $Re\left(\langle (A + BK)S^K(t)y_0, S^K(t)y_0\rangle\right) = \frac{1}{2}\frac{d}{dt}||S^K(t)y_0||^2$, we deduce that

$$\int_0^t ||\chi_\Gamma S^K(s)y_0||^2_{L^2(\Gamma)} ds \geq \beta ||S^K(t)y_0||^2 - \beta ||y_0||^2$$

then $||S^K(t)y_0|| \leq M(y_0) := \frac{1}{\beta}\langle P y_0, y_0\rangle + ||y_0||^2, \; t \geq 0$.

By the density of $\mathcal{D}(A)$ and the continuity of the mapping $y_0 \longrightarrow M(y_0)$, we deduce that $||S^K(t)y_0||$ is bounded on Ω. □

Remark 4.14 We can show that, under the condition

$$Re(\langle \chi_\Gamma (A + BK)y, \chi_\Gamma y\rangle_{L^2(\Gamma)}) \geq \; \beta.Re(\langle (A + BK)y, y\rangle) \text{ for all } y \in \mathcal{D}(A), \beta > 0,$$

the state of the system (4.57) is stabilized on Γ and remains bounded on Ω.

Example 4.13 Let a system be defined on the disk $\Omega = \{(x, y) \mid x^2 + y^2 < 1\}$ by

$$\begin{cases} \dfrac{\partial y(t)}{\partial t} = f(x,z)y(t) + v(t)\chi_{D} & Q \\ y(0) = y_0 \end{cases} \tag{4.67}$$

where $D = \{(x,z) \in \mathbb{R}^2 \mid \; 0.9 \le x^2 + z^2 \le 1\}$ and

$$f(x,z) = \begin{cases} 0 & \text{if } (x,z) \in D \\ \dfrac{1}{10}(x^2+z^2-0.9)^2 & \text{otherwise} \end{cases}$$

The solution of system (4.67) with $v = 0$ is given by

$$y(t) = \begin{cases} y_0 & D \\ e^{t(x^2+y^2-0.9)^2} y_0 & \text{otherwise} \end{cases}$$

The system is unstable on $\Gamma = \{(x,z) \mid z \ge 0 \text{ and } x^2+z^2 = 1\}$.

We will see that we can stabilize (4.67) on Γ by a control which minimizes the functional (4.58) with $R = i_{\Gamma}$. The operator $P = i_{\Gamma}$ is solution of (4.60), and the condition (4.61) is verified. The control $v^*(t) = -P_{\Gamma} y^*(t)$ strongly stabilizes (4.67) on Γ and minimizes the functional (4.58).

4.5.2 Internal Approach

In this paragraph we develop an approach that characterizes an optimal control which stabilizes the system (4.57) on a boundary surface Γ, considered as limit to a sequence of internal regions $(\omega_j)_{j\ge1} \subset \Omega$, with $\Gamma \subset \partial\omega_j$ for all $j \ge 1$. The control stabilizing the system on the boundary region Γ is then a limit to a sequence of controls stabilizing (4.57) on each of (ω_j).

We consider the system (4.67) and the problem (4.58) with $\langle Ry, y\rangle = \|\chi_{\Gamma} y\|^2_{L^2(\Gamma)}$. In what follows we assume that the Eq. (4.60) has a unique positive solution P.

Let $(\omega_j)_{j\ge1}$ be a sequence of neighborhoods of Γ defined by:

$$\omega_j = \left\{x \in \Omega \mid \text{dist}(x,\Gamma) < \frac{1}{j}\right\} \tag{4.68}$$

where "dist" indicates the distance given by: $\text{dist}(x,\Gamma) = \inf_{\xi\in\Gamma} |x-\xi|_{\mathbb{R}^n}$.

The sequence $(\omega_j)_{j\ge1}$ is decreasing and converges to $\bigcap_{j\ge1} \overline{\omega_j} = \Gamma$.

Consider the minimization problem

$$\begin{cases} \min_v q_j(v) = \int_0^{+\infty} \langle R_j y(t), y(t)\rangle dt + \int_0^{+\infty} \|v(t)\|_V^2 dt \\ v \in V_j = \{v \in L^2(0, +\infty; V);\ q_j(v) < \infty\} \end{cases} \tag{4.69}$$

where $\langle R_j y, y\rangle = j\|\chi_{\omega_j} y\|^2$. Then we have

Theorem 4.2 *If the system (4.57) is exponentially stabilized on an internal region $\omega \subset \Omega$ containing Γ on its boundary ($\Gamma \subset \partial\omega$) and if the condition (4.61) is satisfied, then the sequence of controls $v_j^*(t)$ solutions of* (4.69) *converges to the control $v^*(t) = -B^* P y^*(t)$ solution of (4.58), and strongly stabilizes the system (4.57) on the boundary region Γ.*

Proof Since (4.57) is exponentially stabilizable on ω, it is stabilizable on $\omega_j \subset \omega$. We deduce that $q_j(v) < +\infty$, for all $j \geq 1$.

On the other hand, the problem (4.69) has a unique solution given by $v_j^*(t) = -B^* P_j y^*(t)$, where $P_j \in \mathcal{L}(X)$ is a positive and self-adjoint operator verifying the Riccati equation:

$$\langle Ay, P_j y\rangle + \langle P_j y, Ay\rangle - \langle B^* P_j y, B^* P_j y\rangle + \langle R_j y, y\rangle = 0 \tag{4.70}$$

and we have $q_j(v_j^*) = \langle P_j y_0, y_0\rangle$, (see [6]).

For $y \in \mathcal{D}(A)$, we have $j\|\chi_{\omega_j} y\|^2 \longrightarrow \|\chi_\Gamma y\|^2_{L^2(\Gamma)}$ when $j \longrightarrow +\infty$ (Chilov 1967), it follows $\lim_{j\longrightarrow+\infty} \langle R_j y, y\rangle = \|\chi_\Gamma y\|^2_{L^2(\Gamma)}$, for all $y \in D(A)$.

Now, for j large enough, we have

$$\langle R_j y, y\rangle \leq \frac{3}{2} \lim_{j\longrightarrow+\infty} \langle R_j y, y\rangle = \frac{3}{2}\|\chi_\Gamma y\|^2_{L^2(\Gamma)}$$

Then for all $v \in V_a d \subset V$

$$q_j(v) \leq q(v) + \frac{1}{2}\int_0^{+\infty} \|\chi_\Gamma y(t)\|^2_{L^2(\Gamma)} dt$$

For $v = v_j^*$, we obtain

$$q_j(v_j^*) \leq q(v_j^*) + \frac{1}{2}\int_0^{+\infty} \|\chi_\Gamma y_j^*(t)\|^2_{L^2(\Gamma)} dt$$

where $y_j^*(t)$ is the associated solution to $v_j^*(t)$. Therefore

$$q(v_j^*) \leq q(v^*) + \frac{1}{2} q_j(v_j^*),$$

and consequently $\langle P_j y_0, y_0\rangle \leq 2q(u^*)$. This shows that the sequence $\langle P_j y_0, y_0\rangle$ is bounded, and according to the Banach–Steinhaus theorem, the sequence $(P_j)_{n\geq 1}$ is uniformly bounded.

Let $P_{\ell(j)}$ be a subsequence which weakly converges (therefore converge strongly) to a positive and self-adjoint operator P verifying the Riccati equation

$$\langle Ay, Py\rangle + \langle Py, Ay\rangle - \langle B^*Py, B^*Py\rangle + \lim_{j\to+\infty} \langle R_j y, y\rangle = 0 \tag{4.71}$$

The operator P is then the unique solution of the Eq. (4.60), therefore P_j converges strongly to P, and consequently the sequence $v_j^*(t)$ converges to $v^*(t) = -B^*Py^*(t)$, as $j \to \infty$ which is the solution of the problem (4.58). □

Example 4.14 We take again the system (4.67) and we will give an approximation of the optimal control $v^*(t) = -i_\Gamma y^*(t)$ stabilizing the system (4.67) on the boundary region $\Gamma = \{(x, z) \mid z \geq 0, x^2 + z^2 = 1\}$.

Lets consider a sequence of regions ω_j $j \geq 0$ neighborhoods of Γ defined by

$$\omega_j = \{(x, z) \in \Omega \mid z \geq 0 \text{ and } (1 - \frac{1}{j})^2 < x^2 + z^2 < 1\}$$

We propose to stabilize the system (4.67) on every ω_j, $j \geq 1$ by minimizing the functional

$$q_j(u) = j\int_0^{+\infty} ||\chi_{\omega_j} y(t)||^2 dt + \int_0^{+\infty} ||v(t)||_V^2 dt \tag{4.72}$$

The control $v_j^*(t) = -\sqrt{j} i_{\omega_j} y_j^*(t)$ minimizes (4.72) and strongly stabilizes (4.67) on ω_j and the corresponding solution to $y_0 = 1$ is given by

$$y_j^*(t) = \begin{cases} e^{-t} & \text{on } \omega_j \\ e^{t(x^2+z^2-0.9)^2} & \text{otherwise} \end{cases}$$

Now by substituting $y_j^*(t)$ in $v_j^*(t)$ we obtain $v_j^*(t) = \sqrt{j} e^{-t} P_{\omega_j}$ which is a sequence of controls that converges to $v^*(t)$. Therefore the sequence of optimal controls $v_j^*(t)$ converges to $v^*(t)$ the optimal control on Γ.

References

1. Datko, R.: Extending a theorem of A. M. Liapunov to Hilbert space. J. Math. Anal. Appl. **32**, 610–616 (1970)
2. Triggiani, R.: On the stabilizability problem in Banach space. J. Math. Anal. Appl. **52**, 383–403 (1975)
3. Curtain, R.F., Zwart, H.J.: An Introduction to Infinite Dimensional Linear Systems Theory. Springer, New York (1995)

4. Weiss, G.: Admissibility of unbounded control operators. SIAM J. Control Optim. **27**, 527–545 (1989)
5. Brézis, H.: Analyse fonctionnelle, théorie et application. Masson, Paris (1992)
6. Zerrik, E., Ouzahra, M.: Regional stabilization for infinite-dimensional systems. Int. J. Control **76**(1), 73–81 (2003)

Chapter 5
Regional Stabilization of Infinite Dimensional Bilinear Systems

The notion of regional stabilization of infinite dimensional systems consists in studying the asymptotic behaviour of such a system only within a subregion ω interior or in the boundary of the system evolution domain Ω. This notion includes the classical one and enables us to analyse the behaviour of a distributed system in any subregion of its spatial domain. Also, it makes sense for the usual concept of stabilization taking into account the spatial variable and then becomes closer to real applications, where one wishes to stabilize a system in a critical subregion.

The aim of this chapter is to study regional exponential, strong and weak stabilization of the bilinear system defined on a bounded and regular spatial domain $\Omega \subset \mathbb{R}^n$ is described by

$$\begin{cases} \dfrac{dy(t)}{dt} = Ay(t) + u(t)By(t) \\ y(0) = y_0 \end{cases} \tag{5.1}$$

where the state space is $L^2(\Omega)$ endowed with its usual inner product, denoted by $\langle ... \rangle$ and $\|.\|$ the associated norm, the operator A generates a semigroup of contractions $S(t)$ on $L^2(\Omega)$, $u(t) \in L^2(0, +\infty)$ is a scalar valued control and B is a linear, bounded and positive operator mapping $L^2(\Omega)$ into itself.

5.1 Regional Stabilization

Consider system (5.1) evolving on an open and regular set Ω of $\mathbb{R}^n$ and let ω be a subregion of Ω, $\chi_\omega : L^2(\Omega) \longrightarrow L^2(\omega)$ be the restriction operator to ω, while χ_ω^* denotes the adjoint operator of χ_ω and given by

E. H. Zerrik and O. Castillo, *Stabilization of Infinite Dimensional Systems*,
Studies in Systems, Decision and Control 355,
https://doi.org/10.1007/978-3-030-68600-0_5

$$(\chi_\omega^* y)(x) = \begin{cases} y(x), & x \in \omega \\ 0, & otherwise \end{cases} \tag{5.2}$$

Assume that $i_\omega B$ is positive where $i_\omega = \chi_\omega^* \chi_\omega$.

Regional stabilization consists in stabilizing the output $z(t) = \chi_\omega y(t)$.

Let us recall a definition of regional stabilization for system (5.1).

Definition 5.1 System (5.1) is said to be regionally weakly (respectively strongly, exponentially) stabilizable on ω, if for any initial condition $y_0 \in L^2(\Omega)$ the corresponding solution $y(t)$ of (5.1) is unique and $\chi_\omega y(t)$ weakly (respectively strongly, exponentially) converges to 0 as $t \longrightarrow +\infty$.

Remark 5.1
1. The concept is general because when $\omega = \Omega$, we obtain the classical definition of stability.
2. Note that regional exponential stabilization $\Rightarrow$ regional strong stabilization $\Rightarrow$ regional weak stabilization.

In what follows, we will discuss regional exponential, strong and weak stabilization of system (5.1) considering the controls

$$u(t) = \begin{cases} -\dfrac{\langle i_\omega B y(t), y(t)\rangle}{\|y(t)\|^2}, & y(t) \neq 0 \\ 0, & y(t) = 0 \end{cases} \tag{5.3}$$

and

$$u(t) = -\langle i_\omega B y(t), y(t)\rangle \tag{5.4}$$

Let us denote $g(y) = \dfrac{\langle i_\omega B y, y\rangle}{\|y\|^2}$ for all $y \neq 0$, $g(0) = 0$ and assume that

$$\langle i_\omega A y, y\rangle \leq 0, \quad \forall y \in \mathcal{D}(A). \tag{5.5}$$

5.1.1 Regional Exponential Stabilization

The following result gives sufficient conditions for regional exponential stabilization of system (5.1).

Theorem 5.1 *Let A generate a semigroup of contractions $(S(t))_{t\geq 0}$ and B satisfy*

$$\int_0^T |\langle i_\omega B S(s) y, S(s) y\rangle| ds \geq \delta \|\chi_\omega y\|_{L^2(\omega)}, \ (for\ some\ T, \delta > 0) \tag{5.6}$$

then control (5.3) *regionally exponentially stabilizes system* (5.1) *on* ω.

Proof System (5.1) has a unique mild solution $y(t)$ on $L^2(\Omega)$ (see [1]) defined on a maximal interval $[0, t_{\max}[$ by the variation of constants formula

$$y(t) = S(t)y_0 - \int_0^t g(y(s))S(t-s)By(s)ds \tag{5.7}$$

Let us consider the nonlinear semigroup $\Gamma(t)y_0 := y(t)$.

Since $S(t)$ is a semigroup of contractions (A is dissipative), then

$$\frac{d}{dt}\|\Gamma(t)y_0\|^2 \leq -2g(\Gamma(t)y_0)\langle B\Gamma(t)y_0, \Gamma(t)y_0\rangle$$

Integrating this inequality, we get

$$\|\Gamma(t)y_0\|^2 \leq \|y_0\|^2 - 2\int_0^t g(\Gamma(s)y_0)\langle B\Gamma(s)y_0, \Gamma(s)y_0\rangle ds, \ \forall t \in [0, t_{max}[$$

Since B and $i_\omega B$ are positive, then

$$\|\Gamma(t)y_0\| \leq \|y_0\|. \tag{5.8}$$

For all $y_0 \in L^2(\Omega)$ and $t \geq 0$, we have

$$\begin{aligned}\langle i_\omega BS(t)y_0, S(t)y_0\rangle &= \langle i_\omega B(S(t)y_0 - \Gamma(t)y_0), S(t)y_0\rangle \\ &- \langle i_\omega B\Gamma(t)y_0, \Gamma(t)y_0 - S(t)y_0\rangle \\ &+ \langle i_\omega B\Gamma(t)y_0, \Gamma(t)y_0\rangle\end{aligned}$$

Since χ_ω is continuous, then there exists $\alpha > 0$ such that

$$\begin{aligned}|\langle i_\omega BS(t)y_0, S(t)y_0\rangle| &\leq 2\alpha\|B\|\|\Gamma(t)y_0 - S(t)y_0\|\|y_0\| \\ &+ |\langle i_\omega B\Gamma(t)y_0, \Gamma(t)y_0\rangle|\end{aligned} \tag{5.9}$$

It follows from (5.7), that

$$\|\Gamma(t)y_0 - S(t)y_0\| \leq \|B\|\int_0^t |g(\Gamma(s)y_0)|\|\Gamma(s)y_0\|ds$$

For a fixed $T \in]0, t_{max}[$, Schwartz's inequality yields

$$\|\Gamma(t)y_0 - S(t)y_0\| \leq \|B\|\left\{T\int_0^T |g(\Gamma(s)y_0)|^2\|\Gamma(s)y_0\|^2 ds\right\}^{\frac{1}{2}}, \ \forall t \in [0, T] \tag{5.10}$$

Using (5.8), we get

$$|\langle i_\omega B\Gamma(t)y_0, \Gamma(t)y_0\rangle| \leq |g(\Gamma(t)y_0)|\|\Gamma(t)y_0\|\|y_0\|, \ \forall t \in [0, T] \tag{5.11}$$

Integrating (5.9) over the interval $[0, T]$ and taking into account (5.10) and (5.11), we obtain

$$\int_0^T |\langle i_\omega BS(s)y_0, S(s)y_0\rangle| ds \le 2\alpha T^{\frac{3}{2}} \|B\|^2 \|y_0\| \Big\{ \int_0^T |g(\Gamma(s)y_0)|^2 \|\Gamma(s)y_0\|^2 ds \Big\}^{\frac{1}{2}} + \sqrt{T} \|y_0\| \Big\{ \int_0^T |g(\Gamma(s)y_0)|^2 \|\Gamma(s)y_0\|^2 ds \Big\}^{\frac{1}{2}} \tag{5.12}$$

Replacing y_0 by $\Gamma(t)y_0$ in (5.12), using a variable change and the superposition property of the semigroup $\Gamma(t)$, we deduce that

$$\begin{aligned}\int_0^T |\langle i_\omega BS(s)\Gamma(t)y_0, S(s)\Gamma(t)y_0\rangle| ds \le\ & 2\alpha T^{\frac{3}{2}} \|B\|^2 \|\Gamma(t)y_0\| \\ & \times \Big\{ \int_t^{t+T} |g(\Gamma(s)y_0)|^2 \|\Gamma(s)y_0\|^2 ds \Big\}^{\frac{1}{2}} \\ & + \sqrt{T} \|\Gamma(t)y_0\| \\ & \times \Big\{ \int_t^{t+T} |g(\Gamma(s)y_0)|^2 \|\Gamma(s)y_0\|^2 ds \Big\}^{\frac{1}{2}}\end{aligned}$$

Using (5.6) and (5.8), we obtain

$$\delta \|\chi_\omega \Gamma(t)y_0\|_{L^2(\omega)} \le \beta \Big\{ \int_t^{t+T} |g(\Gamma(s)y_0)|^2 \|\Gamma(s)y_0\|^2 ds \Big\}^{\frac{1}{2}}$$

where $\beta = (2\alpha T \|B\|^2 + 1)\sqrt{T} \|y_0\|$.

Then

$$\delta^2 \|\chi_\omega \Gamma(t)y_0\|^2_{L^2(\omega)} \le \beta^2 \int_t^{t+T} |g(\Gamma(s)y_0)|^2 \|\Gamma(s)y_0\|^2 ds \tag{5.13}$$

Using (5.5), we get

$$\frac{d}{dt} \|\chi_\omega \Gamma(t)y_0\|^2_{L^2(\omega)} \le -2|g(\Gamma(t)y_0)|^2 \|\Gamma(t)y_0\|^2 \tag{5.14}$$

integrating this inequality from kT to $(k+1)T$, $(k \in \mathbb{N})$, we obtain

$$\|\chi_\omega \Gamma(kT)y_0\|^2_{L^2(\omega)} - \|\chi_\omega \Gamma((k+1)T)y_0\|^2_{L^2(\omega)} \ge 2 \int_{kT}^{(k+1)T} |g(\Gamma(s)y_0)|^2 \|\Gamma(s)y_0\|^2 ds$$

Using (5.13) and (5.14), we deduce

$$\Big(1 + 2\Big(\frac{\delta}{\beta}\Big)^2\Big) \|\chi_\omega \Gamma((k+1)T)y_0\|^2_{L^2(\omega)} \le \|\chi_\omega \Gamma(kT)y_0\|^2_{L^2(\omega)}$$

Then

$$\|\chi_\omega \Gamma((k+1)T)y_0\|_{L^2(\omega)} \le \rho \|\chi_\omega \Gamma(kT)y_0\|_{L^2(\omega)}$$

where $\rho = \dfrac{1}{(1+2(\frac{\delta}{\beta})^2)^{\frac{1}{2}}}$.

By recurrence, we show that

$$\|\chi_\omega \Gamma(kT)y_0\|_{L^2(\omega)} \le \rho^k \|\chi_\omega y_0\|_{L^2(\omega)}$$

Taking $k = E(\frac{t}{T})$ the integer part of $\frac{t}{T}$, we conclude that

$$\|\chi_\omega \Gamma(t)y_0\|_{L^2(\omega)} \le K e^{-\sigma t}\|y_0\|, \ \forall t \ge 0$$

where $K = \alpha(1+2(\frac{\delta}{\beta})^2)^{\frac{1}{2}}$ and $\sigma = \dfrac{\ln(1+2(\frac{\delta}{\beta})^2)}{2T}$, which shows the regional exponential stability of system (5.1) on ω. ■

5.1.2 Regional Weak Stabilization

Here, we give sufficient conditions that allow regional weak stabilization of system (5.1) using control (5.3).

Theorem 5.2 *Let A generate a semigroup of contractions $(S(t))_{t\ge 0}$ and B be compact such that*

$$\langle i_\omega BS(t)z, S(t)z\rangle = 0, \ \forall t \ge 0 \Longrightarrow \chi_\omega z = 0 \tag{5.15}$$

then control (5.3) *regionally weakly stabilizes system* (5.1) *on* ω.

Proof Let us consider the nonlinear semigroup $\Gamma(t)y_0 = y(t)$ and let (t_n) be a sequence of real numbers such that $t_n \longrightarrow +\infty$ as $n \longrightarrow +\infty$.

By (5.8), $\Gamma(t_n)y_0$ is bounded in the Hilbert space $L^2(\Omega)$, then there exists a subsequence $(t_{\phi(n)})$ of (t_n) and $z \in L^2(\Omega)$ such that

$$\Gamma(t_{\phi(n)})y_0 \rightharpoonup z \quad as \quad n \longrightarrow +\infty$$

For all $n \ge 0$, we set

$$\lambda(t_{\phi(n)}) = \int_{t_{\phi(n)}}^{t_{\phi(n)}+T} |g(\Gamma(s)y_0)|^2 \|\Gamma(s)y_0\|^2 ds.$$

Integrating (5.14), we get

$$2\int_0^t |g(\Gamma(s)y_0)|^2 \|\Gamma(s)y_0\|^2 ds \le \|\chi_\omega y_0\|^2_{L^2(\omega)} - \|\chi_\omega \Gamma(t)y_0\|^2_{L^2(\omega)}.$$

Then

$$\int_0^t |g(\Gamma(s)y_0)|^2 \|\Gamma(s)y_0\|^2 ds \leq \frac{1}{2}\|\chi_\omega y_0\|^2_{L^2(\omega)}.$$

It follows that

$$\int_0^{+\infty} |g(\Gamma(s)y_0)|^2 \|\Gamma(s)y_0\|^2 ds < +\infty.$$

We deduce that $\lambda(t_{\phi(n)}) \longrightarrow 0$ as $n \longrightarrow +\infty$.

Replacing y_0 by $\Gamma(t_{\phi(n)})y_0$ in (5.12) and using the superposition property of the semigroup $\Gamma(t_{\phi(n)})$, we deduce that

$$\int_0^T |\langle i_\omega BS(s)\Gamma(t_{\phi(n)})y_0, S(s)\Gamma(t_{\phi(n)})y_0\rangle| ds \leq (2\alpha\|B\|^2 T + 1)\|y_0\|\sqrt{T\lambda(t_{\phi(n)})} \tag{5.16}$$

Since $\lambda(t_{\phi(n)}) \longrightarrow 0$ as $n \longrightarrow +\infty$, then

$$\lim_{n\longrightarrow+\infty} \int_0^T \langle i_\omega BS(s)\Gamma(t_{\phi(n)})y_0, S(s)\Gamma(t_{\phi(n)})y_0\rangle ds = 0$$

Since B is compact and $S(s)$ is continuous $\forall s \geq 0$, we have

$$\lim_{n\longrightarrow+\infty} \langle i_\omega BS(s)\Gamma(t_{\phi(n)})y_0, S(s)\Gamma(t_{\phi(n)})y_0\rangle = \langle i_\omega BS(s)z, S(s)z\rangle$$

By dominated convergence theorem, we get

$$\int_0^T |\langle i_\omega BS(s)z, S(s)z\rangle| ds = 0$$

It follows that

$$\langle i_\omega BS(s)z, S(s)z\rangle = 0, \quad \forall s \in [0, T]$$

Using (5.15), we deduce that

$$\chi_\omega\Gamma(t_{\phi(n)})y_0 \rightharpoonup 0 \quad as \quad n \longrightarrow +\infty$$

Which implies that $\forall\phi \in L^2(\Omega)$, $\langle \chi_\omega\Gamma(t_n)y_0, \phi\rangle \longrightarrow 0$ as $n \longrightarrow +\infty$, and hence

$$\chi_\omega\Gamma(t)y_0 \rightharpoonup 0 \quad as \quad t \longrightarrow +\infty.$$

In other words $\chi_\omega y(t)$ converges weakly to 0 as $t \longrightarrow +\infty$, and then system (5.1) is regionally weakly stabilizable on ω. ∎

Remark 5.2 In the case $\omega = \Omega$, we retrieve the result established in [2] about the weak stabilization of system (5.1) on the whole domain Ω.

Example 5.1 On $\Omega =]0, 1[$, we consider the heat equation defined by

$$\begin{cases} \dfrac{\partial y(x,t)}{\partial t} = \dfrac{\partial^2 y(x,t)}{\partial x^2} + u(t)By(x,t), & \Omega\times]0,+\infty[\\ y(x,0) = y_0, & \Omega \\ y'(0,t) = y'(1,t) = 0, & \partial\Omega\times]0,+\infty[\end{cases} \tag{5.17}$$

Let us denote $Ay = \Delta y$, $\forall y \in \mathcal{D}(A) = \{y \in H^2(\Omega) |\ y'(0,t) = y'(1,t) = 0\}$.

The eigenvalues and corresponding eigenfunctions of A are given by $\lambda_k = -(k\pi)^2$ and $\varphi_k(x) = \sqrt{2}\cos(k\pi x)$ for all $k \in \mathbb{N}$.

The operator of control defined by $By = \sum_{k=1}^{+\infty} \frac{1}{k^2}\langle i_\omega y, \varphi_k\rangle\varphi_k$ is linear and compact.

The operator A generates the semigroup $S(t)y = \sum_{k=0}^{+\infty} e^{\lambda_k t}\langle y, \varphi_k\rangle\varphi_k$.

For $\omega =]0, \frac{1}{2}[$, we have

$$\langle i_\omega BS(t)y, S(t)y\rangle = 0,\ \forall t \geq 0 \Longrightarrow \chi_\omega y = 0.$$

So (5.15) holds.

Consequently, system (5.17) is regionally weakly stabilizable on ω by the control

$$u(t) = -\frac{\sum_{k=1}^{+\infty} \frac{1}{k^2}|\langle i_\omega y(t), \varphi_k\rangle|^2}{\|y(t)\|^2}.$$

5.1.3 Regional Strong Stabilization

In this subsection we give sufficient conditions for regional strong stabilization.

Theorem 5.3 *Suppose that A generates a semigroup of contractions $(S(t))_{t\geq 0}$ and B satisfies*

$$\int_0^T |\langle i_\omega BS(s)y, S(s)y\rangle| ds \geq \delta\|\chi_\omega y\|^2_{L^2(\omega)},\ (for\ some\ T, \delta > 0) \tag{5.18}$$

then control (5.4) *regionally strongly stabilizes system (5.1) on ω.*

Proof Using (5.5), we obtain

$$\frac{d}{dt}\|\chi_\omega y(t)\|^2_{L^2(\omega)} \leq -2|\langle i_\omega By(t), y(t)\rangle|^2,\ \forall t \geq 0.$$

Integrating this inequality, we get

$$\int_0^t |\langle i_\omega By(s), y(s)\rangle|^2 ds \leq \frac{1}{2}\|\chi_\omega y_0\|^2_{L^2(\omega)}, \ \forall t \geq 0.$$

Hence

$$\int_0^{+\infty} |\langle i_\omega By(s), y(s)\rangle|^2 ds < +\infty. \tag{5.19}$$

Letting $\Phi(t) = \int_0^t S(t-s)u(s)By(s)ds$, we obtain the following relation

$$\begin{aligned}\langle i_\omega BS(t)y_0, S(t)y_0\rangle = &\langle i_\omega B(S(t)y_0 - y(t)), S(t)y_0\rangle - \langle i_\omega By(t), \Phi(t)\rangle \\ &+\langle i_\omega By(t), y(t)\rangle\end{aligned}$$

Since χ_ω is continuous, then there exists $\alpha > 0$ such that

$$\begin{aligned}|\langle i_\omega BS(t)y_0, S(t)y_0\rangle| \leq &\alpha\|B\|\|\Phi(t)\|(\|S(t)y_0\| + \|y(t)\|) \\ &+|\langle i_\omega By(t), y(t)\rangle|\end{aligned} \tag{5.20}$$

Then

$$|\langle i_\omega BS(t)y_0, S(t)y_0\rangle| \leq 2\alpha\|B\|\|\Phi(t)\|\|y_0\| + |\langle i_\omega By(t), y(t)\rangle|, \forall t \in [0, T].$$

Moreover, we have

$$\|\Phi(t)\| \leq \|B\|\|y_0\|\sqrt{T}\Big(\int_0^T |\langle i_\omega By(s), y(s)\rangle|^2 ds\Big)^{\frac{1}{2}} \tag{5.21}$$

Integrating (5.20) over the interval $[0, T]$ and taking into account (5.21), we obtain

$$\begin{aligned}\int_0^T |\langle i_\omega BS(s)y_0, S(s)y_0\rangle| ds \leq &2\alpha T^{\frac{3}{2}}\|y_0\|^2\|B\|\Big(\int_0^T |\langle i_\omega By(s), y(s)\rangle|^2 ds\Big)^{\frac{1}{2}} \\ &+\sqrt{T}\|y_0\|\Big(\int_0^T |\langle i_\omega By(s), y(s)\rangle|^2 ds\Big)^{\frac{1}{2}}.\end{aligned}$$

Replacing y_0 by $y(t)$ and using the superposition property of the solution, we get

$$\begin{aligned}\int_0^T |\langle i_\omega BS(s)y(t), S(s)y(t)\rangle| ds \leq &2\alpha T^{\frac{3}{2}}\|y_0\|\|B\|\Big(\int_t^{t+T} |\langle i_\omega By(s), y(s)\rangle|^2 ds\Big)^{\frac{1}{2}} \\ &+\sqrt{T}\|y_0\|\Big(\int_t^{t+T} |\langle i_\omega By(s), y(s)\rangle|^2 ds\Big)^{\frac{1}{2}}.\end{aligned}$$

From (5.19), we have

$$\int_t^{t+T} |\langle i_\omega By(s), y(s)\rangle|^2 ds \longrightarrow 0, \; as \; t \longrightarrow +\infty.$$

It follows that

$$\int_0^T |\langle i_\omega BS(s)y(t), S(s)y(t)\rangle| ds \longrightarrow 0, \; as \; t \longrightarrow +\infty. \tag{5.22}$$

From (5.18) and (5.22), we deduce that $\|\chi_\omega y(t)\|_{L^2(\omega)} \longrightarrow 0, \; as \; t \longrightarrow +\infty$, which completes the proof. ■

Remark 5.3 When $\omega = \Omega$, we retrieve the result established in [3] about the strong stabilization of system (5.1).

Example 5.2 On $\Omega =]0, 1[$, we consider the following beam equation

$$\begin{cases} \dfrac{\partial^2 y}{\partial t^2}(x,t) = -\dfrac{\partial^4 y}{\partial x^4}(x,t) + u(t)\dfrac{\partial y}{\partial t}(x,t) \quad on \; \Omega\times]0, +\infty[\\ y(\xi, t) \quad = \dfrac{\partial^2 y}{\partial x^2}(\xi, t) = 0, \; \xi = 0, 1 \quad on \;]0, +\infty[\end{cases} \tag{5.23}$$

Let $P = \dfrac{\partial^4 y}{\partial x^4}$ with domain $\mathcal{D}(P) = \{y \in L^2(0,1)/\dfrac{\partial^4 y}{\partial x^4} \in L^2(0,1), y(\xi,t) = \dfrac{\partial^2 y}{\partial x^2}(\xi, t) = 0, \; \xi = 0, 1\}$.

The set $\mathcal{D}(P^{\frac{1}{2}})$ forms a Hilbert space under the inner product $\langle y_1, y_2\rangle_P = \langle P^{\frac{1}{2}} y_1, P^{\frac{1}{2}} y_2\rangle_{L^2(0,1)}$.

The eigenvalues of P are $\lambda_j = (j\pi)^4$, corresponding to eigenfunctions $\varphi_j(x) = \sqrt{2}\sin(j\pi x) \; \forall j \in \mathbb{N}^*$.

To write system (5.23) in the form (5.1) we set $H = (H^2(0,1) \cap H_0^1(0,1)) \times L^2(0,1)$ endowed with the inner product $\langle (y_1, z_1), (y_2, z_2)\rangle = \langle y_1, y_2\rangle_P + \langle z_1, z_2\rangle_{L^2(0,1)}$ and the operators

$$A = \begin{pmatrix} 0 & I \\ -P & 0 \end{pmatrix} \; and \; B = \begin{pmatrix} 0 & 0 \\ 0 & I \end{pmatrix} \; .$$

A is skew-adjoint, and B satisfies condition (5.18), indeed, for $y \in H$, we have $y = \sum_{j=1}^{\infty} \begin{pmatrix} \alpha_j \\ \lambda_j^{\frac{1}{2}} \beta_j \end{pmatrix} \varphi_j$, where $(\alpha_j, \beta_j) \in \mathbb{R}^2 \; \forall j \geq 1$. Then

$$S(s)y = \sum_{j=1}^{\infty} \begin{pmatrix} \alpha_j \cos(\lambda_j^{\frac{1}{2}} s) + \beta_j \sin(\lambda_j^{\frac{1}{2}} s) \\ \beta_j \lambda_j^{\frac{1}{2}} \cos(\lambda_j^{\frac{1}{2}} s) - \alpha_j \lambda_j^{\frac{1}{2}} \sin(\lambda_j^{\frac{1}{2}} s) \end{pmatrix} \varphi_j, \quad \forall s \geq 0.$$

It follows that

$$\begin{aligned}\langle i_\omega BS(s)y, S(s)y\rangle &= \sum_{j=1}^{\infty} \lambda_j \{\alpha_j \sin(\lambda_j^{\frac{1}{2}} s) - \beta_j \cos(\lambda_j^{\frac{1}{2}} s)\} \\ &= \sum_{j=1}^{\infty} \lambda_j \{\alpha_j^2 \sin^2(j\pi s) - \beta_j^2 \cos^2(j\pi s) - \sin(2j\pi s)\beta_j \alpha_j\}.\end{aligned}$$

Integrating this relation over the time interval [0, 2], we obtain

$$\int_0^2 |\langle i_\omega BS(s)y, S(s)y\rangle| ds = \sum_{j=1}^{\infty} \lambda_j (\alpha_j^2 + \beta_j^2) = \|\chi_\omega y\|_{L^2(\omega)}^2$$

then (5.18) holds.

We conclude that the control

$$u(t) = -\|\chi_\omega \partial_t y(., t)\|_{L^2(\omega)}^2$$

regionally strongly stabilizes system (5.23). ∎

5.2 Regional Stabilization Problem

The aim of this section is to characterize a control that stabilizes regionally system (5.1) and minimizes a given performance cost. We study the case when $i_\omega A$ generates a semigroup of isometries and the one when A generates a semigroup of contractions.

5.2.1 Case of Isometries Semigroup

Consider the problem

$$\begin{cases} \min J(u) = \displaystyle\int_0^{+\infty} \frac{\langle i_\omega By(t), y(t)\rangle^2}{\|y(t)\|^2} dt + \int_0^{+\infty} \|y(t)\|^2 |u(t)|^2 dt \\ u \in \mathcal{U}_{ad} = \{u \in L^2(0, +\infty) | y(t) \; is \; a \; global \; solution \; and \; J(u) < +\infty\}. \end{cases} \tag{5.24}$$

Theorem 5.4 *Let $i_\omega A$ generate a semigroup of isometries $(S(t))_{t\geq 0}$ and assume that*

$$\int_0^T |\langle i_\omega BS(t)y, S(t)y\rangle| dt \geq \delta \|\chi_\omega y\|_{L^2(\omega)}, \; (for \; some \; T, \delta > 0) \tag{5.25}$$

then the control

$$u^*(t) = -\frac{\langle i_\omega By(t), y(t)\rangle}{\|y(t)\|^2} \tag{5.26}$$

is the unique solution of (5.24) *and regionally exponentially stabilizes system* (5.1) *on* ω.

Proof Since $S(t)$ is a semigroup of isometries, then

$$\frac{d}{dt}\|\chi_\omega y(t)\|^2_{L^2(\omega)} = -2\frac{|\langle i_\omega By(t), y(t)\rangle|^2}{\|y(t)\|^2}, \quad \forall t \geq 0 \tag{5.27}$$

Integrating (5.27), we obtain

$$\int_0^t \frac{|\langle i_\omega By(s), y(s)\rangle|^2}{\|y(s)\|^2}ds \leq \frac{1}{2}\|\chi_\omega y_0\|^2_{L^2(\omega)}, \quad \forall t \geq 0 \tag{5.28}$$

which is verified $\forall y_0 \in L^2(\Omega)$, then $J(u^*) < +\infty$ and $u^* \in \mathcal{U}_{ad}$.

Let $y_0 \in D(A)$, from (5.27), we deduce that

$$\frac{d}{dt}\|\chi_\omega y(t)\|^2_{L^2(\omega)} = \|y(t)\|^2(|\frac{\langle i_\omega By(t), y(t)\rangle}{\|y(t)\|^2} + u(t)|^2 - \frac{\langle i_\omega By(t), y(t)\rangle^2}{\|y(t)\|^4} - u(t)^2),$$

Then for all $t \geq 0$, we have

$$\begin{aligned}\int_0^t \|y(s)\|^2(\frac{\langle i_\omega By(s), y(s)\rangle^2}{\|y(s)\|^4} + u(s)^2)ds + \|\chi_\omega y(t)\|^2_{L^2(\omega)} - \|\chi_\omega y_0\|^2_{L^2(\omega)} \\ = \int_0^t \|y(s)\|^2|\frac{\langle i_\omega By(s), y(s)\rangle}{\|y(s)\|^2} + u(s)|^2 ds.\end{aligned} \tag{5.29}$$

The solution $y(.)$ is continuous with respect to initial condition (see [1]) and $D(A)$ is dense in $L^2(\Omega)$, then (5.29) holds $\forall y_0 \in L^2(\Omega)$.

From Theorem 5.1, we deduce that control (5.26) regionally exponentially stabilizes system (5.1) on ω.

Thus, as $t \longrightarrow +\infty$ in (5.29), we obtain

$$J(u) = \|\chi_\omega y_0\|^2_{L^2(\omega)} + \int_0^{+\infty} \|y(s)\|^2|\frac{\langle i_\omega By(s), y(s)\rangle}{\|y(s)\|^2} + u(s)|^2 ds \tag{5.30}$$

Then $J(u) \geq \|\chi_\omega y_0\|^2_{L^2(\omega)} = J(u^*)$, which implies that (5.26) is the unique solution of (5.24). ■

For the next result, we consider this problem

$$\begin{cases} \min J(u) = \int_0^{+\infty} \langle i_\omega By(t), y(t)\rangle^2 dt + \int_0^{+\infty} |u(t)|^2 dt \\ u \in \mathcal{U}_{ad} = \{u \in L^2(0, +\infty) | y(t) \; is \; a \; global \; solution \; and \; J(u) < +\infty\}. \end{cases} \tag{5.31}$$

Proposition 5.1 *Suppose that* $i_\omega A$ *generates a semigroup of isometries* $(S(t))_{t\geq 0}$ *and*

$$\int_0^T |\langle i_\omega BS(s)y, S(s)y\rangle| ds \geq \delta \|\chi_\omega y\|^2_{L^2(\omega)}, \ (for\ some\ T, \delta > 0) \tag{5.32}$$

then the control

$$u^*(t) = -\langle i_\omega By(t), y(t)\rangle \tag{5.33}$$

is the unique solution of (5.31) *and regionally strongly stabilizes system* (5.1) *on* ω.

Proof Since $S(t)$ is a semigroup of isometries, then

$$\frac{d}{dt}\|\chi_\omega y(t)\|^2_{L^2(\omega)} = -2|\langle i_\omega By(t), y(t)\rangle|^2, \quad \forall t \geq 0 \tag{5.34}$$

Integrating (5.34), we obtain

$$\int_0^t |\langle i_\omega By(s), y(s)\rangle|^2 ds \leq \frac{1}{2}\|\chi_\omega y_0\|^2_{L^2(\omega)}, \quad \forall t \geq 0$$

which is verified $\forall y_0 \in L^2(\Omega)$, then $J(u^*) < +\infty$ and $u^* \in \mathcal{U}_{ad}$.

Let $y_0 \in D(A)$, from (5.34), we deduce that

$$\frac{d}{dt}\|\chi_\omega y(t)\|^2_{L^2(\omega)} = |\langle i_\omega By(t), y(t)\rangle + u(t)|^2 - \langle i_\omega By(t), y(t)\rangle^2 - u(t)^2,$$

Then for all $t \geq 0$, we have

$$\begin{aligned}\int_0^t (\langle i_\omega By(s), y(s)\rangle^2 + u(s)^2) ds + \|\chi_\omega y(t)\|^2_{L^2(\omega)} - \|\chi_\omega y_0\|^2_{L^2(\omega)} \\ = \int_0^t |\langle i_\omega By(s), y(s)\rangle + u(s)|^2 ds.\end{aligned} \tag{5.35}$$

The solution $y(.)$ is continuous with respect to initial condition (see [1]) and $D(A)$ is dense in $L^2(\Omega)$, then (5.35) holds $\forall y_0 \in L^2(\Omega)$.

To prove that (5.33) is optimal, let $u \in \mathcal{U}_{ad}$, $t \geq 0$ and define

$$z(\tau) = y(\tau) - S(\tau - t)y(t), \quad \forall \tau \in [t, t+T].$$

Applying the variation of constants formula with $y(t)$ as initial state,

$$y(\tau) = S(\tau - t)y(t) + \int_t^\tau u(s)S(\tau - s)By(s)ds, \quad \forall \tau \in [t, t+T]. \tag{5.36}$$

Therefore, $\forall \tau \in [t, t+T]$, we have

$$z(\tau) = \int_t^\tau u(s)S(\tau - s)By(s)ds$$

and using Schwartz's inequality, we obtain

$$\|z(\tau)\| \leq \|B\| \|y(t)\| \sqrt{T\mu(t)} \quad \forall \tau \in [t, t+T], \tag{5.37}$$

where $\mu(t) := \int_t^{t+T} |u(\tau)|^2 d\tau$.

Moreover, using (5.36) and Gronwall's inequality, there exists $u_{\max} \geq |u(\tau)|$ such that

$$\|y(\tau)\| \leq \|y(t)\| e^{\|B\| T u_{\max}} \quad \forall \tau \in [t, t+T], \tag{5.38}$$

Schwartz's inequality and (5.38) give

$$\int_t^{t+T} |\langle i_\omega By(\tau), y(\tau)\rangle| d\tau \leq M \|y(t)\| \sqrt{\gamma(t)}, \tag{5.39}$$

where $\gamma(t) = \int_t^{t+T} |\langle i_\omega By(\tau), y(\tau)\rangle|^2 d\tau$ and $M = \sqrt{T} e^{\|B\| T u_{\max}}$.

We have

$$\begin{aligned}\langle i_\omega BS(\tau - t)y(t), S(\tau - t)y(t)\rangle &= \langle i_\omega B(S(\tau - t)y(t) - y(\tau)), S(\tau - t)y(t)\rangle \\ &\quad - \langle i_\omega Bz(\tau), y(\tau)\rangle + \langle i_\omega By(\tau), y(\tau)\rangle\end{aligned}$$

Then there exists $\alpha > 0$ such that

$$\begin{aligned}|\langle i_\omega BS(\tau - t)y(t), S(\tau - t)y(t)\rangle| &\leq \alpha \|B\| \|y(t)\| \|z(\tau)\| + \alpha \|B\| \|y(\tau)\| \|z(\tau)\| \\ &\quad + |\langle i_\omega By(\tau), y(\tau)\rangle|.\end{aligned}$$

Using (5.37) and (5.38), we deduce

$$\begin{aligned}|\langle i_\omega BS(\tau - t)y(t), S(\tau - t)y(t)\rangle| &\leq \alpha(1 + e^{\|B\| T u_{\max}}) \|y(t)\| \|B\|^2 \sqrt{T\mu(t)} \\ &\quad + |\langle i_\omega By(\tau), y(\tau)\rangle|.\end{aligned}$$

Integrating this inequality over [0, T] and using (5.39), we get

$$\int_0^T |\langle i_\omega BS(\tau)y(t), S(\tau)y(t)\rangle| d\tau \leq N \|y(t)\| \sqrt{\mu(t)} + M \|y(t)\| \sqrt{\gamma(t)}, \tag{5.40}$$

where $N = \alpha(1 + e^{\|B\| T u_{\max}}) \|B\|^2 T \sqrt{T}$.

It follows from (5.32) and (5.40) that

$$\delta \|\chi_\omega y(t)\|^2_{L^2(\omega)} \leq \|y_0\| (N \sqrt{\mu(t)} + M \sqrt{\gamma(t)}),$$

Hence

$$\|\chi_\omega y(t)\|_{L^2(\omega)} \longrightarrow 0 \quad as \quad t \longrightarrow +\infty.$$

Then control (5.33) regionally strongly stabilizes system (5.1) on ω.

Thus, as $t \longrightarrow +\infty$ in (5.35), we obtain

$$J(u) = \|\chi_\omega y_0\|^2_{L^2(\omega)} + \int_0^{+\infty} |\langle i_\omega By(s), y(s)\rangle + u(s)|^2 ds$$

Then $J(u) \geq \|\chi_\omega y_0\|^2_{L^2(\omega)} = J(u^*)$, which implies that control (5.33) is the unique solution of (5.31). ■

5.2.2 Case of Contractions Semigroup

Here, we consider the problem

$$\begin{cases} \min J(u) = \displaystyle\int_0^{+\infty} \frac{\langle P_\omega By(t), y(t)\rangle^2}{\|y(t)\|^2} dt + \int_0^{+\infty} \|y(t)\|^2 |u(t)|^2 dt \\ u \in \mathcal{U}_{ad} = \{u \in L^2(0, +\infty) | y(t) \text{ is a global solution and } J(u) < +\infty\}. \end{cases} \tag{5.41}$$

where $P_\omega = i_\omega P i_\omega$ with P is a self-adjoint, positive and bounded operator which satisfies the equation

$$\langle P_\omega Ay, y\rangle + \langle y, P_\omega Ay\rangle = 0, \; y \in \mathcal{D}(A) \tag{5.42}$$

Theorem 5.5 *Let A generate a semigroup of contractions $(S(t))_{t\geq 0}$ and assume that*

$$\int_0^T |\langle P_\omega BS(t)y, S(t)y\rangle| dt \geq \delta \|\chi_\omega y\|_{L^2(\omega)}, \; (for\ some\ T, \delta > 0). \tag{5.43}$$

and there exists $\eta > 0$ such that

$$\langle P_\omega By, y\rangle \leq \eta \langle i_\omega By, y\rangle \tag{5.44}$$

then the control

$$u^*(t) = -\frac{\langle P_\omega By(t), y(t)\rangle}{\|y(t)\|^2} \tag{5.45}$$

is the unique solution of (5.41) *and regionally exponentially stabilizes system* (5.1) *on ω.*

Proof Let us define the function $V(y) = \langle P_\omega y, y\rangle, \forall y \in L^2(\Omega)$.

For all $y_0 \in \mathcal{D}(A)$ and $t \geq 0$, using (5.42), we obtain

$$\frac{dV(y(t))}{dt} = -2\frac{\langle P_\omega By(t), y(t)\rangle^2}{\|y(t)\|^2}$$

Integrating this relation, we get

$$\int_0^t \frac{\langle P_\omega By(s), y(s)\rangle^2}{\|y(s)\|^2} ds \leq \frac{1}{2} V(y_0), \ t \geq 0 \tag{5.46}$$

The solution $y(.)$ is continuous with respect to the initial condition y_0 (see [1]) and $D(A)$ is dense in $L^2(\Omega)$, then (5.46) holds for all $y_0 \in L^2(\Omega)$ so $J(u^*)$ is finite for all $y_0 \in L^2(\Omega)$.

Now, let us consider the nonlinear semigroup $\Gamma(t)y_0 := y(t)$.
For all $y_0 \in L^2(\Omega)$ and $t \geq 0$, we have

$$\begin{aligned}\langle P_\omega BS(t)y_0, S(t)y_0\rangle &= \langle P_\omega B(S(t)y_0 - \Gamma(t)y_0), S(t)y_0\rangle \\ &- \langle P_\omega B\Gamma(t)y_0, \Gamma(t)y_0 - S(t)y_0\rangle \\ &+ \langle P_\omega B\Gamma(t)y_0, \Gamma(t)y_0\rangle\end{aligned}$$

Since χ_ω is continuous, there exists $\alpha > 0$ such that

$$\begin{aligned}|\langle P_\omega BS(t)y_0, S(t)y_0\rangle| &\leq 2\alpha\|P\|\|B\|\|\Gamma(t)y_0 - S(t)y_0\|\|y_0\| \\ &+ |\langle P_\omega B\Gamma(t)y_0, \Gamma(t)y_0\rangle|\end{aligned} \tag{5.47}$$

Moreover, we have

$$\|\Gamma(t)y_0 - S(t)y_0\| \leq \|B\| \int_0^T \frac{|\langle P_\omega B\Gamma(s)y_0, \Gamma(s)y_0\rangle|}{\|\Gamma(s)y_0\|} ds$$

Schwartz's inequality yields

$$\|\Gamma(t)y_0 - S(t)y_0\| \leq \|B\| \left\{T \int_0^T \frac{|\langle P_\omega B\Gamma(s)y_0, \Gamma(s)y_0\rangle|^2}{\|\Gamma(s)y_0\|^2} ds\right\}^{\frac{1}{2}}, \ \forall t \in [0, T] \tag{5.48}$$

Using (5.8), we get

$$|\langle P_\omega B\Gamma(t)y_0, \Gamma(t)y_0\rangle| \leq \frac{|\langle P_\omega B\Gamma(t)y_0, \Gamma(t)y_0\rangle|}{\|\Gamma(t)y_0\|}\|y_0\|, \ \forall t \in [0, T]$$

Integrating (5.47) over the interval $[0, T]$ and taking into account (5.48), we obtain

$$\int_0^T |\langle P_\omega BS(s)y_0, S(s)y_0\rangle| ds \le 2\alpha\|P\|\|B\|^2 T^{\frac{3}{2}}\|y_0\| \times\left\{\int_0^T \frac{|\langle P_\omega B\Gamma(s)y_0, \Gamma(s)y_0\rangle|^2}{\|\Gamma(s)y_0\|^2}ds\right\}^{\frac{1}{2}} + \sqrt{T}\|y_0\|\left\{\int_0^T \frac{|\langle P_\omega B\Gamma(s)y_0, \Gamma(s)y_0\rangle|^2}{\|\Gamma(s)y_0\|^2}ds\right\}^{\frac{1}{2}} \tag{5.49}$$

Replacing y_0 by $\Gamma(t)y_0$ in (5.49) and using the superposition property of the semi-group $\Gamma(t)$, we deduce that

$$\int_0^T |\langle P_\omega BS(s)\Gamma(t)y_0, S(s)\Gamma(t)y_0\rangle| ds \le 2\alpha\|P\|\|B\|^2 T^{\frac{3}{2}}\|\Gamma(t)y_0\| \times\left\{\int_t^{t+T} \frac{|\langle P_\omega B\Gamma(s)y_0, \Gamma(s)y_0\rangle|^2}{\|\Gamma(s)y_0\|^2}ds\right\}^{\frac{1}{2}} + \sqrt{T}\|\Gamma(t)y_0\|\left\{\int_t^{t+T} \frac{|\langle P_\omega B\Gamma(s)y_0, \Gamma(s)y_0\rangle|^2}{\|\Gamma(s)y_0\|^2}ds\right\}^{\frac{1}{2}}$$

Using (5.43) and (5.8), we obtain

$$\delta\|\chi_\omega\Gamma(t)y_0\|_{L^2(\omega)} \le \beta\left\{\int_t^{t+T} \frac{|\langle P_\omega B\Gamma(s)y_0, \Gamma(s)y_0\rangle|^2}{\|\Gamma(s)y_0\|^2}ds\right\}^{\frac{1}{2}}$$

with $\beta = (2\alpha\|P\|\|B\|^2T + 1)\sqrt{T}\|y_0\|$.

Then

$$\delta^2\|\chi_\omega\Gamma(t)y_0\|^2_{L^2(\omega)} \le \beta^2 \int_t^{t+T} \frac{|\langle P_\omega B\Gamma(s)y_0, \Gamma(s)y_0\rangle|^2}{\|\Gamma(s)y_0\|^2}ds \tag{5.50}$$

Using (5.44) and (5.5), we get

$$\frac{d}{dt}\|\chi_\omega\Gamma(t)y_0\|^2_{L^2(\omega)} \le -\frac{2}{\eta^2}\frac{|\langle P_\omega B\Gamma(t)y_0, \Gamma(t)y_0\rangle|^2}{\|\Gamma(t)y_0\|^2}$$

Integrating this inequality from kT to $(k+1)T$, $(k \in \mathbb{N})$, we obtain

$$\|\chi_\omega\Gamma(kT)y_0\|^2_{L^2(\omega)} - \|\chi_\omega\Gamma((k+1)T)y_0\|^2_{L^2(\omega)} \ge \frac{2}{\eta^2}\int_{kT}^{(k+1)T} \frac{|\langle P_\omega B\Gamma(s)y_0, \Gamma(s)y_0\rangle|^2}{\|\Gamma(s)y_0\|^2}ds$$

Using (5.50) and since $\|\chi_\omega\Gamma(t)y_0\|_{L^2(\omega)}$ decreases, we deduce

$$\left(1 + 2\left(\frac{\delta}{\eta\beta}\right)^2\right)\|\chi_\omega\Gamma((k+1)T)y_0\|^2_{L^2(\omega)} \le \|\chi_\omega\Gamma(kT)y_0\|^2_{L^2(\omega)}$$

Then

$$\|\chi_\omega\Gamma((k+1)T)y_0\|_{L^2(\omega)} \le \rho\|\chi_\omega\Gamma(kT)y_0\|_{L^2(\omega)}$$

where $\rho = \dfrac{1}{(1 + 2(\frac{\delta}{\eta\beta})^2)^{\frac{1}{2}}}$.

By recurrence, we show that

$$\|\chi_\omega \Gamma(kT) y_0\|_{L^2(\omega)} \leq \rho^k \|\chi_\omega y_0\|_{L^2(\omega)}$$

Taking $k = E(\frac{t}{T})$ the integer part of $\frac{t}{T}$, we conclude that

$$\|\chi_\omega \Gamma(t) y_0\|_{L^2(\omega)} \leq K e^{-\sigma t} \|y_0\|, \ \forall t \geq 0 \tag{5.51}$$

where $K = \alpha(1 + 2(\frac{\delta}{\eta\beta})^2)^{\frac{1}{2}}$ and $\sigma = \dfrac{\ln(1 + 2(\frac{\delta}{\eta\beta})^2)}{2T}$, then control (5.45) regionally exponentially stabilizes system (5.1) on ω.

Now, to prove that (5.45) is the unique solution of (5.41), we have $V(y(t)) \leq \alpha \|P\| \|\chi_\omega y(t)\|^2_{L^2(\omega)}$, it follows from (5.51) that $V(y(t)) \longrightarrow 0$ as $t \to +\infty$.

Let $y_0 \in \mathcal{D}(A)$, integrating the relation

$$\frac{dV(y(t))}{dt} = \|y(t)\|^2([\frac{\langle P_\omega B y(t), y(t)\rangle}{\|y(t)\|^2} + u(t)]^2 - \frac{\langle P_\omega B y(t), y(t)\rangle^2}{\|y(t)\|^4} - u^2(t)) \tag{5.52}$$

we have

$$J(u) = V(y_0) + \int_0^{+\infty} \|y(s)\|^2 [\frac{\langle P_\omega B y(s), y(s)\rangle}{\|y(s)\|^2} + u(s)]^2 ds$$

then $J(u) \geq V(y_0)$.

Setting $u = u^*$, we obtain $J(u^*) = V(y_0)$.

Let $y_0 \in L^2(\Omega)$, and a sequence $y_{0n} \subset \mathcal{D}(A)$ such that $y_{0n} \longrightarrow y_0$ as $n \longrightarrow +\infty$, we have

$$J(u) = V(y_{0n}) + \int_0^{+\infty} \|y_n(s)\|^2 [\frac{\langle P_\omega B y_n(s), y_n(s)\rangle}{\|y_n(s)\|^2} + u(s)]^2 ds$$

Thus $J(u) \geq V(y_{0n})$.

We deduce that $J(u) \geq V(y_0) = J(u^*)$. Hence control (5.45) is the unique solution of the problem (5.41). ■

Proposition 5.2 *Suppose that A generates a semigroup of contractions $(S(t))_{t\geq 0}$, P and B are compact such that*

$$\langle P_\omega B S(t) z, S(t) z\rangle = 0, \ \forall t \geq 0 \Longrightarrow \chi_\omega z = 0 \tag{5.53}$$

then control (5.45) *is the unique solution of* (5.41) *and regionally weakly stabilizes system* (5.1) *on ω.*

Proof Let us consider the nonlinear semigroup $\Gamma(t)y_0 = y(t)$ and let (t_n) be a sequence of real numbers such that $t_n \longrightarrow +\infty$ as $n \longrightarrow +\infty$.

From (5.8), $\Gamma(t_n)y_0$ is bounded in the Hilbert space $L^2(\Omega)$, then there exists a subsequence $t_{\phi(n)}$ of (t_n) and $z \in L^2(\Omega)$ such that

$$\Gamma(t_{\phi(n)})y_0 \rightharpoonup z \quad as \quad n \longrightarrow +\infty.$$

For all $n \geq 0$, we set

$$f(t_{\phi(n)}) = \int_{t_{\phi(n)}}^{t_{\phi(n)}+T} \frac{|\langle P_\omega By(s), y(s)\rangle|^2}{\|y(s)\|^2} ds.$$

From (5.46), we have

$$\int_0^{+\infty} \frac{\langle P_\omega By(s), y(s)\rangle^2}{\|y(s)\|^2} ds < +\infty$$

We deduce that $f(t_{\phi(n)}) \longrightarrow 0$ as $n \longrightarrow +\infty$.

Replacing y_0 by $\Gamma(t_{\phi(n)})y_0$ in (5.49) and using the superposition property of the semigroup $\Gamma(t)$, we deduce that

$$\int_0^T |\langle P_\omega BS(s)\Gamma(t_{\phi(n)})y_0, S(s)\Gamma(t_{\phi(n)})y_0\rangle| ds \leq (2\alpha\|P\|\|B\|^2 T + 1)\|y_0\|\sqrt{Tf(t_{\phi(n)})} \tag{5.54}$$

Since $f(t_{\phi(n)}) \longrightarrow 0$ as $n \longrightarrow +\infty$, then

$$\lim_{n\longrightarrow+\infty} \int_0^T \langle P_\omega BS(s)\Gamma(t_{\phi(n)})y_0, S(s)\Gamma(t_{\phi(n)})y_0\rangle ds = 0$$

Since B and P are compact and $S(s)$ is continuous $\forall s \geq 0$, then

$$\lim_{n\longrightarrow+\infty} \langle P_\omega BS(s)\Gamma(t_{\phi(n)})y_0, S(s)\Gamma(t_{\phi(n)})y_0\rangle = \langle P_\omega BS(s)z, S(s)z\rangle$$

By dominated convergence theorem, we obtain

$$\int_0^T |\langle P_\omega BS(s)z, S(s)z\rangle| ds = 0 \text{ andthen } \langle P_\omega BS(s)z, S(s)z\rangle = 0, \quad \forall s \in [0, T]$$

Using (5.53), we deduce that

$$\chi_\omega\Gamma(t_{\phi(n)})y_0 \rightharpoonup 0 \quad as \quad n \longrightarrow +\infty$$

Which implies that $\forall\phi \in L^2(\Omega)$, $\langle\chi_\omega\Gamma(t_n)y_0, \phi\rangle \longrightarrow 0$ as $n \longrightarrow +\infty$, and hence

$$\chi_\omega \Gamma(t) y_0 \rightharpoonup 0 \quad as \quad t \longrightarrow +\infty.$$

In means that $\chi_\omega y(t)$ converges weakly to 0 as $t \longrightarrow +\infty$, and then system (5.1) is regionally weakly stabilizable on ω.

Now, let us prove that (5.45) is the unique solution of (5.41).

Since P is compact, it follows that $V(y(t)) \longrightarrow 0$ as $t \to +\infty$.

Let $y_0 \in \mathcal{D}(A)$, integrating the relation (5.52), we obtain

$$J(u) = V(y_0) + \int_0^{+\infty} \|y(s)\|^2 [\frac{\langle P_\omega B y(s), y(s)\rangle}{\|y(s)\|^2} + u(s)]^2 ds$$

then

$$J(u) \geq V(y_0), \ \forall u \in \mathcal{U}_{ad}.$$

For $u = u^*$, we obtain $J(u^*) = V(y_0)$.

Let $y_0 \in L^2(\Omega)$, and a sequence $y_{0n} \subset \mathcal{D}(A)$ such that $y_{0n} \longrightarrow y_0$ as $n \longrightarrow +\infty$, we have

$$J(u) = V(y_{0n}) + \int_0^{+\infty} \|y_n(s)\|^2 [\frac{\langle P_\omega B y_n(s), y_n(s)\rangle}{\|y_n(s)\|^2} + u(s)]^2 ds$$

Thus $J(u) \geq V(y_{0n})$.

We deduce that $J(u) \geq V(y_0) = J(u^*)$, which implies that (5.45) is the unique solution of problem (5.41). ∎

For the next result, we consider the problem

$$\begin{cases} \min J(u) = \int_0^{+\infty} \langle P_\omega B y(t), y(t)\rangle^2 dt + \int_0^{+\infty} |u(t)|^2 dt \\ u \in \mathcal{U}_{ad} = \{u \in L^2(0, +\infty) | y(t) \ is \ a \ global \ solution \ and \ J(u) < +\infty\}. \end{cases} \tag{5.55}$$

Proposition 5.3 *Let A generate a semigroup of contractions $(S(t))_{t\geq 0}$ and assume that*

$$\int_0^T |\langle P_\omega B S(t) y, S(t) y\rangle| dt \geq \delta \|\chi_\omega y\|^2_{L^2(\omega)}, \ (for \ some \ T, \delta > 0) \tag{5.56}$$

then the control

$$u^*(t) = -\langle P_\omega B y(t), y(t)\rangle \tag{5.57}$$

is the unique solution of (5.55) *and regionally strongly stabilizes system* (5.1) *on ω.*

Proof Let us define the function $V(y) = \langle P_\omega y, y\rangle$, $\forall y \in L^2(\Omega)$.

For all $y_0 \in \mathcal{D}(A)$ and $t \geq 0$, using (5.42), we obtain

$$\frac{dV(y(t))}{dt} = -2\langle P_\omega By(t), y(t)\rangle^2$$

Integrating this relation, we get

$$\int_0^t \langle P_\omega By(s), y(s)\rangle^2 ds \leq \frac{1}{2}V(y_0),\ t \geq 0 \tag{5.58}$$

The solution $y(.)$ is continuous with respect to the initial condition y_0 (see [1]) and $D(A)$ is dense in $L^2(\Omega)$, then (5.58) holds for all $y_0 \in L^2(\Omega)$ so $J(u^*)$ is finite for all $y_0 \in L^2(\Omega)$.

For all $y_0 \in L^2(\Omega)$ and $t \geq 0$, we have

$$\begin{aligned}\langle P_\omega BS(t)y_0, S(t)y_0\rangle &= \langle P_\omega B(S(t)y_0 - y(t)), S(t)y_0\rangle \\ &\quad -\langle P_\omega By(t), \Phi(t)\rangle + \langle P_\omega By(t), y(t)\rangle\end{aligned}$$

Since χ_ω is continuous, there exists $\alpha > 0$ such that

$$\begin{aligned}|\langle P_\omega BS(t)y_0, S(t)y_0\rangle| &\leq \alpha\|P\|\|B\|\|\Phi(t)\|(\|S(t)y_0\| + \|y(t)\|) \\ &\quad +|\langle P_\omega By(t), y(t)\rangle|.\end{aligned} \tag{5.59}$$

where $\Phi(t) = \displaystyle\int_0^t S(t-s)\langle P_\omega By(s), y(s)\rangle By(s)ds.$

Moreover, we have

$$\|\Phi(t)\| \leq \|B\|\|y_0\|\sqrt{T}\Big(\int_0^T |\langle P_\omega By(s), y(s)\rangle|^2 ds\Big)^{\frac{1}{2}} \tag{5.60}$$

Integrating (5.59) over the interval $[0, T]$ and taking into account (5.60), we obtain

$$\begin{aligned}\int_0^T |\langle P_\omega BS(s)y_0, S(s)y_0\rangle| ds &\leq 2\alpha T^{\frac{3}{2}}\|y_0\|^2\|B\|\Big(\int_0^T |\langle P_\omega By(s), y(s)\rangle|^2 ds\Big)^{\frac{1}{2}} \\ &\quad +\sqrt{T}\|y_0\|\Big(\int_0^T |\langle P_\omega By(s), y(s)\rangle|^2 ds\Big)^{\frac{1}{2}}.\end{aligned}$$

Replacing y_0 by $y(t)$ and using the superposition property of the solution, we get

$$\begin{aligned}\int_0^T |\langle P_\omega BS(s)y(t), S(s)y(t)\rangle| ds &\leq 2\alpha T^{\frac{3}{2}}\|y_0\|\|B\|\Big(\int_t^{t+T} |\langle P_\omega By(s), y(s)\rangle|^2 ds\Big)^{\frac{1}{2}} \\ &\quad +\sqrt{T}\|y_0\|\Big(\int_t^{t+T} |\langle P_\omega By(s), y(s)\rangle|^2 ds\Big)^{\frac{1}{2}}.\end{aligned}$$

From (5.58), we have

$$\int_t^{t+T} |\langle P_\omega By(s), y(s)\rangle|^2 ds \longrightarrow 0, \; as \; t \longrightarrow +\infty.$$

It follows that

$$\int_0^T |\langle P_\omega BS(s)y(t), S(s)y(t)\rangle| ds \longrightarrow 0, \; as \; t \longrightarrow +\infty. \tag{5.61}$$

From (5.56) and (5.61), we deduce that $\|\chi_\omega y(t)\|_{L^2(\omega)} \longrightarrow 0, \; as \; t \longrightarrow +\infty$.

Then the control (5.57) regionally strongly stabilizes system (5.1) on ω.

Now, let us show that control (5.57) is the unique solution of (5.55).

We have $V(y(t)) \leq \alpha \|P\| \|\chi_\omega y(t)\|^2_{L^2(\omega)}$, then $V(y(t)) \longrightarrow 0$ as $t \to +\infty$.

Let $y_0 \in \mathcal{D}(A)$, integrating the relation

$$\frac{dV(y(t))}{dt} = [\langle P_\omega By(t), y(t)\rangle + u(t)]^2 - \langle P_\omega By(t), y(t)\rangle^2 - u^2(t)$$

we obtain

$$J(u) = V(y_0) + \int_0^{+\infty} [\langle P_\omega By(s), y(s)\rangle + u(s)]^2 ds$$

Setting $u = u^*$, we get $J(u^*) = V(y_0)$.

Let $y_0 \in L^2(\Omega)$, and a sequence $y_{0n} \subset \mathcal{D}(A)$ such that $y_{0n} \longrightarrow y_0$ as $n \longrightarrow +\infty$, we have

$$J(u) = V(y_{0n}) + \int_0^{+\infty} [\langle P_\omega By_n(s), y_n(s)\rangle + u(s)]^2 ds$$

thus $J(u) \geq V(y_{0n})$.

We deduce that $J(u) \geq V(y_0) = J(u^*)$, which implies that the unique solution of (5.55) is given by the control (5.57). ■

In order to illustrate the previous results numerically, we perform the following algorithm:

Step 1: *Initial data: initial condition* y_0 *and subregion* ω;
Step 2: *Solve the equation* (5.42) *using Bartels-Stewart method* [4];
Step 3: *Apply the control given by* (5.3) *or* (5.45);
Step 4: *Solve system* (5.1) *using Petrov-Galerkin method given in* [5];
Step 5: *Calculate the stabilization cost* J.

5.3 Simulation Results

On $\Omega =]0, 1[$, we consider the system

$$\begin{cases} \pi^2 \dfrac{\partial y}{\partial t}(x,t) = 0.01\Delta y(x,t) + u(t)By(x,t) & \Omega\times]0,+\infty[\\ \dfrac{\partial y}{\partial x}(1,t) = \exp(-t), \quad y(0,t) = 0 & \\ y(x,0) = x(2x^2 - 3x + 1) & \end{cases} \tag{5.62}$$

where $By = \sum_{j=1}^{+\infty} \frac{1}{j^2}\langle i_\omega y, \varphi_j\rangle\varphi_j$ with $\varphi_j(x) = \sqrt{\frac{2}{1+(j\pi)^2}}\cos(j\pi x)$.

1. Regional stabilizing control.
The operator B is linear, compact and verifies the assumption (5.15), then control (5.3) regionally weakly stabilizes (5.62) on ω.

Let $\omega =]0, \frac{1}{4}[$ and applying the above algorithm, we obtain the following figures

Figure 5.1 shows that system (5.62) is stabilized on the subregion ω, with a stabilization error equals to 1.02×10^{-4} (Fig. 5.2).

For $\omega =]0, \frac{1}{2}[$, we obtain

Figure 5.3, shows that system (5.62) is stabilized on the subregion ω, with stabilization error 4.23×10^{-4} (Fig. 5.4).

For $\omega =]0, 1[$, we obtain

Figure 5.5 shows that system (5.62) is stabilized on Ω, with stabilization error 1.98×10^{-3} (Fig. 5.6).

2. Regional optimal stabilizing control.
Consider system (5.62) and problem (5.41) such that P_ω is the unique solution of the equation (5.42). Then control (5.45) is the unique solution of (5.41) and regionally

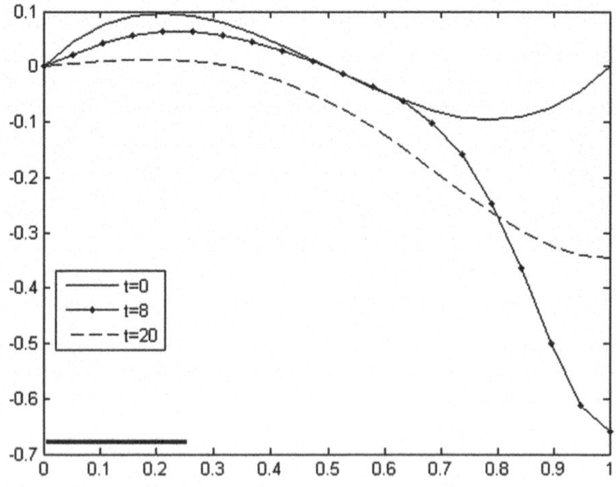

Fig. 5.1 Evolution of the state

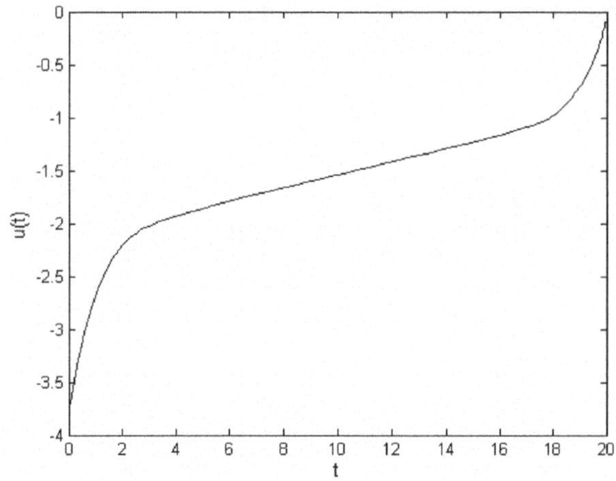

Fig. 5.2 Evolution of the control

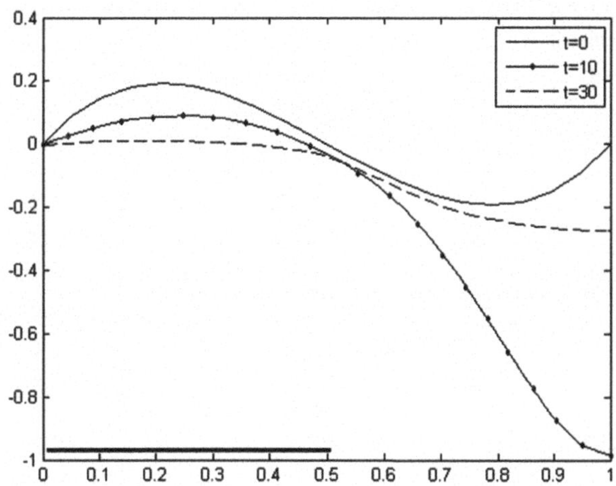

Fig. 5.3 Evolution of the state

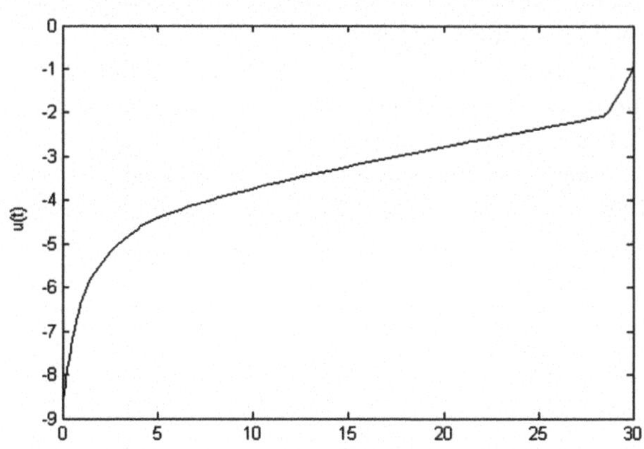

Fig. 5.4 Evolution of the control

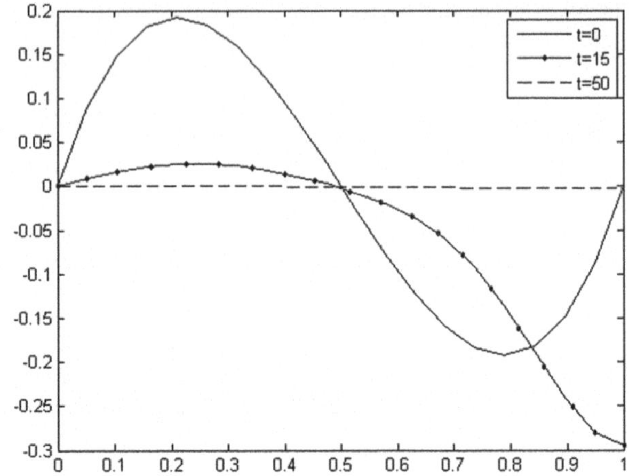

Fig. 5.5 Evolution of the state

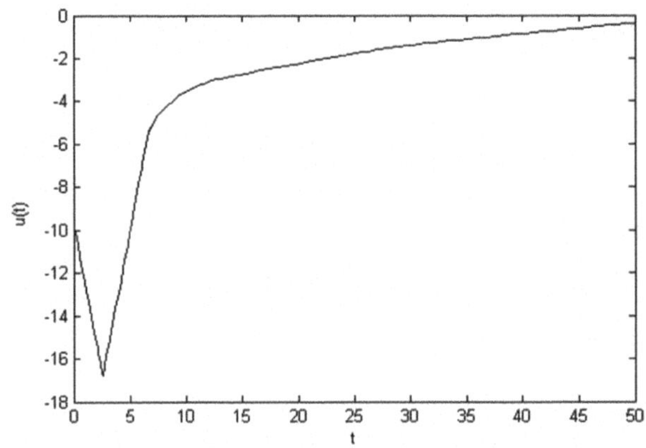

Fig. 5.6 Evolution of the control

weakly stabilizes system (5.62). Let $\omega =]0, \frac{1}{4}[$, applying the above algorithm, we obtain the following figures

Figure 5.7 shows how system (5.62) is stabilized by the control (5.45) on ω with a stabilization error equals to $8.87\ 10^{-5}$ and the cost $J(u^*) = 3.36 \times 10^{-3}$ (Fig. 5.8).

For $\omega = \Omega$,

Figure 5.9 shows that system (5.62) is stabilized on Ω, with stabilization error 7.96×10^{-4} and the cost $J(u^*) = 1.64 \times 10^{-2}$ (Fig. 5.10).

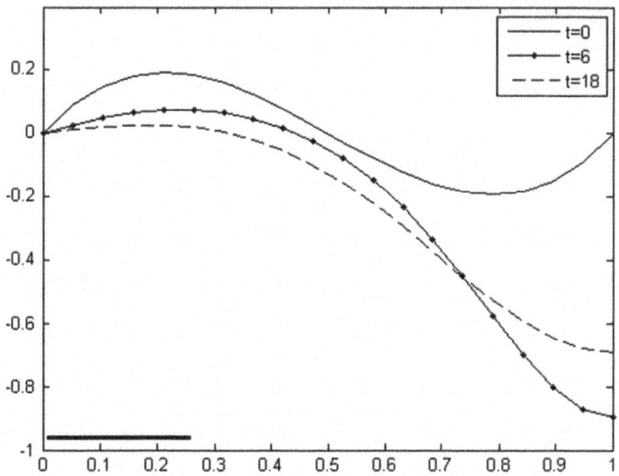

Fig. 5.7 Evolution of the state

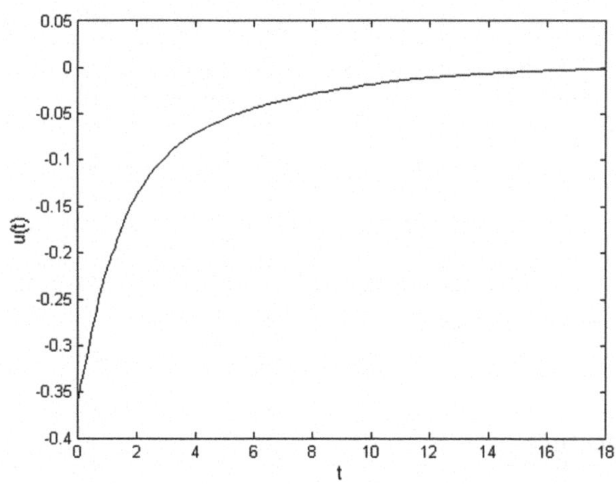

Fig. 5.8 Evolution of the control

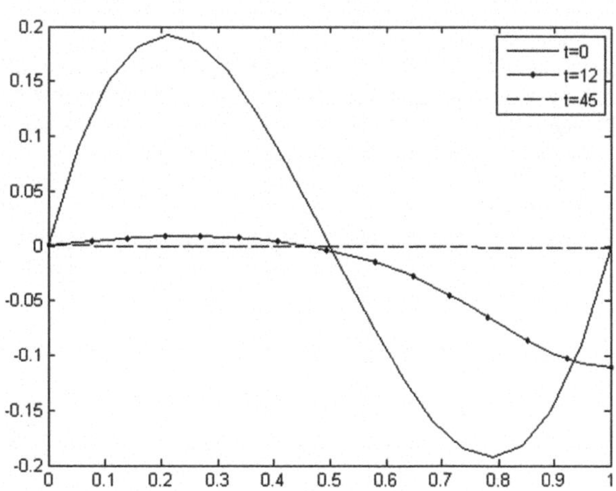

Fig. 5.9 Evolution of the state

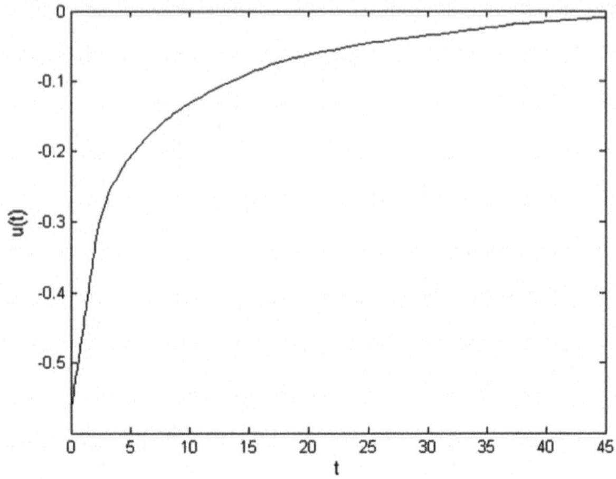

Fig. 5.10 Evolution of the control

The following table shows that there exists a relation between the area of target subregion ω, the cost and the error of stabilization.

ω	$]0, 0.3[$	$]0, 0.5[$	$]0, 0.6[$	$]0, 0.8[$	$]0, 1[$
Error	9.72×10^{-5}	1.565×10^{-4}	3.589×10^{-4}	6.426×10^{-4}	7.98×10^{-4}
Cost	4.528×10^{-3}	5.07×10^{-3}	7.05×10^{-3}	9.33×10^{-3}	1.66×10^{-2}

More the area of the subregion increases, more both the stabilization error and the cost increase.

References

1. Pazy, A.: Semigroups of linear operators and applications to partial differential equations. Springer Verlag, New York (1983)
2. Ouzahra, M.: Exponential and weak stabilization of constrained bilinear systems. SIAM Journal on Control and Optimization **48**(6), 3962–3974 (2010)
3. Berrahmoune, L. *Asymptotic stabilization and decay estimate for distributed bilinear systems*, Recerche di Matematica. 2001, fasc. 1, pp. 89-103
4. Penzl, T.: Numerical solution of generalized Lyapunov equations. Advances in Comp. Math **8**, 33–48 (1998)
5. Skeel, R.D., Berzins, M.: A Method for the spatial discretization of parabolic equations in one space variable. SIAM J. ScI. Stat. Comput. **11**, 1–32 (1990)

Chapter 6
Regional Stabilization of Infinite Dimensional Semilinear Systems

Semilinear systems is an important subclass of nonlinear systems and numerous real-world problems have a semilinear structure. They include applications in nuclear, thermal, chemical, social processes, etc [1].

Let $\Omega \subset \mathbb{R}^n$ be an open bounded domain with regular boundary $\partial\Omega$, we consider the system

$$\begin{cases} \dfrac{dy(t)}{dt} = Ay(t) + u(t)By(t) \\ y(0) = y_0 \end{cases} \tag{6.1}$$

where the operator A generates a semigroup of contractions $(S(t))_{t\geq 0}$ on the Hilbert space $L^2(\Omega)$ with inner product $\langle .,. \rangle$ and norm $\|.\|$, $u(t) \in L^2(0, +\infty)$ is a scalar valued control and B is a nonlinear operator mapping $L^2(\Omega)$ into itself such that $B(0) = 0$.

In this chapter, we study regional exponential, strong and weak stabilization of system (6.1). Moreover, we give a control that stabilizes regionally system (6.1) on ω minimizing a functional cost.

6.1 Regional Stabilization

Here we deal with regional exponential, strong and weak stabilization of system (6.1). Let ω be an open subregion of Ω and denote by $\chi_\omega : L^2(\Omega) \longrightarrow L^2(\omega)$ the restriction operator to ω. Let us denote $i_\omega = \chi_\omega^* \chi_\omega$ with χ_ω^* is the adjoint operator.

Assume that

$$\langle i_\omega By, y\rangle\langle By, y\rangle \geq 0, \quad \forall y \in L^2(\Omega). \tag{6.2}$$

E. H. Zerrik and O. Castillo, *Stabilization of Infinite Dimensional Systems*, Studies in Systems, Decision and Control 355,
https://doi.org/10.1007/978-3-030-68600-0_6

and

$$\langle i_\omega A y, y\rangle \le 0, \quad \forall y \in \mathcal{D}(A). \tag{6.3}$$

6.1.1 Regional Exponential Stabilization

The following result gives sufficient conditions for regional exponential stabilization of system (6.1).

Theorem 6.1 *Suppose that B is locally Lipschitz such that*

$$\int_0^T |\langle i_\omega BS(s)y, S(s)y\rangle| ds \ge \alpha \|\chi_\omega y\|_{L^2(\omega)}, \ (for\ some\ T, \alpha > 0) \tag{6.4}$$

then the control

$$u(t) = \begin{cases} -\dfrac{\langle i_\omega By(t), y(t)\rangle}{\|y(t)\|^2}, & y(t) \ne 0 \\ 0, & y(t) = 0 \end{cases} \tag{6.5}$$

regionally exponentially stabilizes system (6.1) *on ω.*

Proof First, let us show that the map $h : y \mapsto \dfrac{\langle i_\omega By, y\rangle}{\|y\|^2} By$ is locally Lipschitz.

Since B is locally Lipshitz, for each $R > 0$ there exists a constant $K > 0$ such that

$$\|By - Bz\| \le K\|y - z\|, \ \forall y, z \in L^2(\Omega) : 0 < \|y\| \le \|z\| \le R. \tag{6.6}$$

Thus

$$h(z)Bz - h(y)By = \frac{\|y\|^2(\langle i_\omega Bz, z\rangle Bz - \langle i_\omega By, y\rangle By)}{\|z\|^2\|y\|^2} - \frac{(\|z\|^2 - \|y\|^2)\langle i_\omega By, y\rangle By}{\|z\|^2\|y\|^2}.$$

Since χ_ω is continuous, there exists $\delta > 0$ such that

$$\begin{aligned}
\|h(z)Bz - h(y)By\| &\le \frac{\|\langle i_\omega Bz, z\rangle Bz - \langle i_\omega By, y\rangle By\|}{\|z\|^2} + K^2\delta|\|z\|^2 - \|y\|^2|.\frac{\|y\|}{\|z\|^2} \\
&\le \frac{\|\langle i_\omega Bz, z\rangle(Bz - By) + (\langle i_\omega Bz, z - y\rangle}{\|z\|^2} \\
&\quad \frac{\langle i_\omega By - i_\omega Bz, y\rangle)By\|}{\|z\|^2} + 2K^2\delta\|z - y\| \\
&\le K^2\delta\|z\|^2\|z - y\| + K^2\delta\|y\|\|z - y\|(\|y\| + \|z\|) + 2K^2\delta\|z - y\| \\
&\le 3K^2\delta\|z\|^2\|z - y\| + 2K^2\delta\|z - y\|.
\end{aligned}$$

We conclude that for all $y, z \in L^2(\Omega)$, we have $\|h(z) - h(y)\| \le L\|z - y\|$ where $L = 3K^2\delta R^2 + 2K^2\delta$.

It follows that system (6.1) has a unique global mild solution $y(t)$ defined on a maximal interval $[0, t_{\max}[$ (see [2]) and given by the variation of constants formula

$$y(t) = S(t)y_0 + \int_0^t S(t-s)u(s)By(s)ds. \tag{6.7}$$

Let us consider the nonlinear semigroup $N(t)y_0 := y(t)$, we have

$$\begin{aligned}\tfrac{d}{dt}\|N(t)y_0\|^2 &= 2\langle AN(t)y_0, N(t)y_0\rangle \\ &\quad -2\frac{\langle i_\omega BN(t)y_0, N(t)y_0\rangle}{1+|\langle i_\omega BN(t)y_0, N(t)y_0\rangle|}\langle BN(t)y_0, N(t)y_0\rangle.\end{aligned}$$

Since $S(t)$ is a semigroup of contractions, then

$$\frac{d}{dt}\|N(t)y_0\|^2 \leq -2\frac{\langle i_\omega BN(t)y_0, N(t)y_0\rangle}{1+|\langle i_\omega BN(t)y_0, N(t)y_0\rangle|}\langle BN(t)y_0, N(t)y_0\rangle.$$

Integrating this inequality, we get

$$\|N(t)y_0\|^2 \leq \|y_0\|^2 - 2\int_0^t \langle h(N(s)y_0), N(s)y_0\rangle ds, \ \forall t \in [0, t_{max}[.$$

Using (6.2), we obtain

$$\|N(t)y_0\| \leq \|y_0\|. \tag{6.8}$$

For all $y_0 \in L^2(\Omega)$ and $t \geq 0$, we have

$$\begin{aligned}\langle i_\omega BS(t)y_0, S(t)y_0\rangle &= \langle i_\omega BS(t)y_0 - i_\omega BN(t)y_0, S(t)y_0\rangle \\ &\quad - \langle i_\omega BN(t)y_0, N(t)y_0 - S(t)y_0\rangle \\ &\quad + \langle i_\omega BN(t)y_0, N(t)y_0\rangle.\end{aligned}$$

Using (6.6) and the continuity of χ_ω, we obtain

$$\begin{aligned}|\langle i_\omega BS(t)y_0, S(t)y_0\rangle| &\leq 2\delta K\|N(t)y_0 - S(t)y_0\|\|y_0\| \\ &\quad + |\langle i_\omega BN(t)y_0, N(t)y_0\rangle|.\end{aligned} \tag{6.9}$$

Moreover, from (6.7) we have

$$\|N(t)y_0 - S(t)y_0\| \leq K\int_0^t \frac{|\langle i_\omega BN(s)y_0, N(s)y_0\rangle|}{\|N(s)y_0\|}ds.$$

For a fixed $T \in]0, t_{max}[$, Schwarz's inequality yields

$$\|N(t)y_0 - S(t)y_0\| \le K\left\{T\int_0^T \frac{|\langle i_\omega BN(s)y_0, N(s)y_0\rangle|^2}{\|N(s)y_0\|^2}ds\right\}^{\frac{1}{2}}, \ \forall t \in [0, T]. \tag{6.10}$$

Using (6.8), we get

$$|\langle i_\omega BN(t)y_0, N(t)y_0\rangle| \le \frac{|\langle i_\omega BN(t)y_0, N(t)y_0\rangle|}{\|N(t)y_0\|}\|y_0\|, \ \forall t \in [0, T]. \tag{6.11}$$

Integrating (6.9) over the interval $[0, T]$ and using (6.10) and (6.11), we obtain

$$\begin{aligned}\int_0^T |\langle i_\omega BS(s)y_0, S(s)y_0\rangle|ds &\le 2\delta K^2 T^{\frac{3}{2}}\|y_0\|\left\{\int_0^T \frac{|\langle i_\omega BN(s)y_0, N(s)y_0\rangle|^2}{\|N(s)y_0\|^2}ds\right\}^{\frac{1}{2}} \\ &+\sqrt{T}\|y_0\|\left\{\int_0^T \frac{|\langle i_\omega BN(s)y_0, N(s)y_0\rangle|^2}{\|N(s)y_0\|^2}ds\right\}^{\frac{1}{2}}.\end{aligned} \tag{6.12}$$

Replacing y_0 by $N(t)y_0$ in (6.12), we get

$$\begin{aligned}\int_0^T |\langle i_\omega BS(s)N(t)y_0, S(s)N(t)y_0\rangle|ds \le\ & 2\delta K^2 T^{\frac{3}{2}}\|N(t)y_0\| \\ &\times\left\{\int_t^{t+T} \frac{|\langle i_\omega BN(s)y_0, N(s)y_0\rangle|^2}{\|N(s)y_0\|^2}ds\right\}^{\frac{1}{2}} \\ &+\sqrt{T}\|N(t)y_0\| \\ &\times\left\{\int_t^{t+T} \frac{|\langle i_\omega BN(s)y_0, N(s)y_0\rangle|^2}{\|N(s)y_0\|^2}ds\right\}^{\frac{1}{2}}.\end{aligned}$$

By (6.4) and (6.8), we obtain

$$\alpha\|\chi_\omega N(t)y_0\|_{L^2(\omega)} \le \beta\left\{\int_t^{t+T} \frac{|\langle i_\omega BN(s)y_0, N(s)y_0\rangle|^2}{\|N(s)y_0\|^2}ds\right\}^{\frac{1}{2}}$$

where $\beta = (2\delta K^2 T + 1)\sqrt{T}\|y_0\|$.

Then

$$\alpha^2\|\chi_\omega N(t)y_0\|^2_{L^2(\omega)} \le \beta^2\int_t^{t+T} \frac{|\langle i_\omega BN(s)y_0, N(s)y_0\rangle|^2}{\|N(s)y_0\|^2}ds. \tag{6.13}$$

Using (6.3), we get

$$\frac{d}{dt}\|\chi_\omega N(t)y_0\|^2_{L^2(\omega)} \le -2\frac{|\langle i_\omega BN(t)y_0, N(t)y_0\rangle|^2}{\|N(t)y_0\|^2}. \tag{6.14}$$

Integrating (6.14) from mT to $(m+1)T$, $(m \in \mathbb{N})$, we obtain

$$\|\chi_\omega N(mT)y_0\|^2_{L^2(\omega)} - \|\chi_\omega N((m+1)T)y_0\|^2_{L^2(\omega)} \geq 2\int_{mT}^{(m+1)T} \frac{|\langle i_\omega BN(s)y_0, N(s)y_0\rangle|^2}{\|N(s)y_0\|^2} ds.$$

By (6.13) and (6.14), we deduce

$$\left(1+2\left(\frac{\alpha}{\beta}\right)^2\right)\|\chi_\omega N((m+1)T)y_0\|^2_{L^2(\omega)} \leq \|\chi_\omega N(mT)y_0\|^2_{L^2(\omega)}.$$

Then $\|\chi_\omega N((m+1)T)y_0\|_{L^2(\omega)} \leq \rho\|\chi_\omega N(mT)y_0\|_{L^2(\omega)}$ where $\rho = \dfrac{1}{(1+2(\frac{\alpha}{\beta})^2)^{\frac{1}{2}}}$.

By recurrence, we show that $\|\chi_\omega N(mT)y_0\|_{L^2(\omega)} \leq \rho^m\|\chi_\omega y_0\|_{L^2(\omega)}$.

Taking $m = E(\frac{t}{T})$ the integer part of $\frac{t}{T}$ and remarking that $m \geq \frac{t}{T} - 1$, we get

$$\|\chi_\omega N(t)y_0\|_{L^2(\omega)} \leq \rho^{\frac{t}{T}-1}\|\chi_\omega y_0\|_{L^2(\omega)}, \ \forall t \geq 0$$

It follows that

$$\|\chi_\omega N(t)y_0\|_{L^2(\omega)} \leq Me^{-\sigma t}\|y_0\|, \ \forall t \geq 0$$

where $M = \delta(1+2(\frac{\alpha}{\beta})^2)^{\frac{1}{2}}$ and $\sigma = \dfrac{\ln(1+2(\frac{\alpha}{\beta})^2)}{2T}$, which shows the regional exponential stability of system (1) on ω. ■

Example 6.1 Let $\Omega =]0,1[$, we consider the wave equation defined by

$$\begin{cases} \dfrac{\partial^2 y}{\partial t^2}(x,t) = \Delta y(x,t) + u(t)\dfrac{\partial y}{\partial t}(x,t) & on\ \Omega\times]0,+\infty[\\ y(x,0) = y_0,\ \dfrac{\partial y}{\partial t}(x,0) = y_1 & on\ \Omega \\ y(0,t) = y(1,t) = 0 & on\ [0,+\infty[\end{cases} \tag{6.15}$$

This system has the form of (6.1) if we set

$$A = \begin{pmatrix} 0 & I \\ \Delta & 0 \end{pmatrix}\ and\ B = \begin{pmatrix} 0 & 0 \\ 0 & I \end{pmatrix}.$$

Let $\tilde{A} = -\Delta$, with domain $\mathcal{D}(\tilde{A}) = H^2(\Omega) \cap H^1_0(\Omega)$.

Consider $H = H^1_0(\Omega) \times L^2(\Omega)$ the state space endowed with the inner product $\langle (y_1,z_1),(y_2,z_2)\rangle = \langle \tilde{A}^{\frac{1}{2}}y_1, \tilde{A}^{\frac{1}{2}}y_2\rangle_{L^2(\Omega)} + \langle z_1,z_2\rangle_{L^2(\Omega)}$.

The eigenvalues of $\tilde{A}$ are $\lambda_k = (k\pi)^2$, corresponding to eigenfunctions $\phi_k(x) = \sqrt{2}\sin(k\pi x)$, $\forall k \in \mathbb{N}^*$.

We set $y = (y_1, y_2) \in H$ with $y_1 = \sum_{k=1}^{\infty} a_k\phi_k$ and $y_2 = \sum_{k=1}^{\infty} \sqrt{\lambda_k}b_k\phi_k$, where $(a_k, b_k) \in \mathbb{R}^2\ \forall k \geq 1$, and the semigroup is given by

$$S(s)y = \sum_{k=1}^{\infty} \begin{pmatrix} a_k \cos(k\pi s) + b_k \sin(k\pi s) \\ b_k k\pi \cos(k\pi s) - a_k k\pi \sin(k\pi s) \end{pmatrix} \phi_k, \quad \forall s \geq 0.$$

For $\omega =]0, \frac{1}{2}[$, we have

$$\langle i_\omega BS(s)y, S(s)y\rangle = \sum_{k=1}^{\infty} \frac{(k\pi)^2}{2}\big(a_k^2 \sin^2(k\pi s) - b_k a_k \sin(2k\pi s) + b_k^2 \cos^2(k\pi s)\big).$$

Taking $T = 2$, we get

$$\int_0^2 |\langle i_\omega BS(s)y, S(s)y\rangle| ds = \sum_{k=1}^{\infty} \frac{(k\pi)^2}{2}(a_k^2 + b_k^2) \geq \alpha \|\chi_\omega y\|_{L^2(\omega)}$$

where $\alpha = \frac{\pi}{2}\sqrt{\eta}$, ($\eta = \min\{(a_k^2 + b_k^2) | k \in \mathbb{N}^*\}$), thus (6.4) holds and the control

$$u(t) = -\frac{\|\chi_\omega \partial_t y\|_{L^2(\omega)}^2}{\|y(t)\|_{H_0^1(\Omega)}^2 + \|\partial_t y(t)\|^2}$$

regionally exponentially stabilizes system (6.15) on ω. ■

6.1.2 Regional Strong Stabilization

In this subsection we give sufficient conditions for regional strong stabilization.

Theorem 6.2 *Assume that B is locally Lipschitz and satisfies*

$$\int_0^T |\langle i_\omega BS(s)y, S(s)y\rangle| ds \geq \alpha \|\chi_\omega y\|_{L^2(\omega)}^2, \ (T, \alpha > 0) \tag{6.16}$$

then the control

$$u(t) = -\frac{\langle i_\omega By(t), y(t)\rangle}{1 + |\langle i_\omega By(t), y(t)\rangle|} \tag{6.17}$$

regionally strongly stabilizes system (6.1) on ω.

Proof By the same arguments using in the proof of Theorem 6.1, we can show that the map $g : y \mapsto \dfrac{\langle i_\omega By, y\rangle}{1 + |\langle i_\omega By, y\rangle|} By$ is locally Lipschitz and we deduce that system (6.1) has a unique global mild solution $y(t)$ (see [2]).

It remains to show that $\chi_\omega y(t) \longrightarrow 0 \ \ as \ \ t \longrightarrow +\infty$.
For all $y_0 \in L^2(\Omega)$, we have

$$\langle i_\omega BS(t)y_0, S(t)y_0\rangle = \langle i_\omega BS(t)y_0 - i_\omega By(t), S(t)y_0\rangle + \langle i_\omega By(t), S(t)y_0\rangle$$

Using formula (6.7), we obtain

$$\begin{aligned}\langle i_\omega BS(t)y_0, S(t)y_0\rangle = &\langle i_\omega BS(t)y_0 - i_\omega By(t), S(t)y_0\rangle - \langle i_\omega By(t), \Phi(t)\rangle \\ &+ \langle i_\omega By(t), y(t)\rangle\end{aligned}$$

where $\Phi(t) = \int_0^t S(t-s)u(s)By(s)ds$.

By (6.6) and the continuity of χ_ω, we deduce that

$$|\langle i_\omega BS(t)y_0, S(t)y_0\rangle| \le \delta K\|\Phi(t)\|(\|S(t)y_0\| + \|y(t)\|) + |\langle i_\omega By(t), y(t)\rangle|. \tag{6.18}$$

Then

$$|\langle i_\omega BS(t)y_0, S(t)y_0\rangle| \le 2\delta K\|\Phi(t)\|\|y_0\| + |\langle i_\omega By(t), y(t)\rangle|, \forall t \in [0, T].$$

Moreover, Schwarz's inequality gives

$$\|\Phi(t)\| \le K\|y_0\|\sqrt{T}\Big(\int_0^T |\langle i_\omega By(s), y(s)\rangle|^2 ds\Big)^{\frac{1}{2}}. \tag{6.19}$$

Integrating (6.18) over the interval $[0, T]$ and using (6.19), we obtain

$$\begin{aligned}\int_0^T |\langle i_\omega BS(s)y_0, S(s)y_0\rangle| ds \le\, &2\delta T^{\frac{3}{2}} K\|y_0\|^2\Big(\int_0^T |\langle i_\omega By(s), y(s)\rangle|^2 ds\Big)^{\frac{1}{2}} \\ &+\sqrt{T}\|y_0\|\Big(\int_0^T |\langle i_\omega By(s), y(s)\rangle|^2 ds\Big)^{\frac{1}{2}}.\end{aligned}$$

Replacing y_0 by $y(t)$, we get

$$\begin{aligned}\int_0^T |\langle i_\omega BS(s)y(t), S(s)y(t)\rangle| ds \le\, &2\delta T^{\frac{3}{2}} K\|y_0\|\Big(\int_t^{t+T} |\langle i_\omega By(s), y(s)\rangle|^2 ds\Big)^{\frac{1}{2}} \\ &+\sqrt{T}\|y_0\|\Big(\int_t^{t+T} |\langle i_\omega By(s), y(s)\rangle|^2 ds\Big)^{\frac{1}{2}}.\end{aligned} \tag{6.20}$$

Furthermore, using (6.3), we obtain

$$\frac{d}{dt}\|\chi_\omega y(t)\|^2_{L^2(\omega)} \le -2\frac{|\langle i_\omega By(t), y(t)\rangle|^2}{1 + |\langle i_\omega By(t), y(t)\rangle|}, \ \forall t \ge 0.$$

Integrating this inequality, we get

$$\int_0^t |\langle i_\omega By(s), y(s)\rangle|^2 ds \le \frac{L_{\|y_0\|}}{2}\|\chi_\omega y_0\|^2_{L^2(\omega)}, \ \forall t \ge 0.$$

with $L_{\|y_0\|} = \sup_{\|y\| \leq \|y_0\|} (1 + |\langle i_\omega By, y\rangle|)$.

Hence

$$\int_0^{+\infty} |\langle i_\omega By(s), y(s)\rangle|^2 ds < +\infty. \tag{6.21}$$

From (6.21) and (6.20), we deduce that

$$\int_0^T |\langle i_\omega BS(s)y(t), S(s)y(t)\rangle| ds \longrightarrow 0,\ as\ t \longrightarrow +\infty. \tag{6.22}$$

It follows from (6.16) and (6.22) that $\|\chi_\omega y(t)\|_{L^2(\omega)} \longrightarrow 0,\ as\ t \longrightarrow +\infty$, which completes the proof. ■

Example 6.2 On $\Omega =]0, 1[$, we consider the following beam equation

$$\begin{cases} \dfrac{\partial^2 y}{\partial t^2}(x,t) = -\dfrac{\partial^4 y}{\partial x^4}(x,t) + u(t)\dfrac{\partial y}{\partial t}(x,t) & on\ \Omega\times]0,+\infty[\\ y(x,0) \quad = y_0,\ \dfrac{\partial y}{\partial t}(x,0) = y_1 & on\ \Omega \\ y(\xi,t) \quad = \dfrac{\partial^2 y}{\partial x^2}(\xi,t) = 0,\ \xi = 0,1 & on\ ]0,+\infty[\end{cases} \tag{6.23}$$

Let $\tilde{A} = \dfrac{\partial^4 y}{\partial x^4}$, with $\mathcal{D}(\tilde{A}) = \{y \in L^2(\Omega)/\dfrac{\partial^4 y}{\partial x^4} \in L^2(\Omega),\ y(\xi,t) = \dfrac{\partial^2 y}{\partial x^2}(\xi,t) = 0,$ $\xi = 0, 1\}$.

Setting $H = (H^2(\Omega) \cap H_0^1(\Omega)) \times L^2(\Omega)$ the state space endowed with the inner product $\langle (y_1, z_1), (y_2, z_2)\rangle = \langle \tilde{A}^{\frac{1}{2}} y_1, \tilde{A}^{\frac{1}{2}} y_2\rangle_{L^2(\Omega)} + \langle z_1, z_2\rangle_{L^2(\Omega)}$.

The eigenvalues of $\tilde{A}$ are $\lambda_j = (j\pi)^4$, corresponding to eigenfunctions $\varphi_j(x) = \sqrt{2}\sin((j\pi)^2 x),\ \forall j \in \mathbb{N}^*$.

System (6.23) has the form of (6.1) if we take

$$A = \begin{pmatrix} 0 & I \\ -\tilde{A} & 0 \end{pmatrix} \ and\ B = \begin{pmatrix} 0 & 0 \\ 0 & I \end{pmatrix}.$$

B satisfies condition (6.16), indeed, for $y \in H$, we have $y = \sum_{j=1}^{\infty} \begin{pmatrix} \alpha_j \\ \lambda_j^{\frac{1}{2}} \beta_j \end{pmatrix} \varphi_j$, where $(\alpha_j, \beta_j) \in \mathbb{R}^2\ \forall j \geq 1$ and the semigroup is given by

$$S(s)y = \sum_{j=1}^{\infty} \begin{pmatrix} \alpha_j \cos(\lambda_j^{\frac{1}{2}} s) + \beta_j \sin(\lambda_j^{\frac{1}{2}} s) \\ \beta_j \lambda_j^{\frac{1}{2}} \cos(\lambda_j^{\frac{1}{2}} s) - \alpha_j \lambda_j^{\frac{1}{2}} \sin(\lambda_j^{\frac{1}{2}} s) \end{pmatrix} \varphi_j, \quad \forall s \geq 0.$$

For $\omega =]0, \frac{1}{2}[$, we have

$$\begin{aligned}\langle i_\omega BS(s)y, S(s)y\rangle &= \sum_{j=1}^{\infty}\frac{\lambda_j}{2}\{\alpha_j \sin(\lambda_j^{\frac{1}{2}}s) - \beta_j \cos(\lambda_j^{\frac{1}{2}}s)\}^2\\ &= \sum_{j=1}^{\infty}\frac{\lambda_j}{2}\{\alpha_j^2 \sin^2(j\pi s) + \beta_j^2 \cos^2(j\pi s) - \alpha_j\beta_j \sin(2j\pi s)\}\end{aligned}$$

Integrating this relation over the time interval [0, 2], we obtain

$$\int_0^2 |\langle i_\omega BS(s)y, S(s)y\rangle| ds = \sum_{j=1}^{\infty}\frac{\lambda_j}{2}(\alpha_j^2 + \beta_j^2) = \|\chi_\omega y\|^2$$

then (6.16) holds.

We conclude that the control

$$u(t) = -\frac{\|\chi_\omega \partial_t y(.,t)\|^2_{L^2(\omega)}}{1 + \|\chi_\omega \partial_t y(.,t)\|^2_{L^2(\omega)}}$$

regionally strongly stabilizes system (6.23) on ω. ■

6.1.3 *Regional Weak Stabilization*

The following result gives sufficient conditions for regional weak stabilization.

Theorem 6.3 *Let B be locally Lipschitz and weakly sequentially continuous such that*

$$\langle i_\omega BS(t)z, S(t)z\rangle = 0,\ \forall t \geq 0 \Longrightarrow \chi_\omega z = 0 \tag{6.24}$$

then control (6.17) *regionally weakly stabilizes system* (6.1) *on* ω.

Proof Consider the nonlinear semigroup $N(t)y_0 = y(t)$ and let (t_k) be a sequence of real numbers such that $t_k \longrightarrow +\infty$ as $k \longrightarrow +\infty$.

From (6.8), $N(t_k)y_0$ is bounded in $L^2(\Omega)$, then there exists a subsequence $(t_{\phi(k)})$ of (t_k) and $z \in L^2(\Omega)$ such that $N(t_{\phi(k)})y_0 \rightharpoonup z \quad as \quad k \longrightarrow +\infty$.

For all $y_0 \in L^2(\Omega)$ and $t \geq 0$, we have

$$\begin{aligned}\langle i_\omega BS(t)y_0, S(t)y_0\rangle = &\langle i_\omega BN(t)y_0, N(t)y_0\rangle - \langle i_\omega BN(t)y_0, N(t)y_0 - S(t)y_0\rangle\\ &+ \langle i_\omega BS(t)y_0 - i_\omega BN(t)y_0, S(t)y_0\rangle.\end{aligned}$$

Using (6.6) and the continuity of χ_ω, we get

$$\begin{aligned}|\langle i_\omega BS(t)y_0, S(t)y_0\rangle| \leq &|\langle i_\omega BN(t)y_0, N(t)y_0\rangle|\\ &+2\delta K\|N(t)y_0 - S(t)y_0\|\|y_0\|, \text{ (for some } \delta > 0).\end{aligned} \tag{6.25}$$

From (6.7) and Schwarz's inequality, we have

$$\|N(t)y_0 - S(t)y_0\| \leq K\|y_0\|\sqrt{Tf(0)}, \quad \forall t \in [0, T] \tag{6.26}$$

where $f(t) = \displaystyle\int_t^{t+T} |\langle i_\omega BN(s)y_0, N(s)y_0\rangle|^2 ds$.

Integrating (6.25) over the interval $[0, T]$ and using (6.26), we obtain

$$\int_0^T |\langle i_\omega BS(s)y_0, S(s)y_0\rangle| ds \leq \beta\|y_0\|\sqrt{f(0)} \tag{6.27}$$

with $\beta = (2\delta K^2 T\|y_0\| + 1)\sqrt{T}$.

Replacing y_0 by $N(t_{\phi(k)})y_0$ in (6.27), we get

$$\int_0^T |\langle i_\omega BS(s)N(t_{\phi(k)})y_0, S(s)N(t_{\phi(k)})y_0\rangle| ds \leq \beta\|y_0\|\sqrt{f(t_{\phi(k)})}.$$

It follows that

$$\int_0^T |\langle i_\omega BS(s)N(t_{\phi(k)})y_0, S(s)N(t_{\phi(k)})y_0\rangle| ds \longrightarrow 0 \quad as \quad k \longrightarrow 0.$$

Since B is weakly sequentially continuous and using the dominated convergence theorem, we deduce that $\langle i_\omega BS(s)z, S(s)z\rangle = 0, \quad \forall s \in [0, T]$.

From (6.24), it follows that $\chi_\omega N(t_{\phi(k)})y_0 \rightharpoonup 0 \quad as \quad k \longrightarrow +\infty$, and then $\forall \phi \in L^2(\Omega)$, $\langle \chi_\omega N(t_k)y_0, \phi\rangle \longrightarrow 0$ as $k \longrightarrow +\infty$, hence $\chi_\omega N(t)y_0 \rightharpoonup 0 \quad as \quad t \longrightarrow +\infty$.

In other words $\chi_\omega y(t)$ converges weakly to 0 as $t \longrightarrow +\infty$, and then system (6.1) is regionally weakly stabilizable on ω. ■

Example 6.3 On $\Omega =]0, 1[$, we consider the following equation

$$\begin{cases} \dfrac{\partial^2 y}{\partial t^2}(x,t) = -\dfrac{\partial^4 y}{\partial x^4}(x,t) + u(t)\dfrac{\partial y}{\partial t}(x,t) & on\ \Omega\times]0,+\infty[\\ y(x,0) = y_0,\ \dfrac{\partial y}{\partial t}(x,0) = y_1 & on\ \Omega \\ y(\xi,t) = \dfrac{\partial^2 y}{\partial x^2}(\xi,t) = 0,\ \xi = 0,1 & on\ ]0,+\infty[\end{cases} \tag{6.28}$$

The state space is $H = (H^2(\Omega) \cap H_0^1(\Omega)) \times L^2(\Omega)$, this system has the form of (6.1) if we set

$$A = \begin{pmatrix} 0 & I \\ -\Delta^2 & 0 \end{pmatrix} \ and\ B = \begin{pmatrix} 0 & 0 \\ 0 & I \end{pmatrix} .$$

For $\omega =]0, \frac{1}{2}[$, by the same arguments using in Example 6.2, we show that

$$\int_0^2 |\langle i_\omega BS(s)y, S(s)y\rangle| ds = \|\chi_\omega y\|^2$$

It follows that

$$\langle i_\omega BS(t)y, S(t)y\rangle = 0 \Longrightarrow \chi_\omega y = 0$$

Then, condition (6.24) holds. We conclude that the control

$$u(t) = -\frac{\|\chi_\omega \partial_t y(., t)\|^2_{L^2(\omega)}}{1 + \|\chi_\omega \partial_t y(., t)\|^2_{L^2(\omega)}}$$

regionally weakly stabilizes system (6.28) on ω. ■

6.2 Regional Stabilization Problem

This section deals with regional stabilization of system (6.1) by considering the following minimization problem

$$\begin{cases} \min J(u) = \displaystyle\int_0^{+\infty} \frac{\langle P_\omega By(t), y(t)\rangle^2}{\|y(t)\|^2} dt + \int_0^{+\infty} \langle i_\omega Ry(t), y(t)\rangle dt + \int_0^{+\infty} \|y(t)\|^2 |u(t)|^2 dt \\ u \in \mathcal{U}_{ad} = \{u \in L^2(0, +\infty) | y(t) \text{ is a global solution and } J(u) < +\infty\}. \end{cases} \tag{6.29}$$

where B is bounded, A satisfies (6.3) and $P_\omega = i_\omega P i_\omega$ with $P \in \mathcal{L}(L^2(\Omega))$ is a positive and bounded operator satisfying the equation

$$\langle P_\omega Ay, y\rangle + \langle y, P_\omega Ay\rangle + \langle i_\omega Ry, y\rangle = 0, \quad y \in \mathcal{D}(A) \tag{6.30}$$

where $R \in \mathcal{L}(L^2(\Omega))$ is a positive operator.

Theorem 6.4 *Suppose that B is locally Lipschitz and P is compact such that*

$$\langle P_\omega BS(t)z, S(t)z\rangle = 0, \ \forall t \geq 0 \Longrightarrow \chi_\omega z = 0 \tag{6.31}$$

then the control

$$u^*(t) = -\frac{\langle P_\omega By(t), y(t)\rangle}{\|y(t)\|^2} \tag{6.32}$$

is the unique solution of (6.29) *and regionally weakly stabilizes system* (6.1) *on ω.*

Proof Let us define the function $V(y) = \langle P_\omega y, y\rangle$, $y \in L^2(\Omega)$.
Using (6.30), for all $y_0 \in \mathcal{D}(A)$ and $t \geq 0$, we have

$$\frac{dV(y(t))}{dt} = -2\frac{\langle P_\omega By(t), y(t)\rangle^2}{\|y(t)\|^2} - \langle i_\omega Ry(t), y(t)\rangle. \tag{6.33}$$

Integrating this relation, we get

$$\int_0^t \frac{\langle P_\omega By(s), y(s)\rangle^2}{\|y(s)\|^2} ds \le \frac{1}{2}V(y_0), \ t \ge 0. \tag{6.34}$$

The solution $y(.)$ is continuous with respect to the initial condition y_0 (see [2]) and $D(A)$ is dense in $L^2(\Omega)$, then (6.34) holds for all $y_0 \in L^2(\Omega)$ so $J(u^*)$ is finite for all $y_0 \in L^2(\Omega)$.

Let (t_k) be a sequence of real numbers such that $t_k \longrightarrow +\infty$ as $k \longrightarrow +\infty$.

By (6.8), $y(t_k)$ is bounded in $L^2(\Omega)$, then there exists a subsequence $(t_{\phi(k)})$ of (t_k) and $z \in L^2(\Omega)$ such that

$$y(t_{\phi(k)}) \rightharpoonup z \quad as \quad k \longrightarrow +\infty.$$

For all $y_0 \in L^2(\Omega)$ and $t \ge 0$, we have

$$\begin{aligned}\langle P_\omega BS(t)y_0, S(t)y_0\rangle &= \langle P_\omega BS(t)y_0 - P_\omega By(t), S(t)y_0\rangle \\ &\quad - \langle P_\omega By(t), y(t) - S(t)y_0\rangle + \langle P_\omega By(t), y(t)\rangle.\end{aligned}$$

Thus

$$|\langle P_\omega BS(t)y_0, S(t)y_0\rangle| \le 2\delta^2\|P\|K\|y(t) - S(t)y_0\|\|y_0\| + |\langle P_\omega By(t), y(t)\rangle|. \tag{6.35}$$

Moreover, we have

$$\|y(t) - S(t)y_0\| \le K\|y_0\| \int_0^t |\langle P_\omega By(s), y(s)\rangle| ds.$$

Schwarz's inequality yields

$$\|y(t) - S(t)y_0\| \le K\sqrt{T\lambda(0)}, \quad \forall t \in [0, T] \tag{6.36}$$

where $\lambda(t) = \displaystyle\int_t^{t+T} |\langle P_\omega By(s), y(s)\rangle|^2 ds.$

Integrating (6.35) over the interval $[0, T]$ and with (6.36), we obtain

$$\int_0^T |\langle P_\omega BS(s)y_0, S(s)y_0\rangle| ds \le (2\delta^2K^2T\|P\| + 1)\|y_0\|\sqrt{T\lambda(0)}. \tag{6.37}$$

Replacing y_0 by $y(t_{\phi(k)})$ in (6.37), we get

$$\int_0^T |\langle P_\omega BS(s)y(t_{\phi(k)}), S(s)y(t_{\phi(k)})\rangle| ds \le (2\delta^2K^2T\|P\| + 1)\|y_0\|\sqrt{T\lambda(t_{\phi(k)})}. \tag{6.38}$$

By (6.38), we get $\lim_{k \longrightarrow +\infty} \int_0^T \langle P_\omega BS(s)y(t_{\phi(k)}), S(s)y(t_{\phi(k)})\rangle ds = 0.$

Since P is compact and $S(s)$ is continuous $\forall s \geq 0$, we have

$$\lim_{k \longrightarrow +\infty} \langle P_\omega BS(s)y(t_{\phi(k)}), S(s)y(t_{\phi(k)})\rangle = \langle P_\omega BS(s)z, S(s)z\rangle.$$

By dominated convergence theorem, we obtain

$$\int_0^T |\langle P_\omega BS(s)z, S(s)z\rangle| ds = 0 \text{ and then } \langle P_\omega BS(s)z, S(s)z\rangle = 0, \quad \forall s \in [0, T]$$

.

Using (6.31), we deduce that $\chi_\omega y(t_{\phi(k)}) \rightharpoonup 0 \ \ as \ \ k \longrightarrow +\infty$.

It follows that $\chi_\omega y(t)$ converges weakly to 0 as $t \longrightarrow +\infty$, and system (6.1) is regionally weakly stabilizable on ω.

Now, let us prove that (6.32) is the unique solution of (6.29).

Since P is compact, it follows that $V(y(t)) \longrightarrow 0$ as $t \to +\infty$.

Let $y_0 \in \mathcal{D}(A)$, formula (6.33) may be written as

$$\begin{aligned}\frac{dV(y(t))}{dt} &= \|y(t)\|^2 \left(\left[\frac{\langle P_\omega By(t), y(t)\rangle}{\|y(t)\|^2} + u(t)\right]^2 - \frac{\langle P_\omega By(t), y(t)\rangle^2}{\|y(t)\|^2} - u^2(t)\right) \\ &-\langle i_\omega Ry(t), y(t)\rangle\end{aligned}$$

integrating this relation, we obtain

$$J(u) = V(y_0) + \int_0^{+\infty} \|y(s)\|^2 \left[\frac{\langle P_\omega By(s), y(s)\rangle}{\|y(s)\|^2} + u(s)\right]^2 ds.$$

Then

$$J(u) \geq V(y_0), \ \forall u \in \mathcal{U}_{ad}.$$

For $u = u^*$, we get $J(u^*) = V(y_0)$.

Let $y_0 \in L^2(\Omega)$, and a sequence $(y_{0k}) \subset \mathcal{D}(A)$ such that $y_{0k} \longrightarrow y_0$ as $k \longrightarrow +\infty$, we have

$$\begin{aligned}J(u) = V(y_{0k}) &+ \int_0^{+\infty} \|y_k(s)\|^2 \left[\frac{\langle P_\omega By_k(s), y_k(s)\rangle}{\|y_k(s)\|^2} + u(s)\right]^2 ds \\ &- \int_0^{+\infty} \langle i_\omega Ry_k(s), y_k(s)\rangle ds.\end{aligned}$$

Thus $J(u) \geq V(y_{0k})$.

We deduce that $J(u) \geq V(y_0) = J(u^*)$, so (6.32) is the unique solution of problem (6.29). ∎

Proposition 6.1 *Suppose that B is locally Lipschitz and satisfies*

$$\int_0^T |\langle P_\omega BS(s)y, S(s)y\rangle|ds \geq \alpha \|\chi_\omega y\|^2_{L^2(\omega)}, \quad (T, \alpha > 0) \tag{6.39}$$

then control (6.32) *is the unique solution of* (6.29) *and regionally strongly stabilizes system (6.1) on* ω.

Proof For all $y_0 \in L^2(\Omega)$, we have

$$\langle P_\omega BS(t)y_0, S(t)y_0\rangle = \langle P_\omega BS(t)y_0 - P_\omega By(t), S(t)y_0\rangle + \langle P_\omega By(t), S(t)y_0\rangle$$

Using (6.7), we obtain the following relation

$$\langle P_\omega BS(t)y_0, S(t)y_0\rangle = \langle P_\omega BS(t)y_0 - P_\omega By(t), S(t)y_0\rangle - \langle P_\omega By(t), \Psi(t)\rangle + \langle P_\omega By(t), y(t)\rangle$$

where $\Psi(t) = \displaystyle\int_0^t S(t-s)u(s)By(s)ds$.

Since χ_ω is continuous, then there exists $\delta > 0$ such that

$$|\langle P_\omega BS(t)y_0, S(t)y_0\rangle| \leq \delta^2 K\|P\|\|\Psi(t)\|(\|S(t)y_0\| + \|y(t)\|) + |\langle P_\omega By(t), y(t)\rangle|. \tag{6.40}$$

Then

$$|\langle P_\omega BS(t)y_0, S(t)y_0\rangle| \leq 2\delta^2 K\|P\|\|\Psi(t)\|\|y_0\| + |\langle P_\omega By(t), y(t)\rangle|, \forall t \in [0, T].$$

Moreover, we have

$$\|\Psi(t)\| \leq K\|y_0\|\sqrt{T}\Big(\int_0^T |\langle P_\omega By(s), y(s)\rangle|^2 ds\Big)^{\frac{1}{2}}. \tag{6.41}$$

Integrating (6.40) over the interval $[0, T]$ and using (6.41), we obtain

$$\int_0^T |\langle P_\omega BS(s)y_0, S(s)y_0\rangle|ds \leq 2\delta^2 T^{\frac{3}{2}} K\|P\|\|y_0\|^2\Big(\int_0^T |\langle P_\omega By(s), y(s)\rangle|^2 ds\Big)^{\frac{1}{2}} + \sqrt{T}\|y_0\|\Big(\int_0^T |\langle P_\omega By(s), y(s)\rangle|^2 ds\Big)^{\frac{1}{2}}.$$

Replacing y_0 by $y(t)$, we get

$$\int_0^T |\langle P_\omega BS(s)y(t), S(s)y(t)\rangle|ds \leq 2\delta^2 T^{\frac{3}{2}} K\|P\|\|y_0\|\Big(\int_t^{t+T} |\langle P_\omega By(s), y(s)\rangle|^2 ds\Big)^{\frac{1}{2}} + \sqrt{T}\|y_0\|\Big(\int_t^{t+T} |\langle P_\omega By(s), y(s)\rangle|^2 ds\Big)^{\frac{1}{2}}. \tag{6.42}$$

From (6.34) and (6.42), we deduce that

$$\int_0^T |\langle P_\omega BS(s)y(t), S(s)y(t)\rangle| ds \longrightarrow 0, \ as\ t \longrightarrow +\infty. \tag{6.43}$$

From (6.39) and (6.43), we deduce that $\|\chi_\omega y(t)\|_{L^2(\omega)} \longrightarrow 0,\ as\ t \longrightarrow +\infty$, which completes the proof. ■

Proposition 6.2 *Assume that B is locally Lipschitz such that*

$$\int_0^T |\langle P_\omega BS(t)y, S(t)y\rangle| dt \geq \delta \|\chi_\omega y\|_{L^2(\omega)}, \ (for\ some\ T, \delta > 0) \tag{6.44}$$

and there exists $\mu > 0$ such that

$$\langle P_\omega By, y\rangle \leq \mu \langle i_\omega By, y\rangle \tag{6.45}$$

then control (6.32) is the unique solution of (6.29) and regionally exponentially stabilizes system (6.1) on ω.

Proof Let us define the function $F(y) = \langle P_\omega y, y\rangle$, $\forall y \in L^2(\Omega)$.
For all $y_0 \in \mathcal{D}(A)$ and $t \geq 0$, using (6.30), we obtain

$$\frac{dF(y(t))}{dt} = -2\frac{\langle P_\omega By(t), y(t)\rangle^2}{\|y(t)\|^2} - \langle i_\omega Ry(t), y(t)\rangle.$$

Integrating this relation, we get

$$\int_0^t \frac{\langle P_\omega By(s), y(s)\rangle^2}{\|y(s)\|^2} ds \leq \frac{1}{2} F(y_0),\ t \geq 0. \tag{6.46}$$

By density, we deduce that $J(u^*)$ is finite for all $y_0 \in L^2(\Omega)$.

For $y_0 \in L^2(\Omega)$ and $t \geq 0$, we have

$$\begin{aligned}\langle P_\omega BS(t)y_0, S(t)y_0\rangle &= \langle P_\omega BS(t)y_0 - P_\omega By(t), S(t)y_0\rangle \\ &\quad - \langle P_\omega By(t), y(t) - S(t)y_0\rangle + \langle P_\omega By(t), y(t)\rangle.\end{aligned}$$

Since χ_ω is continuous, there exists $\alpha > 0$ such that

$$|\langle P_\omega BS(t)y_0, S(t)y_0\rangle| \leq 2\alpha K\|P\|\|y(t) - S(t)y_0\|\|y_0\| + |\langle P_\omega By(t), y(t)\rangle|. \tag{6.47}$$

Moreover, we have

$$\|y(t) - S(t)y_0\| \leq K \int_0^T \frac{|\langle P_\omega By(s), y(s)\rangle|}{\|y(s)\|} ds.$$

Schwarz's inequality yields

$$\|y(t) - S(t)y_0\| \leq K\Big\{T \int_0^T \frac{|\langle P_\omega By(s), y(s)\rangle|^2}{\|y(s)\|^2} ds\Big\}^{\frac{1}{2}}, \ \forall t \in [0, T]. \tag{6.48}$$

Using (6.8), we get

$$|\langle P_\omega By(t), y(t)\rangle| \leq \frac{|\langle P_\omega By(t), y(t)\rangle|}{\|y(t)\|}\|y_0\|, \ \forall t \in [0, T].$$

Integrating (6.47) over the interval $[0, T]$ and taking into account (6.48), we obtain

$$\int_0^T |\langle P_\omega BS(s)y_0, S(s)y_0\rangle| ds \leq (2\alpha K^2 T\|P\| + 1)\sqrt{T}\|y_0\|\sqrt{f(0)} \tag{6.49}$$

where $f(t) = \displaystyle\int_t^{t+T} \frac{|\langle P_\omega By(s), y(s)\rangle|^2}{\|y(s)\|^2} ds.$

Replacing y_0 by $y(t)$ in (6.49), we get

$$\int_0^T |\langle P_\omega BS(s)y(t), S(s)y(t)\rangle| ds \leq (2\alpha\|P\|K^2 T + 1)\sqrt{T}\|y(t)\|\sqrt{f(t)}.$$

Using (6.44) and (6.8), we obtain $\delta\|\chi_\omega y(t)\|_{L^2(\omega)} \leq \beta\sqrt{f(t)}$ with $\beta = (2\alpha K^2 T\|P\| + 1)\sqrt{T}\|y_0\|$.

Then

$$\delta^2\|\chi_\omega y(t)\|^2_{L^2(\omega)} \leq \beta^2\sqrt{f(t)}. \tag{6.50}$$

Using (6.45) and (6.3), we get

$$\frac{d}{dt}\|\chi_\omega y(t)\|^2_{L^2(\omega)} \leq -\frac{2}{\mu^2}\frac{|\langle P_\omega By(t), y(t)\rangle|^2}{\|y(t)\|^2}.$$

Integrating this inequality from mT to $(m+1)T$, ($m \in \mathbb{N}$), we obtain

$$\|\chi_\omega y(mT)\|^2_{L^2(\omega)} - \|\chi_\omega y((m+1)T)\|^2_{L^2(\omega)} \geq \frac{2}{\mu^2}\int_{mT}^{(m+1)T} \frac{|\langle P_\omega By(s), y(s)\rangle|^2}{\|y(s)\|^2} ds.$$

Using (6.50) and since $\|\chi_\omega y(t)\|_{L^2(\omega)}$ decreases, we deduce

$$\Big(1 + 2\Big(\frac{\delta}{\mu\beta}\Big)^2\Big)\|\chi_\omega y((m+1)T)\|^2_{L^2(\omega)} \leq \|\chi_\omega y(mT)\|^2_{L^2(\omega)}.$$

Then

$$\|\chi_\omega y((m+1)T)\|_{L^2(\omega)} \leq \sigma\|\chi_\omega y(mT)\|_{L^2(\omega)}$$

where $\sigma = \dfrac{1}{(1 + 2(\frac{\delta}{\mu\beta})^2)^{\frac{1}{2}}}.$

This implies that

$$\|\chi_\omega y(mT)\|_{L^2(\omega)} \leq \sigma^m \|\chi_\omega y_0\|_{L^2(\omega)}.$$

Taking $m = E(\frac{t}{T})$ the integer part of $\frac{t}{T}$ and remarking that $m \geq \frac{t}{T} - 1$, we obtain

$$\|\chi_\omega y(t)\|_{L^2(\omega)} \leq \sigma^{\frac{t}{T}-1} \|\chi_\omega y_0\|_{L^2(\omega)}.$$

We deduce that

$$\|\chi_\omega y(t)\|_{L^2(\omega)} \leq F e^{-\rho t} \|y_0\|, \ \forall t \geq 0 \tag{6.51}$$

where $F = \alpha(1 + 2(\frac{\delta}{\mu\beta})^2)^{\frac{1}{2}}$ and $\rho = \dfrac{\ln(1 + 2(\frac{\delta}{\mu\beta})^2)}{2T}$, then control (6.32) regionally exponentially stabilizes system (6.1) on ω.

It remains to show that (6.32) is the unique solution of (6.29).

Remarking that $F(y(t)) \leq \alpha \|P\| \|\chi_\omega y(t)\|^2_{L^2(\omega)}$, it follows from (6.51) that $F(y(t)) \longrightarrow 0$ as $t \rightarrow +\infty$.

Let $y_0 \in \mathcal{D}(A)$, integrating the relation

$$\frac{dF(y(t))}{dt} = \|y(t)\|^2 \left(\left[\frac{\langle P_\omega B y(t), y(t)\rangle}{\|y(t)\|^2} + u(t) \right]^2 - \frac{\langle P_\omega B y(t), y(t)\rangle^2}{\|y(t)\|^4} - u^2(t) \right)$$
$$-\langle i_\omega R y(t), y(t)\rangle \tag{6.52}$$

we have

$$J(u) = F(y_0) + \int_0^{+\infty} \|y(s)\|^2 \left[\frac{\langle P_\omega B y(s), y(s)\rangle}{\|y(s)\|^2} + u(s) \right]^2 ds$$

then $J(u) \geq F(y_0)$.

Setting $u = u^*$, we obtain $J(u^*) = F(y_0)$.

Let $y_0 \in L^2(\Omega)$, and a sequence $y_{0k} \subset \mathcal{D}(A)$ such that $y_{0k} \longrightarrow y_0$ as $k \longrightarrow +\infty$, we have

$$J(u) = F(y_{0k}) + \int_0^{+\infty} \|y_k(s)\|^2 \left[\frac{\langle P_\omega B y_k(s), y_k(s)\rangle}{\|y_k(s)\|^2} + u(s) \right]^2 ds.$$

Thus $J(u) \geq F(y_0) = J(u^*)$.

Hence control (6.32) is the unique solution of the problem (6.29). ■

In order to illustrate the previous results, we perform the following algorithm:
Step 1: Initial data : initial condition y_0 and subregion ω;
Step 2: Solve equation (6.30) using Bartels-Stewart method given in [3];
Step 3: Apply the control given by (6.17) or (6.32);
Step 4: Solve system (6.1) using Petrov-Galerkin method;

6.3 Simulation Results

Let $\Omega =]0, 1[$, we consider the system

$$\begin{cases} \frac{\partial y}{\partial t}(x,t) = 0.01\Delta y(x,t) + u(t)\frac{y(x,t)}{1+\|y(x,t)\|}, & \Omega\times]0,+\infty[\\ y(0,t) \;\; = y(1,t) = 0, &]0,+\infty[\\ y(x,0) \;\; = 2\pi\sin(2\pi x), & \Omega \end{cases} \tag{6.53}$$

where the state space is $L^2(\Omega)$.

1. Regional stabilizing control.
Let $\omega =]0, \frac{1}{2}[$, we perform the above algorithm applying the control (6.17) to system (6.53), we obtain the following figures.

Figure 6.1, shows that system (6.53) is stabilized on the subregion ω, with stabilization error equals to 3.84×10^{-4} (Fig. 6.2).

For $\omega =]0, 1[$, we obtain.

Figure 6.3 shows that system (6.53) is stabilized on Ω, with stabilization error equals to 9.83×10^{-4} (Fig. 6.4).

2. Regional optimal stabilizing control.
Consider system (6.53) and problem (6.29) with $R = I$ such that P_ω is the unique solution of the equation (6.30).
Let $\omega =]0, \frac{1}{2}[$, we perform the above algorithm applying the control (6.32) to system (6.53), we obtain the figures below.

Figure 6.5 shows how system (6.53) is stabilized by the control (6.32) on ω with a stabilization error equals to $2.1\ 10^{-4}$ and the cost $J(u^*) = 4.9 \times 10^{-2}$ (Fig. 6.6).

For $\omega = \Omega$.

Figure 6.7 shows that system (6.53) is stabilized on Ω, with stabilization error equals to 8.23×10^{-4} and the cost $J(u^*) = 9.75 \times 10^{-2}$ (Fig. 6.8).

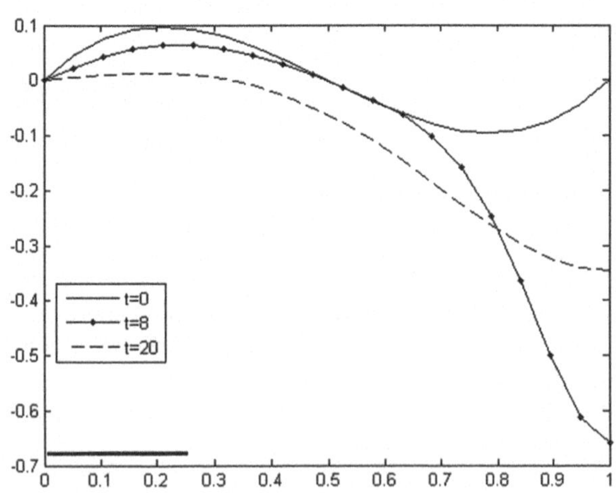

Fig. 6.1 Evolution of the state

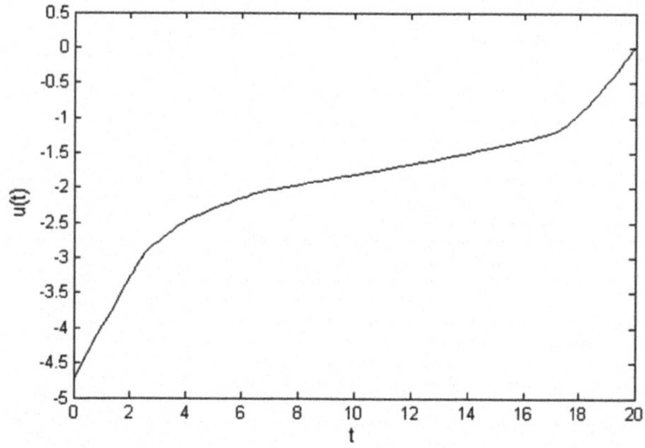

Fig. 6.2 Evolution of the control

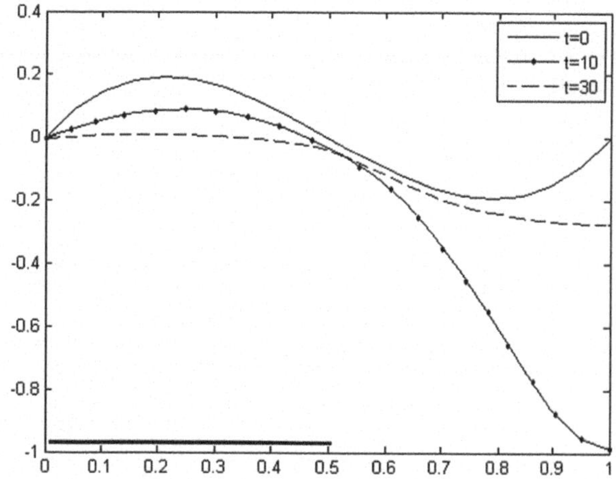

Fig. 6.3 Evolution of the state

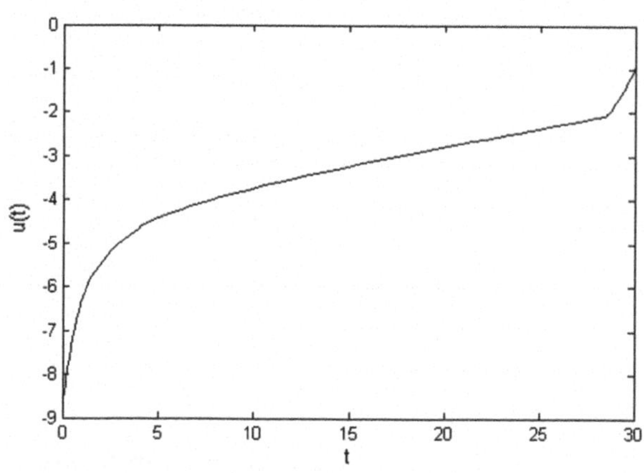

Fig. 6.4 Evolution of the control

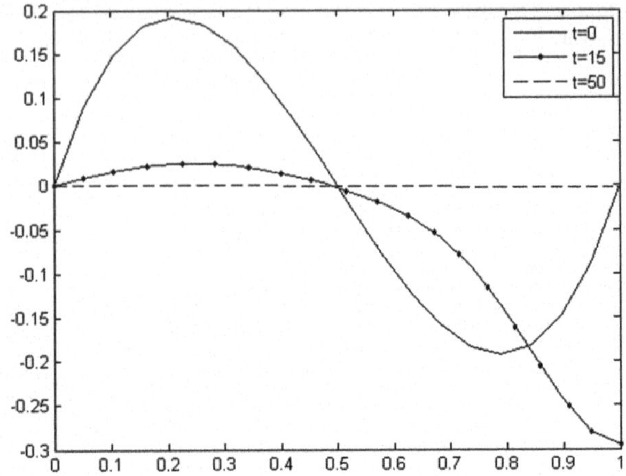

Fig. 6.5 Evolution of the state

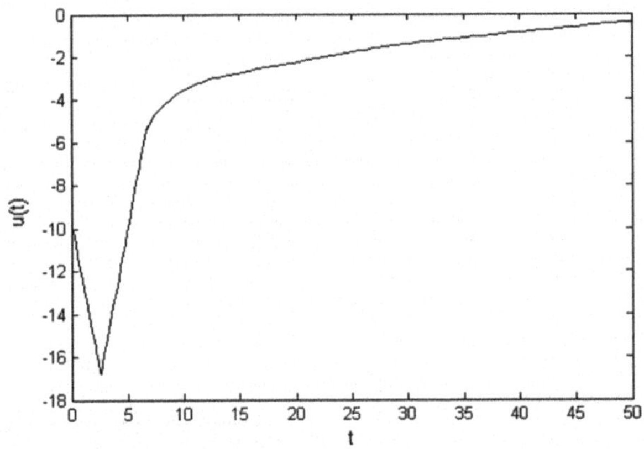

Fig. 6.6 Evolution of the control

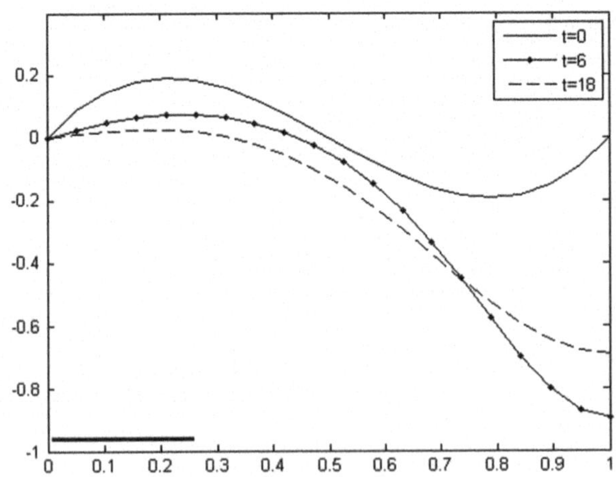

Fig. 6.7 Evolution of the state

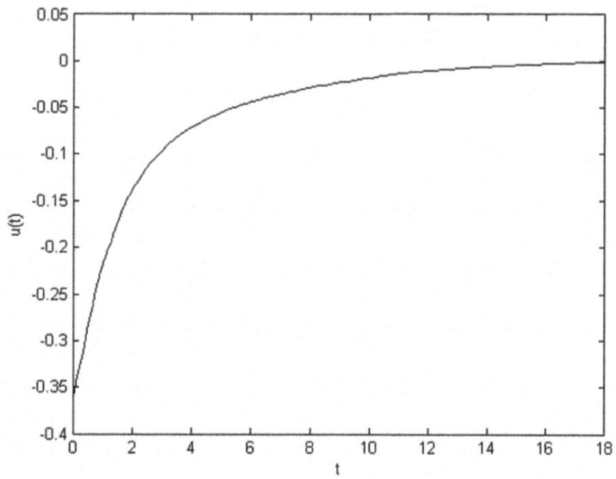

Fig. 6.8 Evolution of the control

The table below shows that there exists a relation between the area of subregion ω, the cost and the error of stabilization. Moreover, we remark that more the area of ω increases more the stabilization cost increases.

ω	$]0, 0.2[$	$]0, 0.4[$	$]0, 0.6[$	$]0, 0.8[$	$]0, 1[$
Error	1.33×10^{-4}	1.73×10^{-4}	3.95×10^{-4}	6.526×10^{-4}	8.23×10^{-4}
Cost	2.38×10^{-2}	3.84×10^{-2}	5.63×10^{-2}	7.56×10^{-2}	9.75×10^{-2}

References

1. Mohler, R.R.: Bilinear Control Processes with Application to Engineering, Ecology and Medicine. Academic, New York (1973)
2. Pazy, A.: Semigroups of Linear Operators and Applications to Partial Differential Equations. Springer, New York (1983)
3. Penzl, T.: Numerical solution of generalized Lyapunov equations. Adv. Comput. Math **8**, 33–48 (1998)

Chapter 7
Output Stabilization of Infinite Dimensional Semilinear Systems

This chapter considers the output stabilization for a class of distributed bilinear and semilinear systems with stabilizing controls that depend on the output operator. We give sufficient conditions for exponential, strong and weak output stabilization of such system. The obtained results are illustrated by many examples.

7.1 Output Stabilization of Semilinear Systems

We consider the following semilinear system

$$\begin{cases} \dfrac{dy(t)}{dt} = Ay(t) + u(t)By(t), \\ y(0) = y_0, \end{cases} \tag{7.1}$$

where $A : \mathcal{D}(A) \subset X \to X$ generates a strongly continuous semigroup of contractions $(S(t))_{t\geq 0}$ on a Hilbert space X endowed with a norm and an inner product denoted, respectively, by $\|.\|$ and $\langle ., . \rangle$, $u(.) \in U_{ad}$ is a scalar valued control, where U_{ad} is the admissible controls set, and B is a nonlinear operator from X to X with $B(0) = 0$ so that the origin be an equilibrium state of system (7.1).

System (7.1) is augmented with the output

$$z(t) := Cy(t). \tag{7.2}$$

E. H. Zerrik and O. Castillo, *Stabilization of Infinite Dimensional Systems*,
Studies in Systems, Decision and Control 355,
https://doi.org/10.1007/978-3-030-68600-0_7

Definition 7.1 The output (7.2) is said to be

1. exponentially stabilizable, if there exists a control $u(t) = f(y(t))$ such that for any initial condition $y_0 \in X$, the corresponding solution $y(t)$ of system (7.1) is global and there exist $\alpha, \beta > 0$ such that

$$\|Cy(t)\|_Y \leq \alpha e^{-\beta t}\|y_0\|, \quad \forall t > 0,$$

2. strongly stabilizable, if there exists a control $u(t) = f(y(t))$ such that for any initial condition $y_0 \in X$, the corresponding solution $y(t)$ of system (7.1) is global and verifies

$$\|Cy(t)\|_Y \to 0, \quad \text{as } t \to \infty,$$

3. weakly stabilizable, if there exists a control $u(t) = f(y(t))$ such that for any initial condition $y_0 \in X$, the corresponding solution $y(t)$ of system (7.1) is global and satisfies

$$\langle Cy(t), \psi\rangle_Y \to 0, \quad \forall \psi \in Y, \quad \text{as } t \to \infty.$$

Remark 7.1 It is clear that exponential stability of (7.2) ⇒ strong stability of (7.2) ⇒ weak stability of (7.2).

7.1.1 Output Exponential Stabilization

In this subsection, we develop sufficient conditions for exponential stabilization of the output (7.2).

The following result concerns the exponential stabilization of (7.2).

Theorem 7.1 *Let A generate a semigroup $(S(t))_{t\geq 0}$ of contractions on X and B be locally Lipschitz. If the conditions*

1. $\langle C^*CA\psi, \psi\rangle \leq 0$, $\forall \psi \in \mathcal{D}(A)$, *where C^* is the adjoint operator of C,*
2. $\langle C^*CB\psi, \psi\rangle\langle B\psi, \psi\rangle \geq 0$, $\forall \psi \in X$,
3. *there exist $T, \gamma > 0$, such that*

$$\int_0^T |\langle C^*CBS(t)\psi, S(t)\psi\rangle| dt \geq \gamma\|C\psi\|_Y^2, \quad \forall \psi \in X, \tag{7.3}$$

hold, then the control

$$u(t) = \begin{cases} -\dfrac{\langle C^*CBy(t), y(t)\rangle}{\|y(t)\|^2}, & \text{if } y(t) \neq 0, \\ 0, & \text{if } y(t) = 0, \end{cases} \tag{7.4}$$

is in $U_{ad} = L^\infty(0, \infty)$, and exponentially stabilizes the output (7.2).

Proof System (7.1) has a unique weak solution $y(t)$ (see [1]) defined on a maximal interval $[0, t_{\max}]$ by

$$y(t) = S(t)y_0 + \int_0^t g(y(s))S(t-s)By(s)ds, \tag{7.5}$$

where

$$g(y(t)) = \begin{cases} -\dfrac{\langle C^*CBy(t), y(t)\rangle}{\|y(t)\|^2}, & \text{if } y(t) \neq 0, \\ 0, & \text{if } y(t) = 0. \end{cases}$$

Since $(S(t))_{t\geq 0}$ is a semigroup of contractions, we have

$$\frac{d}{dt}\|y(t)\|^2 \leq 2g(y(t))\langle By(t), y(t)\rangle.$$

Integrating this inequality, and using hypothesis 2 of Theorem 7.1, it follows that

$$\|y(t)\| \leq \|y_0\|. \tag{7.6}$$

For all $y_0 \in X$ and $t \geq 0$, we have

$$\begin{aligned}\langle C^*CBS(t)y_0, S(t)y_0\rangle &= \langle C^*CBy(t), y(t)\rangle - \langle C^*CBy(t), y(t) - S(t)y_0\rangle \\ &\quad + \langle C^*CBS(t)y_0 - C^*CBy(t), S(t)y_0\rangle.\end{aligned}$$

Since B is locally Lipschitz, there exists a constant positive L that depends on $\|y_0\|$ such that

$$|\langle C^*CBS(t)y_0, S(t)y_0\rangle| \leq |\langle C^*CBy(t), y(t)\rangle| + 2\alpha L\|y(t) - S(t)y_0\|\|y_0\|, \tag{7.7}$$

where α is a positive constant.

Using (7.6), we deduce

$$|\langle C^*CBy(t), y(t)\rangle| \leq |g(y(t))|\|y(t)\|\|y_0\|, \ \forall t \in [0, T]. \tag{7.8}$$

While from (7.5) and using Schwartz's inequality, we obtain

$$\|y(t) - S(t)y_0\| \leq L\left(T\int_0^T |g(y(t))|^2\|y(t)\|^2 dt\right)^{\frac{1}{2}}. \tag{7.9}$$

Integrating (7.7) over the interval $[0, T]$ and taking into account (7.8) and (7.9), we get

$$\int_0^T |\langle C^*CBS(t)y_0, S(t)y_0\rangle| dt \le 2\alpha T^{\frac{3}{2}} L^2 \|y_0\| \left(\int_0^T |g(y(t))|^2 \|y(t)\|^2 dt\right)^{\frac{1}{2}} + T^{\frac{1}{2}} \|y_0\| \left(\int_0^T |g(y(t))|^2 \|y(t)\|^2 dt\right)^{\frac{1}{2}}. \quad (7.10)$$

Now, let us consider the nonlinear semigroup $U(t)y_0 := y(t)$ (see [2]).

Replacing y_0 by $U(t)y_0$ in (7.10), and using the superposition properties of the semigroup $(U(t))_{t\geq 0}$, we deduce that

$$\begin{aligned}\int_0^T |\langle C^*CBS(s)U(t)y_0, S(s)U(t)y_0\rangle| ds \le\ & 2\alpha T^{\frac{3}{2}} L^2 \|U(t)y_0\| \\ & \times \left(\int_t^{t+T} |g(U(s)y_0)|^2 \|U(s)y_0\|^2 ds\right)^{\frac{1}{2}} \\ & + T^{\frac{1}{2}} \|U(t)y_0\| \\ & \times \left(\int_t^{t+T} |g(U(s)y_0)|^2 \|U(s)y_0\|^2 ds\right)^{\frac{1}{2}}\end{aligned} \quad (7.11)$$

Thus, by using (7.3) and (7.11), it follows that

$$\gamma \|CU(t)y_0\|_Y \le M \left(\int_t^{t+T} |g(U(s)y_0)|^2 \|U(s)y_0\|^2 ds\right)^{\frac{1}{2}}, \quad (7.12)$$

where $M = (2\alpha T L^2 + 1)T^{\frac{1}{2}}$ is a non-negative constant depending on $\|y_0\|$ and T.

From hypothesis 1 of Theorem 7.1, we have

$$\frac{d}{dt}\|CU(t)y_0\|_Y^2 \le -2|g(U(t)y_0)|^2 \|U(t)y_0\|^2. \quad (7.13)$$

Integrating (7.13) from nT and $(n+1)T$, $(n \in \mathbb{N})$, we obtain

$$\|CU(nT)y_0\|_Y^2 - \|CU((n+1)T)y_0\|_Y^2 \ge 2 \int_{nT}^{(n+1)T} |g(U(s)y_0)|^2 \|U(s)y_0\|^2 ds.$$

Using (7.12), (7.13) and the fact that $\|CU(t)z_0\|_Y$ decreases, it follows

$$\left(1 + 2\left(\frac{\gamma}{M}\right)^2\right) \|CU((n+1)T)y_0\|_Y^2 \le \|CU(nT)y_0\|_Y^2.$$

Then

$$\|CU((n+1)T)y_0\|_Y \le \beta \|CU(nT)y_0\|_Y,$$

where $\beta = \frac{1}{\left(1+2\left(\frac{\gamma}{M}\right)^2\right)^{\frac{1}{2}}}$.

By recurrence, we show that $\|CU(nT)y_0\|_Y \leq \beta^n \|Cy_0\|_Y$.

Taking $n = E(\frac{t}{T})$ the integer part of $\frac{t}{T}$, we deduce that

$$\|CU(t)y_0\|_Y \leq Re^{-\sigma t}\|y_0\|,$$

where $R = \alpha\left(1 + 2\left(\frac{\gamma}{M}\right)^2\right)^{\frac{1}{2}}$, with $\alpha > 0$ and $\sigma = \frac{\ln\left(1+2\left(\frac{\gamma}{M}\right)^2\right)}{2T} > 0$, which achieves the proof. □

Corollary 7.1 *Let $\Omega \subset \mathbb{R}^n$ be the system evolution domain assumed to be regular and ω be a subregion of Ω, if we take $C = \chi_\omega$, where $\chi_\omega : X = L^2(\Omega) \to Y = L^2(\omega)$ is the restriction operator to ω, and χ_ω^* denotes the adjoint operator of χ_ω given by*

$$(\chi_\omega^* z)(x) = \begin{cases} z(x), & \text{if } x \in \omega, \\ 0, & \text{if } x \in \Omega \backslash \omega. \end{cases}$$

If the assumptions of Theorem 7.1 hold, then the control

$$u(t) = \begin{cases} -\dfrac{\langle \chi_\omega^* \chi_\omega B y(t), y(t)\rangle}{\|y(t)\|^2}, & \text{if } y(t) \neq 0, \\ 0, & \text{if } y(t) = 0, \end{cases}$$

exponentially stabilizes system (7.1) on ω.

Example 7.1 Let us consider the system defined on an open and bounded domain $\Omega \subset \mathbb{R}^n$ with C^∞ boundary $\partial\Omega$ by the equation

$$\begin{cases} \dfrac{\partial y(x,t)}{\partial t} = -i\Delta y(x,t) + u(t)By(x,t), & x \in \Omega,\ t > 0, \\ y(x,t) = 0, & x \in \partial\Omega,\ t > 0, \\ y(x,0) = y_0(x), & x \in \Omega, \end{cases} \tag{7.14}$$

where $X = L^2(\Omega)$ endowed with its natural complex inner product, $A = -i\Delta$, $(i \in \mathbb{C})$ with domain $\mathcal{D}(A) = H^2(\Omega) \cap H_0^1(\Omega)$ and the operator of control given by $By = y + \dfrac{y}{1+\|y\|}$.

Let ω be a subregion of Ω that verifies the geometric control condition (GCC) (see [3]).

System (7.14) is augmented with the output

$$z(t) = \chi_\omega y(t). \tag{7.15}$$

The operator A generates a semigroup of isometries on $L^2(\Omega)$, and

$$\mathcal{R}e\Big(\big\langle \chi_\omega^* \chi_\omega A z, z\big\rangle\Big) = 0, \forall z \in \mathcal{D}(A).$$

For all $y_0 \in L^2(\Omega)$, we have

$$\begin{aligned}\big\langle \chi_\omega^* \chi_\omega B S(t) y_0, S(t) y_0\big\rangle &= \big\langle S(t) y_0, S(t) y_0\big\rangle_{L^2(\omega)} + \frac{1}{1 + \|S(t) y_0\|} \int_\omega |S(t) y_0|^2 dx \\ &\geq \big\langle S(t) y_0, S(t) y_0\big\rangle_{L^2(\omega)}.\end{aligned}$$

Integrating this inequality, we get

$$\int_0^T \big\langle \chi_\omega^* \chi_\omega B S(t) y_0, S(t) y_0\big\rangle dt \geq \int_0^T \big\langle S(t) y_0, S(t) y_0\big\rangle_{L^2(\omega)} dt.$$

Since the subregion ω verifies GCC, the inequality

$$\int_0^T \big\langle S(t) y_0, S(t) y_0\big\rangle_{L^2(\omega)} dt \geq \alpha \|y_0\|^2 \geq \alpha \|\chi_\omega y_0\|^2_{L^2(\omega)}, \text{ for some } \alpha, T > 0,$$

holds (see [3, 4]), we deduce that

$$\int_0^T \big\langle \chi_\omega^* \chi_\omega B S(t) y_0, S(t) y_0\big\rangle dt \geq \alpha \|\chi_\omega y_0\|^2_{L^2(\omega)}.$$

When we apply Theorem 7.1, we conclude that the control

$$u(t) = -\frac{\|\chi_\omega y(.,t)\|^2_{L^2(\omega)}}{\|y(.,t)\|^2} \left(1 + \frac{1}{1 + \|y(.,t)\|}\right) 1\!\!1_{\{t \geq 0,\ y(.,t) \neq 0\}},$$

exponentially stabilizes the output (7.15).

7.1.2 Output Strong Stabilization

The following result provides sufficient conditions for strong stabilisation of the output (7.2).

Theorem 7.2 *Let A generate a semigroup $(S(t))_{t \geq 0}$ of contractions on X, B be locally Lipschitz and let assumptions 1, 2 and 3 of Theorem 7.1 hold, then the control*

$$u(t) = -\langle C^* C B y(t), y(t)\rangle, \tag{7.16}$$

is in $U_{ad} = L^2(0, \infty)$, and strongly stabilizes the output (7.2).

Proof From hypothesis 1 of Theorem 7.1, we obtain

$$\frac{d}{dt}\|Cy(t)\|_Y^2 \leq -2|\langle C^*CBy(t), y(t)\rangle|^2. \tag{7.17}$$

Integrating this inequality, we deduce that

$$2\int_0^t |\langle C^*CBy(s), y(s)\rangle|^2 ds \leq \|Cy(0)\|_Y^2.$$

Hence, we get

$$\int_0^{+\infty} |\langle C^*CBy(s), y(s)\rangle|^2 ds < +\infty, \tag{7.18}$$

so $u \in L^2(0, \infty)$.

From the variation of constants formula and using Schwartz's inequality, we deduce

$$\|y(t) - S(t)y_0\| \leq LT^{\frac{1}{2}} \left(\int_0^T |\langle C^*CBy(s), y(s)\rangle|^2 ds\right)^{\frac{1}{2}}. \tag{7.19}$$

Integrating (7.7) over the interval $[0, T]$ and taking into account (7.19), we obtain

$$\int_0^T |\langle C^*CBS(s)y_0, S(s)y_0\rangle| ds \leq 2\alpha L^2 T^{\frac{3}{2}} \|y_0\|^2 \left(\int_0^T |\langle C^*CBy(s), y(s)\rangle|^2 ds\right)^{\frac{1}{2}} + T^{\frac{1}{2}} \left(\int_0^T |\langle C^*CBy(s), y(s)\rangle|^2 ds\right)^{\frac{1}{2}}.$$

Replacing y_0 by $y(t)$ and using the superposition property of the solution, we get

$$\int_0^T |\langle C^*CBS(s)y(t), S(s)y(t)\rangle| ds \leq 2\alpha L^2 T^{\frac{3}{2}} \|y_0\|^2 \left(\int_t^{t+T} |\langle C^*CBy(s), y(s)\rangle|^2 ds\right)^{\frac{1}{2}} + T^{\frac{1}{2}} \left(\int_t^{t+T} |\langle C^*CBy(s), y(s)\rangle|^2 ds\right)^{\frac{1}{2}}. \tag{7.20}$$

By (7.18), we get

$$\int_t^{t+T} |\langle C^*CBS(s)y(t), S(s)y(t)\rangle| ds \to 0, \text{ as } t \to +\infty. \tag{7.21}$$

From (7.3) and (7.21), we deduce that $\|Cy(t)\|_Y \to 0$, as $t \to +\infty$, which completes the proof. □

Proposition 7.1 *If B is linear bounded and the assumptions 1, 2 and 3 of Theorem 7.1 hold, then the control (7.16) strongly stabilizes the output (7.2) with decay* *estimate*

$$\|Cy(t)\|_Y = O(t^{-\frac{1}{2}}), \ \ as \ t \to +\infty. \qquad (7.22)$$

Proof Since B is bounded and while using (7.20), we get

$$\int_0^T |\langle C^*CBS(s)U(t)y_0, S(s)U(t)y_0\rangle| ds \le \theta\sqrt{\xi(t)}, \qquad (7.23)$$

where $\theta = (2\alpha T\|B\|^2\|y_0\|^2 + 1)T^{\frac{1}{2}}$ and $\xi(t) = \left(\int_t^{t+T} |\langle C^*CBU(s)y_0, U(s)y_0\rangle|^2 ds\right)$.

From (7.3) and (7.23), we deduce that

$$\varrho\sqrt{\xi(nT)} \ge \|CU(nT)y_0\|_Y^2, \quad \forall n \ge 0, \qquad (7.24)$$

where $\varrho = \frac{1}{\gamma}\theta$.

Integrating the inequality

$$\frac{d}{dt}\|CU(t)y_0\|_Y^2 \le -2|\langle C^*CBU(t)y_0, U(t)y_0\rangle|^2,$$

from nT to $(n+1)T$, $(n \in \mathbb{N})$ and using (7.24), we obtain

$$\|CU(nT)y_0\|_Y^2 - \|CU(nT+T)y_0\|_Y^2 \ge 2\xi(nT), \quad \forall n \ge 0.$$

Hence, we get

$$\varrho^2\|CU(nT+T)y_0\|_Y^2 - \varrho^2\|CU(nT)y_0\|_Y^2 \le -2\|CU(nT)y_0\|_Y^4, \quad \forall n \ge 0. \qquad (7.25)$$

Let us introduce the sequence $r_n = \|CU(nT)y_0\|_Y^2, \ \forall n \ge 0$.

Using (7.25), we deduce that

$$\frac{r_n - r_{n+1}}{r_n^2} \ge \frac{2}{\varrho^2}, \quad \forall n \ge 0.$$

Since the sequence (r_n) decreases, we get

$$\frac{r_n - r_{n+1}}{r_n . r_{n+1}} \ge \frac{2}{\varrho^2}, \quad \forall n \ge 0,$$

and also

$$\frac{1}{r_{n+1}} - \frac{1}{r_n} \ge \frac{2}{\varrho^2}, \quad \forall n \ge 0.$$

We deduce that

$$r_n \le \frac{r_0}{\frac{2r_0}{\varrho^2}n + 1}, \quad \forall n \ge 0.$$

Finally, introducing the integer part $n = E(\frac{t}{T})$ and from (7.17), we get $\|CU(t)y_0\|_Y$ decreases. We deduce the estimate

$$\|Cy(t)\|_Y = O(t^{-1/2}), \quad \text{as } t \to +\infty.$$

□

Corollary 7.2 *Let A generate a semigroup of contractions on X and suppose that $\langle C^*CAz, z\rangle = 0$, for all $z \in \mathcal{D}(A)$. If B is locally Lipschitz and condition (7.3) holds, then the control (7.16) strongly stabilizes the output (7.2) and it is the unique solution of problem*

$$\begin{cases} \min Q(u) = \displaystyle\int_0^{+\infty} |\langle C^*CBy(t), y(t)\rangle|^2 dt + \int_0^{+\infty} |u(t)|^2 dt, \\ u \in U_{ad} = \{u \in L^2(0,\infty) \mid y(t) \text{ is a global solution and } Q(u) < +\infty\}. \end{cases} \tag{7.26}$$

Proof For $y_0 \in \mathcal{D}(A)$, we get

$$\frac{d}{dt}\|Cy(t)\|_Y^2 = -2|\langle C^*CBy(t), y(t)\rangle|^2. \tag{7.27}$$

Integrating (7.27), we deduce

$$\int_0^t |\langle C^*CBy(s), y(s)\rangle|^2 ds \le \frac{1}{2}\|Cy_0\|_Y^2, \ \forall t \ge 0.$$

This inequality holds for all $y_0 \in X$, since the correspondence $y_0 \longrightarrow y(.)$ is continuous in $\mathcal{C}([0,t]; X)$, then $Q(u) < +\infty$, for all $y_0 \in X$. In other word $u \in U_{ad}$, so $U_{ad} \neq \emptyset$.

From (7.27), we get

$$\frac{d}{dt}\|Cy(t)\|_Y^2 = |\langle C^*CBy(t), y(t)\rangle + u(t)|^2 - \langle C^*CBy(t), y(t)\rangle^2 - u(t)^2. \tag{7.28}$$

Integrating (7.28) over $(0,t)$, we obtain

$$\int_0^t \left(\langle C^*CBy(s), y(s)\rangle^2 + u(s)^2\right) ds + \|Cy(t)\|_Y^2 - \|Cy_0\|_Y^2 = \int_0^t |\langle C^*CBy(s), y(s)\rangle + u(s)|^2 ds. \tag{7.29}$$

From Theorem 7.2, we get

$$\|Cy(t)\|_Y \to 0, \text{ as } t \to +\infty.$$

Let $t \to +\infty$ in (7.29), we obtain

$$Q(u) = \|Cy_0\|_Y^2 + \int_0^{+\infty} |\langle C^*CBy(s), y(s)\rangle + u(s)|^2 ds, \tag{7.30}$$

which implies that $Q(u) \geq \|Cy_0\|_Y^2$, so (7.16) is an optimal control.

Now, let $vu_i(t), i = 1, 2$, be two solutions of problem (7.26).

From (7.30), we deduce that $u_i(t) = \langle C^*CBy_i(t), y_i(t)\rangle$, where $y_i(t)$ verifies

$$\begin{cases} \dot{y}(t) = Ay(t) - \langle C^*CBy(t), y(t)\rangle By(t), \\ y(0) = y_0, \end{cases}$$

thus $y_i(t) = y(t),\ i = 1, 2$ and $u_1(t) = u_2(t)$. □

Example 7.2 Consider the system defined on $\Omega \subset \mathbb{R}^n$ by

$$\begin{cases} \dfrac{\partial y(x,t)}{\partial t} = -i\Delta y(x,t) + u(t)By(x,t), & x \in \Omega,\ t > 0, \\ y(x,t) = 0, & x \in \partial\Omega,\ \ t > 0, \\ y(x,0) = y_0(x), & x \in \Omega, \end{cases} \tag{7.31}$$

where $X = L^2(\Omega)$ is endowed with its natural complex inner product, $A = -i\Delta$, ($i \in \mathbb{C}$) with domain $\mathcal{D}(A) = H^2(\Omega) \cap H_0^1(\Omega)$, $B = \chi_\omega^*\chi_\omega$ and the control function $u(.) \in L^2(0, +\infty)$.

Let ω be a subregion of Ω that verify GCC, and system (7.31) is augmented with the output

$$z(t) = \chi_\omega y(t). \tag{7.32}$$

The operator A generates a semigroup of isometries on $L^2(\Omega)$, and

$$\mathcal{R}e\Big(\big\langle \chi_\omega^*\chi_\omega Az, z\big\rangle\Big) = 0, \forall z \in \mathcal{D}(A).$$

Since the subregion ω verifies GCC, so that the inequality (7.3) holds (see [3, 4]).

We deduce that the control $u(t) = -\|\chi_\omega y(., t)\|_{L^2(\omega)}^2$, strongly stabilizes the output (7.32) and minimizes the cost

$$Q(u) = \int_0^{+\infty} \|\chi_\omega y(., t)\|_{L^2(\omega)}^2 dt + \int_0^{+\infty} |u(t)|^2 dt.$$

7.1.3 Output Weak Stabilization

The following result discusses the weak stabilization of the output (7.2).

Theorem 7.3 *Let A generate a semigroup $(S(t))_{t\geq 0}$ of contractions on X, B be locally Lipschitz and weakly sequentially continuous. If assumptions 1, 2 of Theorem 7.1 and*

$$\langle C^*CBS(t)\psi, S(t)\psi\rangle = 0, \quad \forall t \geq 0 \Longrightarrow C\psi = 0, \tag{7.33}$$

hold, then the control

$$u(t) = -\langle C^*CBy(t), y(t)\rangle, \tag{7.34}$$

is in $U_{ad} = L^2(0, \infty)$, *and weakly stabilizes the output (7.2).*

Proof Let us consider $\psi \in Y$ and $(t_n) \geq 0$ be a sequence of real numbers such that $t_n \to +\infty$, as $n \to +\infty$.

Using (7.17), we deduce that the sequence $h_n = \langle Cy(t_n), \psi\rangle_Y$ is bounded.

Let $h_{\gamma(n)}$ be an arbitrary convergent subsequence of h_n.

From (7.6), we get $y(t_{\gamma(n)})$ is bounded in H, so we can extract a subsequence still denoted by $y(t_{\gamma(n)})$ such that $y(t_{\gamma(n)}) \rightharpoonup \varphi \in X$, as $n \to +\infty$.

Since C is continuous, B is weakly sequentially continuous and $S(t)$ is continuous $\forall t \geq 0$, we get

$$\lim_{n\to+\infty} \langle C^*CBS(t)y(t_{\gamma(n)}), S(t)y(t_{\gamma(n)})\rangle = \langle C^*CBS(t)\varphi, S(t)\varphi\rangle.$$

From (7.21), we have

$$\int_0^T \langle C^*CBS(s)y(t_{\gamma(n)}), S(s)y(t_{\gamma(n)})\rangle ds \to 0, \text{ as } n \to +\infty.$$

Using the dominated convergence Theorem, we deduce that

$$\langle C^*CBS(t)\varphi, S(t)\varphi\rangle = 0, \text{ for all } t \geq 0,$$

which implies, according to (7.33), that $C\varphi = 0$, and hence $h_n \to 0$, as $t \to +\infty$.

We deduce that $\langle Cy(t), \psi\rangle_Y \to 0$, as $t \to +\infty$. In other words $Cy(t) \rightharpoonup 0$, as $t \to +\infty$, which achieves the proof. □

Remark 7.2 Condition (7.33) may be verified for some subregion ω, but not on the whole domain Ω, as illustrated below.

Let us consider the system defined in $\Omega =]0, +\infty[$ by

$$\begin{cases} \dfrac{\partial y(x,t)}{\partial t} = -\dfrac{\partial y(x,t)}{\partial x} + u(t)By(x,t), & x \in \Omega, \ t > 0, \\ y(x,0) = y_0(x), & x \in \Omega, \end{cases} \tag{7.35}$$

where $X = L^2(\Omega)$, $Ay = -\dfrac{\partial y}{\partial x}$ with domain $\mathcal{D}(A) = \{y \in H^1(\Omega) \mid y(0) = 0, \ y(x) \to 0, \text{ as } x \to +\infty\}$, $By = \left|\int_0^1 y(x)dx\right|$ the control operator and $u(.) \in L^2(0, +\infty)$.

The operator A generates a semigroup of contractions

$$(S(t)y_0)(x) = \begin{cases} y_0(x-t), & \text{if } x \geq t, \\ 0, & \text{if } x < t. \end{cases}$$

Let $\omega =]0, 1[$ be a subregion of Ω and system (7.35) is augmented with the output

$$z(t) = \chi_\omega y(t). \tag{7.36}$$

Thus stabilizing the output (7.36) means the stabilization of system (7.35) on ω.

We have

$$\begin{aligned} \langle \chi_\omega^* \chi_\omega Ay, y \rangle &= -\int_0^1 y'(x)y(x)dx \\ &= -\frac{y^2(1)}{2} \leq 0, \end{aligned}$$

so, the assumption 1 of Theorem 7.1 holds.

The operator B is sequentially continuous and verifies

$$\langle \chi_\omega^* \chi_\omega BS(t)y_0, S(t)y_0 \rangle = |\int_0^{1-t} y_0(x)dx| \int_0^{1-t} y_0(x)dx, \quad 0 \leq t \leq 1.$$

Thus

$$\langle \chi_\omega^* \chi_\omega BS(t)y_0, S(t)y_0 \rangle = 0, \quad \forall t \geq 0 \implies y_0(x) = 0 \text{ a.e } x \in]0, 1[, \quad \text{i.e } \chi_{]0,1[} y_0 = 0.$$

Then, the control

$$u(t) = -|\int_0^1 y(x,t)dx| \int_0^1 y(x,t)dx, \tag{7.37}$$

weakly stabilizes the output (7.36).

7.2 Output Stabilization of Bilinear Systems

We consider the following bilinear system

$$\begin{cases} \dfrac{dy(t)}{dt} = Ay(t) + u(t)By(t) \\ y(0) = y_0, \end{cases} \tag{7.38}$$

where $B : X \to X$ is a linear bounded operator. The system (7.38) is augmented with the output

$$z(t) := Cy(t). \tag{7.39}$$

7.2.1 Output Exponential Stabilization

In this subsection, we develop sufficient conditions for exponential stabilization of the output (7.39).

The following result concerns the exponential stabilization of (7.39).

Theorem 7.4 *Let A generate a semigroup $(S(t))_{t\geq 0}$ of contractions on X and B is a bounded control operator. If the conditions:*
1. $\langle C^*CA\psi, \psi\rangle \leq 0,\ \forall\psi \in \mathcal{D}(A)$, *where C^* is the adjoint of C,*
2. $\langle C^*CB\psi, \psi\rangle\langle B\psi, \psi\rangle \geq 0,\ \forall\psi \in X$,
3. *there exist $T, \gamma > 0$, such that*

$$\int_0^T |\langle C^*CBS(t)\psi, S(t)\psi\rangle| dt \geq \gamma \|C\psi\|_Y^2, \quad \forall\, \psi \in X, \tag{7.40}$$

hold, then the control

$$u(t) = \begin{cases} -\dfrac{\langle C^*CBy(t), y(t)\rangle}{\|y(t)\|^2} & \text{if } y(t) \neq 0 \\ 0 & \text{if } y(t) = 0, \end{cases} \tag{7.41}$$

exponentially stabilizes the output (7.39).

Proof System (7.38) has a unique weak solution $y(t)$ (see [1]) defined on a maximal interval $[0, t_{\max}]$ by

$$y(t) = S(t)y_0 + \int_0^t g(y(s))S(t-s)By(s)ds, \tag{7.42}$$

where

$$g(y(t)) = \begin{cases} -\frac{\langle C^*CBy(t), y(t)\rangle}{\|y(t)\|^2} & \text{if } y(t) \neq 0 \\ 0 & \text{if } y(t) = 0. \end{cases}$$

Since $(S(t))_{t\geq 0}$ is a semigroup of contractions, we deduce

$$\frac{d}{dt}\|y(t)\|^2 \leq 2g(y(t))\langle By(t), y(t)\rangle.$$

Integrating this inequality over the interval $[0, t]$, we have

$$\|y(t)\|^2 - \|y(0)\|^2 \leq 2\int_0^t g(y(s))\langle By(s), y(s)\rangle ds.$$

Using hypothesis 2 of Theorem 7.4, it follows that

$$\|y(t)\| \leq \|y_0\|. \tag{7.43}$$

For all $y_0 \in X$ and $t \geq 0$, we have

$$\begin{aligned}\langle C^*CBS(t)y_0, S(t)y_0\rangle &= \langle C^*CBy(t), y(t)\rangle - \langle C^*CBy(t), y(t) - S(t)y_0\rangle \\ &+ \langle C^*CBS(t)y_0 - C^*CBy(t), S(t)y_0\rangle.\end{aligned}$$

Since B is bounded, then

$$|\langle C^*CBS(t)y_0, S(t)y_0\rangle| \leq |\langle C^*CBy(t), y(t)\rangle| + 2\alpha\|B\|\|y(t) - S(t)y_0\|\|y_0\|, \tag{7.44}$$

where α is a positive constant.

Using (7.43), we deduce

$$|\langle C^*CBy(t), y(t)\rangle| \leq |g(y(t))|\|y(t)\|\|y_0\|, \ \forall t \in [0, T]. \tag{7.45}$$

While from (7.42) and using Schwartz's inequality, we obtain

$$\|y(t) - S(t)y_0\| \leq \|B\| \left(T\int_0^T |g(y(t))|^2\|y(t)\|^2 dt\right)^{\frac{1}{2}}. \tag{7.46}$$

Integrating (7.44) over the interval $[0, T]$ and taking into account (7.45) and (7.46), we have

$$\begin{aligned}\int_0^T |\langle C^*CBS(t)y_0, S(t)y_0\rangle| dt &\leq 2\alpha T^{\frac{3}{2}}\|B\|^2\|y_0\| \left(\int_0^T |g(y(t))|^2\|y(t)\|^2 dt\right)^{\frac{1}{2}} \\ &+ T^{\frac{1}{2}}\|y_0\| \left(\int_0^T |g(y(t))|^2\|y(t)\|^2 dt\right)^{\frac{1}{2}}. \end{aligned}\tag{7.47}$$

Let us consider the nonlinear semigroup $U(t)y_0 := y(t)$. Replacing y_0 by $U(t)y_0$ in (7.47), and using the superposition properties of the semigroup $(U(t))_{t\geq 0}$, we deduce that

$$\begin{aligned}\int_0^T |\langle C^*CBS(s)U(t)y_0, S(s)U(t)y_0\rangle| ds &\leq 2\alpha T^{\frac{3}{2}}\|B\|^2\|U(t)y_0\| \\ &\times \left(\int_t^{t+T} |g(U(s)y_0)|^2\|U(s)y_0\|^2 ds\right)^{\frac{1}{2}} \\ &+ T^{\frac{1}{2}}\|U(t)y_0\| \left(\int_t^{t+T} |g(U(s)y_0)|^2\|U(s)y_0\|^2 ds\right)^{\frac{1}{2}}. \end{aligned}\tag{7.48}$$

Thus, by using (7.40) and (7.48), it follows that

$$\gamma\|CU(t)y_0\|_Y \leq M\left(\int_t^{t+T} |g(U(s)y_0)|^2\|U(s)y_0\|^2 ds\right)^{\frac{1}{2}}, \tag{7.49}$$

where $M = (2\alpha T\|B\|^2 + 1)T^{\frac{1}{2}}$ is a positive constant depending on $\|y_0\|$ and T.

From hypothesis 1 of Theorem 7.1, we have

$$\frac{d}{dt}\|CU(t)y_0\|_Y^2 \leq -2|g(U(t)y_0)|^2\|U(t)y_0\|^2. \tag{7.50}$$

Integrating (7.50) from nT and $(n+1)T$, $(n \in \mathbb{N})$, we obtain

$$\|CU(nT)y_0\|_Y^2 - \|CU((n+1)T)y_0\|_Y^2 \geq 2\int_{nT}^{(n+1)T} |g(U(s)y_0)|^2\|U(s)y_0\|^2 ds.$$

Using (7.49), (7.50) and the fact that $\|CU(t)y_0\|_Y$ decreases, it follows

$$\left(1 + 2\left(\frac{\gamma}{M}\right)^2\right)\|CU((n+1)T)y_0\|_Y^2 \leq \|CU(nT)y_0\|_Y^2.$$

Then

$$\|CU((n+1)T)y_0\|_Y \leq \beta\|CU(nT)y_0\|_Y,$$

where $\beta = \frac{1}{\left(1+2\left(\frac{\gamma}{M}\right)^2\right)^{\frac{1}{2}}}$. By recurrence, we show that $\|CU(nT)y_0\|_Y \leq \beta^n\|Cy_0\|_Y$. Taking $n = E(\frac{t}{T})$ the integer part of $\frac{t}{T}$, we deduce that

$$\|CU(t)y_0\|_Y \leq Re^{-\sigma t}\|y_0\|,$$

where $R = \alpha\left(1 + 2\left(\frac{\gamma}{M}\right)^2\right)^{\frac{1}{2}}$, with $\alpha > 0$ and $\sigma = \frac{\ln\left(1+2\left(\frac{\gamma}{M}\right)^2\right)}{2T} > 0$, which achieves the proof. □

7.2.2 Output Strong Stabilization

The following result will be used to prove strong stabilization of the output (7.39).

Theorem 7.5 *Let A generate a semigroup $(S(t))_{t\geq 0}$ of contractions on X and B : $X \to X$ is a bounded linear operator. If the conditions:*
1. $\langle C^*CA\psi, \psi\rangle \leq 0, \ \forall\psi \in \mathcal{D}(A)$,
2. $\langle C^*CB\psi, \psi\rangle\langle B\psi, \psi\rangle \geq 0, \ \forall\psi \in X$,
hold, then control

$$u(t) = -\frac{\langle C^*CBy(t), y(t)\rangle}{1 + |\langle C^*CBy(t), y(t)\rangle|}, \tag{7.51}$$

allows the estimate

$$\left(\int_0^T |\langle C^*CBS(s)y(t), S(s)y(t)\rangle| ds\right)^2 = \Theta\left(\int_t^{t+T} \frac{|\langle C^*CBy(s), y(s)\rangle|^2}{1+|\langle C^*CBy(s), y(s)\rangle|} ds\right),$$
$$\text{as } t \to +\infty. \tag{7.52}$$

Proof We have

$$\frac{1}{2}\frac{d}{dt}\langle Cy(t), Cy(t)\rangle_Y = \langle CAy(t), Cy(t)\rangle_Y + u(t)\langle CBy(t), Cy(t)\rangle_Y.$$

Then

$$\frac{1}{2}\frac{d}{dt}\langle Cy(t), Cy(t)\rangle_Y = \frac{1}{2}\frac{d}{dt}\|Cy(t)\|_Y^2 = \langle C^*CAy(t), y(t)\rangle + u(t)\langle C^*CBy(t), y(t)\rangle.$$

From hypothesis 1 of Theorem 7.5, we have

$$\frac{1}{2}\frac{d}{dt}\|Cy(t)\|_Y^2 \le u(t)\langle C^*CBy(t), y(t)\rangle.$$

In order to make the function $\frac{1}{2}\|Cy(t)\|_Y^2$ nonincreasing, we consider the control

$$u(t) = -\frac{\langle C^*CBy(t), y(t)\rangle}{1+|\langle C^*CBy(t), y(t)\rangle|},$$

so that the resulting closed-loop system is

$$y'(t) = Ay(t) + f(y(t)), \;\; y(0) = y_0, \tag{7.53}$$

where $f(z) = -\frac{\langle C^*CBz, z\rangle Bz}{1+|\langle C^*CBz, z\rangle|}, \;\; \forall\, z \in X.$

Since f is locally Lipschitz, then system (7.38) has a unique mild solution $y(t)$ (see [1]) defined on a maximal interval $[0, t_{max}]$ by

$$y(t) = S(t)y_0 + \int_0^t S(t-s)f(y(s))ds. \tag{7.54}$$

Because of the contractions of the semigroup (i.e $\langle A\psi, \psi\rangle \le 0, \;\forall\, \psi \in \mathcal{D}(A)$), we have

$$\frac{d}{dt}\|y(t)\|^2 \le -2\frac{\langle C^*CBy(t), y(t)\rangle\langle By(t), y(t)\rangle}{1+|\langle C^*CBy(t), y(t)\rangle|}.$$

Integrating this inequality over the interval $[0, t]$, we deduce

$$\|y(t)\|^2 - \|y(0)\|^2 \le -2\int_0^t \frac{\langle C^*CBy(s), y(s)\rangle\langle By(s), y(s)\rangle}{1+|\langle C^*CBy(s), y(s)\rangle|} ds.$$

Using condition 2 of Theorem 7.5, it follows that

$$\|y(t)\| \leq \|y_0\|. \tag{7.55}$$

From hypothesis 1 of Theorem 7.5, we have

$$\frac{d}{dt}\|Cy(t)\|_Y^2 \leq -2\frac{|\langle C^*CBy(t), y(t)\rangle|^2}{1+|\langle C^*CBy(t), y(t)\rangle|}.$$

Integrating this inequality, we deduce

$$\|Cy(t)\|_Y^2 - \|Cy(0)\|_Y^2 \leq -2\int_0^t \frac{|\langle C^*CBy(s), y(s)\rangle|^2}{1+|\langle C^*CBy(s), y(s)\rangle|}ds. \tag{7.56}$$

While from (7.54) and using Schwartz inequality, we obtain

$$\|y(t) - S(t)y_0\| \leq \|B\|\|y_0\|\left(T\int_0^t \frac{|\langle C^*CBy(s), y(s)\rangle|^2}{1+|\langle C^*CBy(s), y(s)\rangle|}ds\right)^{\frac{1}{2}}, \ \forall t \in [0, T]. \tag{7.57}$$

Since B is bounded and C continuous, we have

$$|\langle C^*CBS(s)y_0, S(s)y_0\rangle| \leq 2K\|B\|\|y(s) - S(s)y_0\|\|y_0\| + |\langle C^*CBy(s), y(s)\rangle|, \tag{7.58}$$

where K is a positive constant. Replacing y_0 by $y(t)$ in (7.57) and (7.58), we deduce

$$\begin{aligned}|\langle C^*CBS(s)y(t), S(s)y(t)\rangle| &\leq 2K\|B\|^2\|y_0\|^2\left(T\int_t^{t+T} \frac{|\langle C^*CBy(s), y(s)\rangle|^2}{1+|\langle C^*CBy(s), y(s)\rangle|}ds\right)^{\frac{1}{2}} \\ &+ |\langle C^*CBy(t+s), y(t+s)\rangle|, \quad \forall t \geq s \geq 0.\end{aligned}$$

Integrating this relation over $[0, T]$ and using Cauchy-Schwartz, we obtain

$$\begin{aligned}\int_0^T |\langle C^*CBS(s)y(t), S(s)y(t)\rangle|ds &\leq \left(2K\|B\|^2T^{\frac{3}{2}} + T\left(1 + K\|B\|\|y_0\|^2\right)\right) \\ &\times \left(\int_t^{t+T} \frac{|\langle C^*CBy(s), y(s)\rangle|^2}{1+|\langle C^*CBy(s), y(s)\rangle|}ds\right)^{\frac{1}{2}},\end{aligned}$$

which achieves the proof. □

The following result gives sufficient conditions for strong stabilization of the output (7.39).

Theorem 7.6 *Let A generate a semigroup $(S(t))_{t\geq 0}$ of contractions on X and B is a bounded linear operator. If the assumptions 1, 2 of Theorem 7.5 and*

$$\int_0^T |\langle C^*CBS(t)\psi, S(t)\psi\rangle| dt \geq \gamma \|C\psi\|_Y^2, \ \ \forall \ \psi \in X, \ (for\ some\ T, \gamma > 0), \tag{7.59}$$

hold, then control (7.51) strongly stabilizes the output (7.39) with decay estimate

$$\|Cy(t)\|_Y = \Theta\left(\frac{1}{\sqrt{t}}\right), \quad as \ t \longrightarrow +\infty. \tag{7.60}$$

Proof Using (7.56), we deduce

$$\|Cy(kT)\|_Y^2 - \|Cy((k+1)T)\|_Y^2 \geq 2\int_{kT}^{k(T+1)} \frac{|\langle C^*CBy(t), y(t)\rangle|^2}{1 + |\langle C^*CBy(t), y(t)\rangle|} dt, \ \ k \geq 0.$$

From (7.52) and (7.59), we have

$$\|Cy(kT)\|_Y^2 - \|Cy((k+1)T)\|_Y^2 \geq \beta \|Cy(kT)\|_Y^4, \tag{7.61}$$

where $\beta = \frac{\gamma^2}{2\left(2K\|B\|^2 T^{\frac{3}{2}} + T\left(1 + K\|B\|\|y_0\|^2\right)\right)^2}$. Taking $s_k = \|Cy(kT)\|_Y^2$, the inequality (7.61) can be written as

$$\beta s_k^2 + s_{k+1} \leq s_k, \ \ \forall k \geq 0.$$

Since $s_{k+1} \leq s_k$, we obtain

$$\beta s_{k+1}^2 + s_{k+1} \leq s_k, \ \ \forall k \geq 0.$$

Taking $p(s) = \beta s^2$ and $q(s) = s - (I + p)^{-1}(s)$ in Lemma 3.3 in [5], we deduce

$$s_k \leq x(k), \ \ k \geq 0,$$

where $x(t)$ is the solution of equation $x'(t) + q(x(t)) = 0, \ \ x(0) = s_0$.

Since $x(k) \geq s_k$ and $x(t)$ decreases give $x(t) \geq 0, \forall t \geq 0$. Furthermore, it is easy to see that $q(s)$ is an increasing function such that

$$0 \leq q(s) \leq p(s), \forall s \geq 0.$$

We obtain $-\beta x(t)^2 \leq x'(t) \leq 0$, which implies that

$$x(t) = \Theta(t^{-1}), \ \text{as } t \to +\infty.$$

Finally the inequality $s_k \leq x(k)$, together with the fact that $\|Cy(t)\|_Y$ decreases, we deduce the estimate

$$\|Cy(t)\|_Y = \Theta\left(\frac{1}{\sqrt{t}}\right), \quad \text{as } t \longrightarrow +\infty.$$

□

7.2.3 Output Weak Stabilization

The following result provides sufficient conditions for weak stabilization of the output (7.39).

Theorem 7.7 *Let A generate a semigroup $(S(t))_{t\geq 0}$ of contractions on X and B is a compact operator. If the conditions:*
1. $\langle C^*CA\psi, \psi\rangle \leq 0, \ \forall \psi \in \mathcal{D}(A)$,
2. $\langle C^*CB\psi, \psi\rangle\langle B\psi, \psi\rangle \geq 0, \ \forall \psi \in X$,
3. $\langle C^*CBS(t)\psi, S(t)\psi\rangle = 0, \ \ \forall t \geq 0 \Longrightarrow C\psi = 0$,
hold, then control (7.51) weakly stabilizes the output (7.39).

Proof Let us consider the nonlinear semigroup $\Gamma(t)y_0 := y(t)$ and let (t_n) be a sequence of real numbers such that $t_n \longrightarrow +\infty$ as $n \longrightarrow +\infty$.

From (7.6), $\Gamma(t_n)y_0$ is bounded in X, then there exists a subsequence $(t_{\phi(n)})$ of (t_n) such that

$$\Gamma(t_{\phi(n)})y_0 \rightharpoonup \psi, \text{ as } n \to \infty.$$

Since B is compact and C continuous, we have

$$\lim_{n\to+\infty} \langle C^*CBS(t)\Gamma(t_{\phi(n)})y_0, S(t)\Gamma(t_{\phi(n)})y_0\rangle = \langle C^*CBS(t)\psi, S(t)\psi\rangle.$$

For all $n \geq$, we set

$$\Lambda_n(t) := \int_{\phi(n)}^{\phi(n)+t} \frac{|\langle C^*CB\Gamma(s)y_0, \Gamma(s)y_0\rangle|^2}{1 + |\langle C^*CB\Gamma(s)y_0, \Gamma(s)y_0\rangle|}ds.$$

It follows that $\forall t \geq 0$, $\Lambda_n(t) \to 0$ as $n \to +\infty$.

Using (7.52), we deduce

$$\lim_{n\to+\infty} \int_0^t |\langle C^*CBS(s)\Gamma(t_{\phi(n)})y_0, S(s)\Gamma(t_{\phi(n)})y_0\rangle|ds = 0.$$

Hence, by the dominated convergence Theorem, we have

$$\int_0^t |\langle C^*CBS(s)\psi, S(s)\psi\rangle|ds = 0.$$

We conclude that

$$\langle C^*CBS(s)\psi, S(s)\psi\rangle = 0, \quad \forall s \in [0, t].$$

Using condition 3 of Theorem 7.7, we deduce that

$$C\Gamma(t_{\phi(n)})y_0 \rightharpoonup 0, \quad \text{as } n \longrightarrow +\infty. \tag{7.62}$$

On the other hand, it is clear that (7.62) holds for each subsequence $(t_{\phi(n)})$ of (t_n) such that $C\Gamma(t_{\phi(n)})y_0$ weakly converges in Y. This implies that $\forall\varphi \in Y$, we have $\langle C\Gamma(t_n)y_0, \varphi\rangle \to 0$ as $n \longrightarrow +\infty$ and hence

$$C\Gamma(t)y_0 \rightharpoonup 0, \quad \text{as } t \longrightarrow +\infty.$$

7.2.4 Examples

Example 7.3 Let Ω denote a bounded open subset of $\mathbb{R}^n$, and consider the following wave equation

$$\begin{cases} \dfrac{\partial^2 y(x,t)}{\partial t^2} - \Delta y(x,t) = u(t)\dfrac{\partial y(x,t)}{\partial t} & \Omega\times]0,+\infty[\\ y(x,t) = 0 & \partial\Omega\times]0,+\infty[\\ y(x,0) = y_0(x), \quad \dfrac{\partial y(x,0)}{\partial t} = y_1(x) & \Omega. \end{cases} \tag{7.63}$$

This system has the form of equation (7.38) if we set $X = H_0^1(\Omega) \times L^2(\Omega)$ with $\langle (y_1, z_1), (y_2, z_2)\rangle = \langle y_1, y_2\rangle_{H^1(\Omega)} + \langle z_1, z_2\rangle_{L^2(\Omega)}$, $A = \begin{pmatrix} 0 & I \\ \Delta & 0 \end{pmatrix}$ and $B = \begin{pmatrix} 0 & 0 \\ 0 & I \end{pmatrix}$. We consider the output operator $C = I$, we have A is skew-adjoint on X and the assumption (7.59) holds. Then the control

$$u(t) = -\frac{\|\dfrac{\partial y(.,t)}{\partial t}\|^2_{L^2(0,1)}}{1 + \|\dfrac{\partial y(.,t)}{\partial t}\|^2_{L^2(0,1)}}, \tag{7.64}$$

strongly stabilises system (7.63) with the decay estimate

$$\|(y(.,t), \frac{\partial y(.,t)}{\partial t})\|_X = \Theta(\frac{1}{\sqrt{t}}), \quad \text{as } t \longrightarrow +\infty.$$

Example 7.4 Let us consider a system defined on $\Omega =]0, 1[$ by

$$\begin{cases} \dfrac{\partial y(x,t)}{\partial t} = Ay(x,t) + u(t)a(x)y(x,t) & \Omega\times]0,+\infty[\\ y(x,0) = y_0(x) & \Omega, \end{cases} \tag{7.65}$$

where $X = L^2(\Omega)$, $Ay = -y$, and $a \in L^\infty(0,1)$ such that $a(x) \geq 0$ a.e on $]0,1[$ and $a(x) \geq c > 0$ on subregion ω of Ω and $u(.) \in L^\infty(0,+\infty)$ is a control function.

System (7.65) is augmented with the output

$$z(t) = \chi_\omega y(t), \tag{7.66}$$

where $\chi_\omega : L^2(\Omega) \longrightarrow L^2(\omega)$, the restriction operator to ω and χ_ω^* is the adjoint operator of χ_ω. The operator A generates a semigroup of contractions on $L^2(\Omega)$ given by $S(t)y_0 = e^{-t}y_0$. For all $y_0 \in L^2(\Omega)$ and $T = 2$, we obtain

$$\begin{aligned} \int_0^2 \langle \chi_\omega^* \chi_\omega BS(t)y_0, S(t)y_0\rangle dt &= \int_0^2 e^{-2t}dt \int_\omega a(x)|y_0|^2 dx \\ &\geq \beta \|\chi_\omega y_0\|^2_{L^2(\omega)}, \end{aligned}$$

with $\beta = c\displaystyle\int_0^2 e^{-2t}dt > 0$.

Then the control

$$u(t) = -\frac{\displaystyle\int_\omega a(x)|y(x,t)|^2 dx}{1 + \displaystyle\int_\omega a(x)|y(x,t)|^2 dx},$$

strongly stabilizes the output (7.15) with decay estimate

$$\|\chi_\omega y(t)\|_{L^2(\omega)} = \Theta\left(\frac{1}{\sqrt{t}}\right), \quad \text{as } t \longrightarrow +\infty.$$

Example 7.5 Consider a system defined in $\Omega =]0,+\infty[$, and described by

$$\begin{cases} \dfrac{\partial y(x,t)}{\partial t} = -\dfrac{\partial y(x,t)}{\partial x} + u(t)By(x,t) & \Omega\times]0,+\infty[\\ y(x,0) = y_0(x) & \Omega, \end{cases} \tag{7.67}$$

where $Ay = -\dfrac{\partial y}{\partial x}$ with domain $\mathcal{D}(A) = \{y \in H^1(\Omega) \mid y(0) = 0,\ y(x) \to 0 \text{ as } x \to +\infty\}$ and $By = \displaystyle\int_0^1 y(x)dx$. The operator A generates a semigroup of contractions

$$(S(t)y_0)(x) = \begin{cases} y_0(x-t) & \text{if } x \geq t \\ 0 & \text{if } x < t. \end{cases}$$

Let $\omega =]0, 1[$ be a subregion of Ω and system (7.67) is augmented with the output

$$z(t) = \chi_\omega y(t). \tag{7.68}$$

We have

$$\begin{aligned}\langle \chi_\omega^* \chi_\omega A y, y \rangle &= -\int_0^1 y'(x) y(x) dx \\ &= -\frac{y^2(1)}{2} \le 0,\end{aligned}$$

so, the Assumption 1 of Theorem 7.7 holds. The operator B is compact and verifies

$$\langle \chi_\omega^* \chi_\omega B S(t) y_0, S(t) y_0 \rangle = \left(\int_0^{1-t} y_0(x) dx \right)^2, \quad 0 \le t \le 1.$$

Thus

$$\langle \chi_\omega^* \chi_\omega B S(t) y_0, S(t) y_0 \rangle = 0, \quad \forall t \ge 0 \implies y_0(x) = 0, \ a.e \ \text{on } \omega.$$

Then, the control

$$u(t) = -\frac{\left(\int_0^1 y(x,t) dx \right)^2}{1 + \left(\int_0^1 y(x,t) dx \right)^2}, \tag{7.69}$$

weakly stabilizes the output (7.68).

7.3 Simulations

Consider system (7.67) with $y(x, 0) = \sin(\pi x)$, and augmented with the output (7.68).

- For $\omega =]0, 2[$, Fig. 7.1 shows that the state is stabilized on ω with error equals 3.4×10^{-4}, and the evolution of control function is given by Fig. 7.2.
- For $\omega =]0, 3[$, Fig. 7.3 shows that the state is stabilized on ω with error equals 7.8×10^{-4} and the evolution of control is given by Fig. 7.4.

Remark 7.3 It is clear that the control (7.69) stabilizes the state on ω, but do not take into account the residual part $\Omega \backslash \omega$.

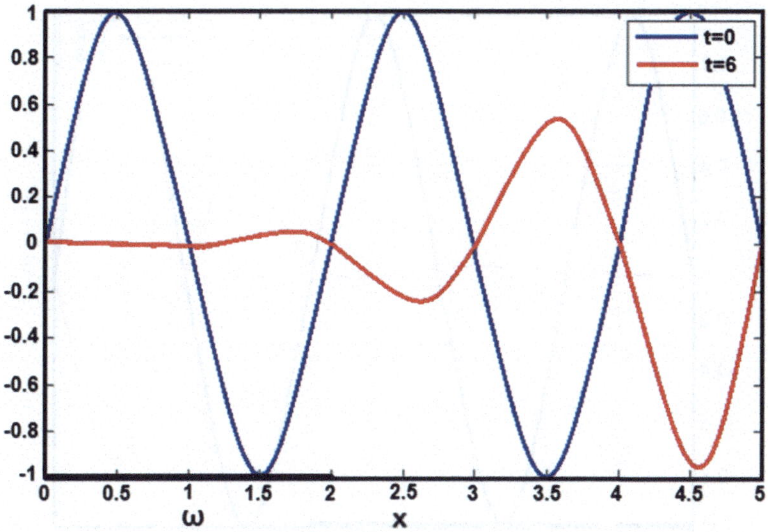

Fig. 7.1 The stabilization of the state on $\omega =]0, 2[$.

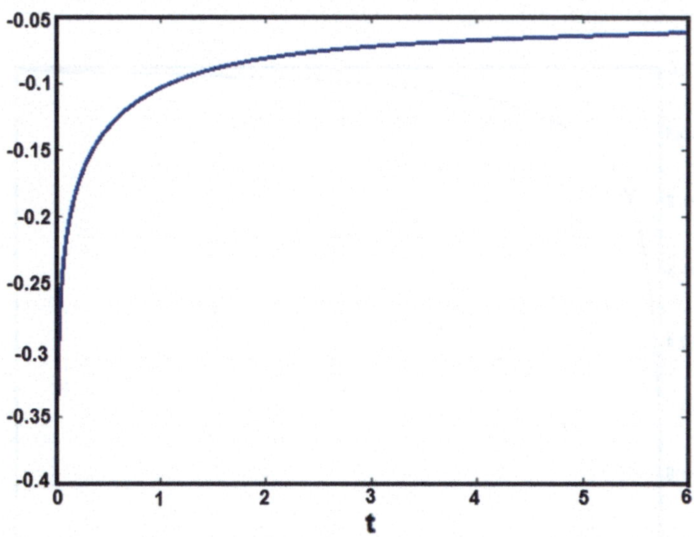

Fig. 7.2 Evolution of control function

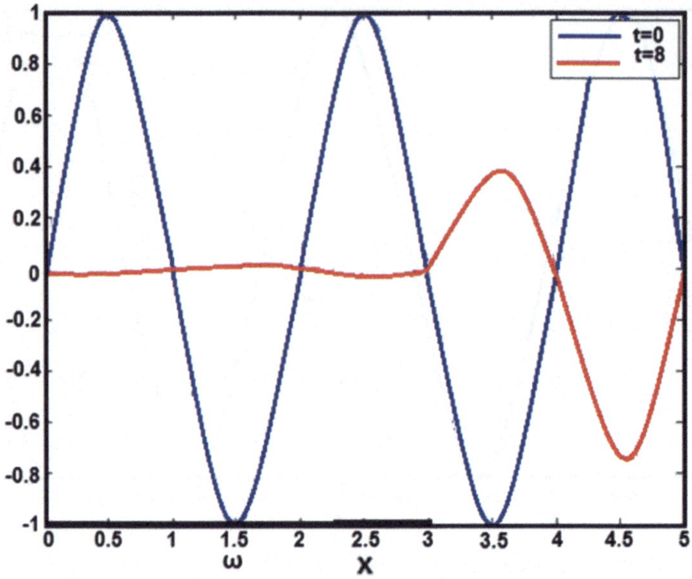

Fig. 7.3 The stabilization of the state on $\omega =]0, 3[$.

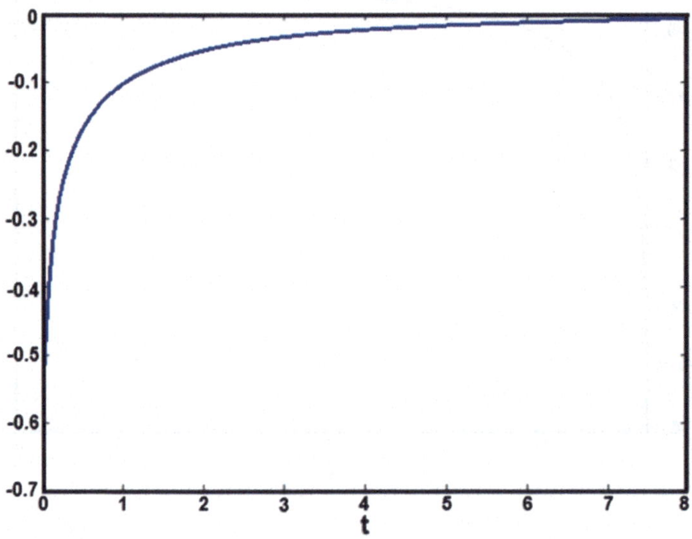

Fig. 7.4 Evolution of control function

References

1. Pazy, A.: Semigroups of Linear Operators and Applications to Partial Differential Equations. Springer, New York (1983)
2. Ball, J.M., Slemrod, M.: Feedback stabilization of distributed semilinear control systems. Appl. Math. Optim. **5**, 169–179 (1979)
3. Bardos, C., Lebeau, G., Rauch, J.: Sharp sufficient conditions for the observation, control and stabilization of waves from the boundary. SIAM J. Control Optim. **30**, 1024–1065 (1992)
4. Lebeau, G.: Contrôle de l'équation de Schrödinger. J. Math. Pures Appl. **71**, 267–291 (1992)
5. Lasiecka, I., Tataru, D.: Uniform boundary stabilisation of semilinear wave equation with nonlinear boundary damping. J. Differ. Integral Equ. **6**, 507–533 (1993)

Chapter 8
Stabilization of Infinite Dimensional Second Order Semilinear Systems

Semilinear are special kinds of nonlinear systems capable of representing a variety of important physical processes.

In this chapter we study strong and exponential stabilization of the following system

$$\begin{cases} y_{tt}(t) = Ay(t) + u(t)By_t(t), \\ y(0) = y_0, \ y_t(0) = y_1, \end{cases} \tag{8.1}$$

where A is an dissipative operator on the Hilbert space H endowed with the inner product $\langle ., . \rangle$ and the corresponding norm $\| \ \|_H$, $B : H \to H$ is a possibly nonlinear and locally Lipschitz operator, u denotes the control function, and $X = V \times H$ is the state space with $V = D((-A)^{\frac{1}{2}})$ endowed with the norm $\|v\|_V$.

8.1 Stabilization of Bilinear Systems

In this section we give controls that ensure strong and exponential stabilization of system (8.1) when B is a linear operator.

The system (8.1) can be written in the form

$$\begin{cases} \frac{d}{dt}\tilde{y}(t) = \tilde{A}\tilde{y}(t) + \tilde{B}\tilde{y}(t) \\ \tilde{y}(0) = \tilde{y}_0 \end{cases} \tag{8.2}$$

E. H. Zerrik and O. Castillo, *Stabilization of Infinite Dimensional Systems*,
Studies in Systems, Decision and Control 355,
https://doi.org/10.1007/978-3-030-68600-0_8

where $\tilde{y}(t) = (y, y_t)$ and $\tilde{A}$ given by

$$\tilde{A}\tilde{y}(t) = \begin{pmatrix} 0 & I \\ A & 0 \end{pmatrix} \tilde{y}(t)$$

and the operator $\tilde{B}$ is given by

$$\tilde{B}\tilde{y}(t) = \begin{pmatrix} 0 \\ u(t)By_t(t) \end{pmatrix}.$$

the system (8.2) has a unique weak solution in $C([0, T], X)$ (see [1]) and given by the variation of constants formula

$$\tilde{y}(t) = S(t)\tilde{y}_0 + \int_0^t S(t-s)\tilde{B}\tilde{y}(s)ds. \tag{8.3}$$

where $S(t)$ is a semigroup generated by the operator $\tilde{A}$.

We consider the following uncontrolled system

$$\begin{cases} z_{tt}(t) = Az(t), \\ z(0) = z_0, \; z_t(0) = z_1, \end{cases} \tag{8.4}$$

8.1.1 Strong Stabilization

In this section we give sufficient conditions for strong stabilization of system (8.1).

Theorem 8.1 *Suppose that the solution z of system (8.4) satisfies*

$$\|z_0\|_V^2 + \|z_1\|_H^2 \leq \alpha \int_0^T |\langle Bz_t(t), z_t(t)\rangle| dt, \; (for \; some \; T, \alpha > 0) \tag{8.5}$$

then the control

$$u(t) = -\langle By_t(t), y_t(t)\rangle \tag{8.6}$$

strongly stabilizes system (8.1).

Proof From the relation

$$\langle Bz_t, z_t\rangle = \langle B(z_t - y_t), z_t\rangle - \langle By_t, y_t - z_t\rangle + \langle By_t, y_t\rangle,$$

and as B is bounded, we have

$$|\langle Bz_t, z_t\rangle| \leq \|B\|_H \|z_t - y_t\|_H \|z_t\|_H + \|B\|_H \|y_t - z_t\|_H \|y_t\|_H + |\langle By_t, y_t\rangle| \tag{8.7}$$

Differentiating the energy $E_y(t) = \frac{1}{2}\{\|y_t\|_H^2 + \|y\|_V^2\}$ and since A is dissipative, we get

$$\frac{d}{dt}E_y(t) \leq -\langle By_t, y_t\rangle^2$$

Then

$$E_y(t) \leq -\int_0^t \langle By_s(s), y_s(s)\rangle^2 ds + E_y(0)$$

Which implies that

$$E_y(t) \leq E_y(0) \tag{8.8}$$

Moreover, we have

$$\|y_t\|_H^2 \leq \|(y, y_t)\|_X^2 \leq 2E_y(t) \quad and \quad \|z_t - y_t\|_H^2 \leq \|(\psi - \phi)\|_X^2 \tag{8.9}$$

where $\psi = (z, z_t)$ and $\phi = (y, y_t)$.

Using (8.7), (8.8) and (8.9), we obtain

$$|\langle Bz_t, z_t\rangle| \leq 2\|B\|_H\|\psi - \phi\|_X\sqrt{E_y(0)} + |\langle By_t, y_t\rangle| \tag{8.10}$$

From (8.3), we have

$$\|\psi - \phi\|_X \leq \sqrt{E_y(0)}\left(\int_0^t |\langle By_s(s), y_s(s)\rangle|^2 ds\right)$$

It follows from Schwartz's inequality, that

$$\|\psi - \phi\|_H \leq \sqrt{TE_y(0)}\left(\int_0^T |\langle By_s(s), y_s(s)\rangle|^2 ds\right)^{\frac{1}{2}} \tag{8.11}$$

Integrating (8.10) over the interval $[0, T]$ and with (8.11), we obtain

$$\int_0^T |\langle Bz_t, z_t\rangle| dt \leq (2E_y(0) + 1)\|B\|_H\sqrt{T}\left(\int_0^T |\langle By_s(s), y_s(s)\rangle|^2 ds\right)^{\frac{1}{2}}$$

Condition (8.5) gives

$$\|z_0\|_V^2 + \|z_1\|_H^2 \leq \alpha(2E_y(0) + 1)\|B\|_H\sqrt{T}\left(\int_0^T |\langle By_s(s), y_s(s)\rangle|^2 ds\right)^{\frac{1}{2}}$$

Taking $(z_0, z_1) = (y_0, y_1)$, we have

$$\|y_0\|_V^2 + \|y_1\|_H^2 \leq \alpha(2E_y(0) + 1)\|B\|_H\sqrt{T}\left(\int_0^T |\langle By_s(s), y_s(s)\rangle|^2 ds\right)^{\frac{1}{2}}$$

Now replacing (y_0, y_1) by (y, y_t), we get

$$\|y(t)\|_V^2 + \|y_t(t)\|_H^2 \leq \beta\left(\int_t^{t+T} |\langle By_s(s), y_s(s)\rangle|^2 ds\right)^{\frac{1}{2}}$$

Then $\|(y, y_t)\|_X \longrightarrow 0, \; as \; t \longrightarrow +\infty$, which completes the proof. □

Example 8.1 We consider on $\Omega =]0, 1[$, the following system

$$\begin{cases} y_{tt}(x,t) = y_{xx}(x,t) + u(t)y_t(x,t), & \Omega\times]0,+\infty[\\ y(x,0) = y_0(x), \;\; y_t(x,0) = y_1(x), & \Omega \\ y(0,t) = y(1,t) = 0, &]0,+\infty[\end{cases} \tag{8.12}$$

where $u(t) \in L^\infty(0, +\infty)$ is a real valued control.

System (8.12) has a unique weak solution on the state space $X = H_0^1(\Omega) \times L^2(\Omega)$ ([6]).

The solution of the uncontrolled system

$$\begin{cases} z_{tt}(x,t) = z_{xx}(x,t), & \Omega\times]0,+\infty[\\ z(x,0) = z_0(x), z_t(x,0) = z_1(x), & \Omega \\ z(0,t) = z(1,t) = 0, &]0,+\infty[\end{cases}$$

is given by

$$z(x,t) = \sum_{k=1}^{\infty}(a_k\cos(k\pi t) + b_k\sin(k\pi t))\sin(k\pi x).$$

The assumption (8.5) is verified, indeed

$$\begin{aligned} \int_0^2 \langle Bz_t(t), z_t(t)\rangle dt &= \int_0^2\int_0^1 |z_t(x,t)|^2\, dxdt \\ &= \sum_{k=1}^{\infty}(k\pi)^2(a_k^2 + b_k^2) \\ &= \|z_t(0)\|_{L^2(\Omega)}^2 + \|z_x(0)\|_{L^2(\Omega)}^2 \end{aligned}$$

It follows that the control

$$u(t) = -\int_0^1 |y_t(x,t)|^2 dx \tag{8.13}$$

strongly stabilizes system (8.12).

8.1.2 Exponential Stabilization

The following result provides sufficient conditions for exponential stabilization of system (8.1).

Theorem 8.2 *Suppose that the solution z of system (8.4) satisfies the condition (8.5), then the control*

$$u(t) = \begin{cases} -\dfrac{\langle By_t(t), y_t(t)\rangle}{\|(y, y_t)\|_X^2}, & (y, y_t) \neq (0, 0) \\ 0, & (y, y_t) = (0, 0) \end{cases} \tag{8.14}$$

stabilizes exponentially system (8.1), in other word there exist $M > 0$ and $\lambda > 0$ such that for every $(y_0, y_1) \in X$

$$\|y(t)\|_V^2 + \|y_t(t)\|_H^2 \leq Me^{-\lambda t}(\|y_0\|_V^2 + \|y_1\|_H^2).$$

Proof Let us consider the solution z of system (8.4) with $z(0) = y_0$, $z_t(0) = y_1$ and the natural energy associated with (8.1)

$$E_y(t) = \frac{1}{2}\{\|y(t)\|_V^2 + \|y_t(t)\|_H^2\}$$

Since A is dissipative, we have

$$\frac{d}{dt}E_y(t) \leq -\frac{\langle By_t, y_t\rangle^2}{\|y(t)\|_V^2 + \|y_t(t)\|_H^2} \tag{8.15}$$

Integrating this inequality, we get

$$E_y(t) \leq -\int_0^t \frac{\langle By_s(s), y_s(s)\rangle^2}{\|y(s)\|_V^2 + \|y_s(s)\|_H^2} ds + E_y(0)$$

Then

$$E_y(t) \leq E_y(0) \tag{8.16}$$

Furthermore, we have

$$\|y_t\|_H^2 \leq \|(y, y_t)\|_X^2 \leq 2E_y(t) \quad \textit{and} \quad \|z_t - y_t\|_H^2 \leq \|(\psi - \phi)\|_X^2 \tag{8.17}$$

where $\psi = (z, z_t)$ and $\phi = (y, y_t)$.

Using (8.7) and combining (8.16) and (8.17), we obtain

$$|\langle Bz_t, z_t\rangle| \leq 2\|B\|_H\|\psi - \phi\|_X\sqrt{2E_y(0)} + |\langle By_t, y_t\rangle| \tag{8.18}$$

Furthermore, from (8.3), we have

$$\|\psi - \phi\|_X \leq \|B\|_H \left\{ \int_0^t \frac{|\langle By_s(s), y_s(s)\rangle|}{2E_y(s)} \sqrt{2E_s(s)} ds \right\}$$

Schwartz's inequality, gives

$$\|\psi - \phi\|_X \leq \|B\|_H \sqrt{T} \left\{ \int_0^T \frac{|\langle By_s(s), y_s(s)\rangle|^2}{2E_y(s)} ds \right\}^{\frac{1}{2}} \tag{8.19}$$

Using (8.16), we have

$$|\langle By_t, y_t\rangle| \leq \frac{|\langle By_t, y_t\rangle|}{2E_y(t)} \sqrt{2E_y(t)} \sqrt{2E_y(0)} \tag{8.20}$$

Integrating (8.18) over the interval $[0, T]$ and taking into account (8.19) and (8.20), we obtain

$$\int_0^T |\langle Bz_t, z_t\rangle| dt \leq (2\|B\|_H + 1)\|B\|_H \sqrt{T 2E_y(0)} \left\{ \int_0^T \frac{|\langle By_s(s), y_s(s)\rangle|^2}{2E_y(s)} ds \right\}^{\frac{1}{2}}$$

It follows from (8.5) that

$$\|z_0\|_V^2 + \|z_1\|_H^2 \leq \alpha(2\|B\|_H + 1)\|B\|_H \sqrt{2TE_y(0)} \left\{ \int_0^T \frac{|\langle By_s(s), y_s(s)\rangle|^2}{2E_y(s)} ds \right\}^{\frac{1}{2}}$$

Taking $(z_0, z_1) = (y_0, y_1)$, we get

$$\|y_0\|_V^2 + \|y_1\|_H^2 \leq (2\|B\|_H + 1)\|B\|_H \sqrt{2TE_y(0)} \left\{ \int_0^T \frac{|\langle By_s(s), y_s(s)\rangle|^2}{2E_y(s)} ds \right\}^{\frac{1}{2}}$$

Now replacing (y_0, y_1) by (y, y_t), we get

$$\|y(t)\|_V^2 + \|y(t)\|_H^2 \leq (2\|B\|_H + 1)\|B\|_H \sqrt{2TE_y(t)} \left\{ \int_t^{t+T} \frac{|\langle By_s(s), y_s(s)\rangle|^2}{2E_y(s)} ds \right\}^{\frac{1}{2}}$$

Then

$$E_y(t) \leq \gamma \int_t^{t+T} \frac{|\langle By_s(s), y_s(s)\rangle|^2}{2E_y(s)} ds \tag{8.21}$$

where $\gamma = (2\|B\|_H + 1)\|B\|_H \sqrt{2T}$.

Integrating inequality (8.15) from nT to $(n+1)T$, and since $E_y(t)$ decreases, we obtain

$$E_y(nT) - E_y((n+1)T) \geq \int_{nT}^{(n+1)T} \frac{|\langle By_s(s), y_s(s)\rangle|^2}{2E_y(s)} ds$$

From (8.21), we get

$$E_y((n+1)T) \le rE_y(nT)$$

where $r = \frac{1}{\gamma}$.

By recurrence, we show that

$$E_y((n+1)T) \le r^n E_y(0)$$

Since $E_y(t)$ decreases and taking n the integer part of $\frac{t}{T}$, we deduce that

$$E_y(t) \le Me^{-\lambda t}E_y(0)$$

where $M = (2\|B\|_H + 1)\|B\|_H\sqrt{T}$ and $\lambda = \dfrac{\ln((2\|B\|_H + 1)\|B\|_H\sqrt{T})}{T}$.

In other word system (8.1) is exponentially stabilizable by the control (8.14). □

Example 8.2 On $\Omega =]0, 1[$, we consider the following system

$$\begin{cases} y_{tt}(x,t) + y_{xxxx}(x,t) = u(t)y_t(x,t), & \Omega\times]0,+\infty[\\ y(x,0) = y_0(x), \ \ y_t(x,0) = y_1(x), & \Omega \\ y(0,t) = y(1,t) = y_{xx}(0,t) = y_{xx}(1,t) = 0, &]0,+\infty[\end{cases} \tag{8.22}$$

where $u(t) \in L^\infty(0,+\infty)$ is a real valued control.

System (8.22) has a unique weak solution on the state space $X = (H^2(\Omega) \cap H_0^1(\Omega)) \times L^2(\Omega)$ ([6]).

Moreover, the solution of the following system

$$\begin{cases} z_{tt}(x,t) + z_{xxxx}(x,t) = 0, & \Omega\times]0,+\infty[\\ z(x,0) = z_0(x), z_t(x,0) = z_1(x), & \Omega \\ z(0,t) = z(1,t) = z_{xx}(0,t) = z_{xx}(1,t) = 0, &]0,+\infty[\end{cases}$$

takes the form

$$z(x,t) = \sum_{m=1}^{\infty}(a_m\cos((m\pi)^2t) + b_m\sin((m\pi)^2t))\sin(m\pi x).$$

The condition (8.5) is verified, indeed

$$\begin{aligned} \int_0^2 \langle Bz_t(t), z_t(t)\rangle dt &= \int_0^2\int_0^1 |z_t(x,t)|^2dxdt \\ &= \sum_{m=1}^{\infty}(m\pi)^4(a_m^2 + b_m^2) \\ &= \|z_t(0)\|^2_{L^2(\Omega)} + \|z_{xx}(0)\|^2_{L^2(\Omega)} \end{aligned}$$

We conclude that the control

$$u(t) = -\frac{\displaystyle\int_0^1 |y_t(x,t)|^2 dx}{\|y_t\|^2_{L^2(\Omega)} + \|y_{xx}\|^2_{L^2(\Omega)}} \tag{8.23}$$

ensures the exponential stabilization of system (8.22).

8.2 Stabilization of Semilinear Systems

In this section we study the strong and exponential stabilization of the semilinear system (8.1) when B is nonlinear.

In other hand, we consider the system

$$\begin{cases} z_{tt}(x,t) = Az(t), \\ z(0) = z_0, \, z_t(0) = z_1, \end{cases} \tag{8.24}$$

8.2.1 *Strong Stabilization*

The following result gives sufficient conditions for strong stabilization of system (8.1).

Theorem 8.3 *If B is locally Lipschitz and there exist $T > 0$ and $\delta > 0$ such that the solution (z, z_t) of system (8.24) verifies the inequality*

$$\int_0^T |\langle Bz_t(t), z_t(t)\rangle| dt \geq \delta\{\|z_t(0)\|^2_H + \|z(0)\|^2_V\}, \tag{8.25}$$

then the control

$$u(t) = -\langle By_t(t), y_t(t)\rangle \tag{8.26}$$

strongly stabilizes system (8.1).

Proof We show first that $h : y_t \mapsto \langle By_t, y_t\rangle By_t$ is locally Lipschitz.

Since B is locally Lipschitz, then for all $R > 0$, there exists $L > 0$ such that

$$\|By_t - Bz_t\|_H \leq L\|y_t - z_t\|_H, \ \forall y_t, z_t \in H : 0 < \|y_t\|_H \leq \|z_t\|_H \leq R. \tag{8.27}$$

Let us remark that

$$\begin{aligned}\langle Bz_t, z_t\rangle Bz_t - \langle By_t, y_t\rangle By_t &= \langle Bz_t, z_t\rangle(Bz_t - By_t) + (\langle Bz_t, z_t\rangle - \langle By_t, y_t\rangle)By_t \\ &= (\langle Bz_t, z_t - y_t\rangle - \langle Bz_t - By_t, y_t\rangle)By_t \\ &+ \langle Bz_t, z_t\rangle(Bz_t - By_t)\end{aligned}$$

Using (8.27), we have

$$\begin{aligned}\|h(z_t) - h(y_t)\|_H &\leq L^2\|z_t\|_H^2\|z_t - y_t\|_H + L^2\|z_t - y_t\|_H(\|z_t\|_H + \|y_t\|_H)\|y_t\|_H \\ &\leq \mathcal{L}\|z_t - y_t\|_H\end{aligned}$$

where $\mathcal{L} = 3LR^2$.

It follows that system (8.1) has a unique weak solution (y, y_t) (see [1]).

Using (8.7) and (8.27), we obtain

$$|\langle Bz_t, z_t\rangle| \leq L\|z_t - y_t\|_H\|z_t\|_H + L\|y_t - z_t\|_H\|y_t\|_H + |\langle By_t, y_t\rangle| \tag{8.28}$$

Now, we differentiate $E_y(t)$

$$\frac{d}{dt}E_y(t) \leq -\langle By_t, y_t\rangle^2$$

Then

$$E_y(t) \leq -\int_0^t \langle By_s(s), y_s(s)\rangle^2 ds + E_y(0)$$

It follows that

$$E_y(t) \leq E_y(0) \tag{8.29}$$

Moreover, we have

$$\|y_t\|_H^2 \leq \|(y, y_t)\|_X^2 \leq 2E_y(t) \quad and \quad \|z_t - y_t\|_H^2 \leq \|(\psi - \phi)\|_X^2 \tag{8.30}$$

where $\psi = (z, z_t)$ and $\phi = (y, y_t)$.

From (8.28) and combining (8.29) with (8.30), we get

$$|\langle Bz_t, z_t\rangle| \leq 2L\|\psi - \phi\|_X\sqrt{2E_y(0)} + |\langle By_t, y_t\rangle| \tag{8.31}$$

From (8.3), we have

$$\|\psi - \phi\|_X \leq \sqrt{2E_y(0)}\int_0^t |\langle By_t, y_t\rangle|^2 ds$$

For a fixed $T > 0$, Schwarz inequality, yields

$$\|\psi - \phi\|_X \leq \sqrt{2TE_y(0)g(0)} \tag{8.32}$$

where $g(t) = \int_t^{t+T} |\langle By_s(s), y_s(s)\rangle|^2 ds$.

Integrating (8.31) over the interval $[0, T]$ and by (8.32), we obtain

$$\int_0^T |\langle Bz_t, z_t\rangle| dt \le (2E_y(0) + 1)L\sqrt{Tg(0)}$$

Condition (8.25) allows

$$\|z_1\|_H^2 + \|z_0\|_V^2 \le \frac{(2E_y(0) + 1)L\sqrt{T}}{\delta}\sqrt{g(0)}$$

Replacing (z_0, z_1) by (y_0, y_1), we get

$$\|y_1\|_H^2 + \|y_0\|_V^2 \le \frac{(2E_y(0) + 1)L\sqrt{T}}{\delta}\sqrt{g(0)}$$

Now, replacing (y_0, y_1) by (y, y_t), we obtain

$$\|y_t\|_H^2 + \|y\|_V^2 \le \beta\sqrt{g(t)}$$

where $\beta = \dfrac{(2E_y(0) + 1)L\sqrt{T}}{\delta}$.

It follows that $\|(y, y_t)\|_X \longrightarrow 0$ as $t \longrightarrow +\infty$, we deduce that the control (8.26) strongly stabilizes system (8.1).

Example 8.3 On $\Omega =]0, 1[$, we consider the following system

$$\begin{cases} y_{tt}(x,t) = y_{xx}(x,t) + u(t)By_t(x,t), & \Omega\times]0, +\infty[\\ y(x,0) = y_0(x), \;\; y_t(x,0) = y_1(x), & \Omega \\ y(0,t) = y(1,t) = 0, &]0, +\infty[\end{cases} \tag{8.33}$$

where the operator $By_t = (\max\{\|y_t\|_{L^2(\Omega)}, 1\})y_t$ and $u(t) \in L^\infty(0, +\infty)$ is a real valued control.

System (8.33) has a unique solution on state space $X = H_0^1(\Omega) \times L^2(\Omega)$ ([6]).
The solution of the uncontrolled system

$$\begin{cases} z_{tt}(x,t) = z_{xx}(x,t), & \Omega\times]0, +\infty[\\ z(x,0) = z_0(x), z_t(x,0) = z_1(x), & \Omega \\ z(0,t) = z(1,t) = 0, &]0, +\infty[\end{cases}$$

is given by

$$z(t,x) = \sum_{k=1}^{\infty} (a_k \cos(k\pi t) + b_k \sin(k\pi t)) \sin(k\pi x).$$

The assumption (8.25) is verified, indeed

$$\begin{aligned}\int_0^2 \langle Bz_t(t), z_t(t)\rangle dt &= (\max\{\|z_t\|_{L^2(\Omega)}, 1\}) \int_0^2 \int_0^1 |z_t(x,t)|^2\, dxdt \\ &\geq \sum_{k=1}^{\infty} (k\pi)^2 (a_k^2 + b_k^2) \\ &\geq \|z_t(0)\|^2_{L^2(\Omega)} + \|z_x(0)\|^2_{L^2(\Omega)}\end{aligned}$$

We conclude that

$$u(t) = -(\max\{\|y_t\|_{L^2(\Omega)}, 1\}) \int_0^1 |y_t(x,t)|^2 dx \tag{8.34}$$

strongly stabilizes system (8.33).

8.2.2 Exponential Stabilization

Theorem 8.4 *Assume that B is locally Lipschitz and there exist $T > 0$ and $C > 0$ such that the solution (z, z_t) of system (8.24) satisfy the following condition*

$$\int_0^T |\langle Bz_t(t), z_t(t)\rangle| dt \geq C\{\|z_t(0)\|_H^2 + \|z(0)\|_V^2\}, \tag{8.35}$$

then the control

$$u(t) = \begin{cases} -\dfrac{\langle By_t(t), y_t(t)\rangle}{\|(y, y_t)\|_X^2}, & (y, y_t) \neq (0,0) \\ 0, & (y, y_t) = (0,0) \end{cases} \tag{8.36}$$

exponentially stabilizes system (8.1).

Proof Let $\phi = (y, y_t)$ et $\psi = (z, z_t)$. We show first that the operator $f : \phi \mapsto \dfrac{\langle By_t, y_t\rangle}{\|(y, y_t)\|_X^2} By_t$ is locally Lipschitz.

Since B is locally Lipschitz, then for all $R > 0$, there exists $K > 0$ such that

$$\|By_t - Bz_t\|_H \leq K\|y_t - z_t\|_H, \ \forall y_t, z_t \in H : 0 < \|y_t\|_H \leq \|z_t\|_H \leq R. \tag{8.37}$$

Then

$$f(\psi) - f(\phi) = \frac{\|\phi\|_X^2(\langle Bz_t, z_t\rangle Bz_t - \langle By_t, y_t\rangle By_t)}{\|\psi\|_X^2\|\phi\|_X^2} - \frac{(\|\psi\|_X^2 - \|\phi\|_X^2)\langle By_t, y_t\rangle By_t}{\|\psi\|_X^2\|\phi\|_X^2}.$$

It follows that

$$\|f(\psi)-f(\phi)\|_X \le \frac{\|\langle Bz_t, z_t\rangle Bz_t - \langle By_t, y_t\rangle By_t\|_H}{\|\psi\|_X^2} + K^2|\|\psi\|_X^2 - \|\phi\|_X^2|.\frac{\|\phi\|_X}{\|\psi\|_X^2}$$

$$\le \frac{\|\langle Bz_t, z_t\rangle(Bz_t - By_t) + (\langle Bz_t, z_t - y_t\rangle}{\|\psi\|_X^2}$$

$$-\frac{\langle By_t - Bz_t, y_t\rangle)By_t\|_H}{\|\psi\|_X^2} + 2K^2\|\psi - \phi\|_X$$

Furthermore, we have

$$\|z_t - y_t\|_H^2 \le \|(\psi - \phi)\|_X^2 \tag{8.38}$$

Using (8.38), we get $\|f(\psi) - f(\phi)\|_X \le L\|\psi - \phi\|_X$ where $L = 5K^2$.

We conclude that system (8.1) has a unique weak solution (y, y_t) (see [1]).

From (8.37) and (8.7), we get

$$|\langle Bz_t, z_t\rangle| \le K\|z_t - y_t\|_H\|z_t\|_H + K\|y_t - z_t\|_H\|y_t\|_H + |\langle By_t, y_t\rangle| \tag{8.39}$$

Moreover, we have

$$\|y_t\|_H^2 \le \|(y, y_t)\|_X^2 \le 2E_y(t)$$

Since A is dissipative, then

$$\frac{d}{dt}E_y(t) \le -\frac{\langle By_t, y_t\rangle^2}{\|y_t\|_H^2 + \|y\|_V^2}$$

Thus

$$E_y(t) = -\int_0^t \frac{\langle By_s(s), y_s(s)\rangle^2}{\|y_s(s)\|_H^2 + \|y(s)\|_V^2}ds + E_y(0)$$

It follows that

$$E_y(t) \le E_y(0) \tag{8.40}$$

Using (8.38), (8.39) and (8.40), we obtain

$$|\langle Bz_t, z_t\rangle| \le 2K\|\psi - \phi\|_X\sqrt{2E_y(0)} + |\langle By_t, y_t\rangle| \tag{8.41}$$

From (8.3), we deduce

$$\|\psi - \phi\|_X \le K\int_0^t |f(s)|\sqrt{2E_y(s)}ds$$

where $f(s) = \dfrac{\langle By_s(s), y_s(s)\rangle}{2E_y(s)}$.

Schwarz inequality, gives

$$\|\psi - \phi\|_X \leq K\sqrt{2T}\left(\int_0^T |f(s)|^2 E_y(s)ds\right)^{\frac{1}{2}} \tag{8.42}$$

Using (8.40), we get

$$|\langle By_t, y_t\rangle| \leq |f(t)|\sqrt{2E_y(t)}\sqrt{2E_y(0)} \tag{8.43}$$

Integrating (8.41) on the interval $[0, T]$ and from (8.42) and (8.43), we obtain

$$\int_0^T |\langle Bz_s(s), z_s(s)\rangle| ds \leq 2(2K^2+1)\sqrt{TE_y(0)}\left(\int_0^T |f(s)|^2 E_y(s)ds\right)^{\frac{1}{2}}$$

It follows from (8.35), that

$$\|z_t(0)\|_H^2 + \|z(0)\|_V^2 \leq \frac{2(2K+1)}{C}\sqrt{TE_y(0)}\left(\int_0^T |f(s)|^2 E_y(s)ds\right)^{\frac{1}{2}}$$

Taking $(z_0, z_1) = (y_0, y_1)$, we have

$$\|y_1\|_H^2 + \|y_0\|_V^2 \leq \frac{2(2K+1)}{C}\sqrt{TE_y(0)}\left(\int_0^T |f(s)|^2 E_y(s)ds\right)^{\frac{1}{2}}$$

Replacing (y_0, y_1) by (y, y_t), we get

$$\|y_t\|_H^2 + \|y\|_V^2 \leq \frac{2(2K+1)}{C}\sqrt{TE_y(t)}\left(\int_t^{t+T} |f(s)|^2 E_y(s)ds\right)^{\frac{1}{2}}$$

It follows that

$$E_y(t) \leq \gamma \int_t^{t+T} |f(s)|^2 E_y(s)ds \tag{8.44}$$

where $\gamma = \dfrac{2(2K+1)}{C}\sqrt{T}$.

Integrating from kT to $(k+1)T$ the inequality

$$\frac{d}{dt}E_y(t) \leq -\frac{\langle By_t, y_t\rangle^2}{2E_y(t)}$$

and since $E_y(t)$ decreases, we obtain

$$E_y(kT) - E_y((k+1)T) \geq \int_{kT}^{(k+1)T} |f(s)|^2 E_y(s)ds$$

From (8.44), we get

$$E_y((k+1)T) \leq r E_y(kT)$$

where $r = \dfrac{1}{\gamma}$.

This implies that

$$E_y((k+1)T) \leq r^k E_y(0) \tag{8.45}$$

Applying (8.45) to the integer part $k = E(\frac{t}{T})$ of $\frac{t}{T}$, we deduce that

$$E_y(t) \leq M e^{-\lambda t} E_y(0)$$

where $M = \frac{2\sqrt{T}}{C}(2K+1)$ and $\lambda = \dfrac{\ln(\frac{2\sqrt{T}}{C}(2K+1))}{T}$.

Then the system (8.1) is exponentially stabilizable by the control (8.36).

Example 8.4 On $\Omega =]0, 1[$, we consider the following system

$$\begin{cases} y_{tt}(x,t) + y_{xxxx}(x,t) = u(t)By_t(x,t), & \Omega\times]0,+\infty[\\ y(x,0) = y_0(x),\ \ y_t(x,0) = y_1(x), & \Omega \\ y(0,t) = y(1,t) = y_{xx}(0,t) = y_{xx}(1,t) = 0, &]0,+\infty[\end{cases} \tag{8.46}$$

where the operator $By_t = (\max\{\|y_t\|_{L^2(\Omega)}, 1\})y_t$ and $u(t) \in L^\infty(0,+\infty)$ is a real valued control.

System (8.46) has a unique solution on the state space $X = (H^2(\Omega) \cap H_0^1(\Omega)) \times L^2(\Omega)$ ([6]).

The condition (8.35) is verified, indeed

$$\begin{aligned} \int_0^2 \langle Bz_t(t), z_t(t)\rangle dt &= (\max\{\|z_t\|_{L^2(\Omega)}, 1\}) \int_0^2 \int_0^1 |z_t(x,t)|^2 dxdt \\ &\geq \sum_{m=1}^{\infty} (m\pi)^4 (a_m^2 + b_m^2) \\ &\geq \|z_t(0)\|^2_{L^2(\Omega)} + \|z_{xx}(0)\|^2_{L^2(\Omega)} \end{aligned}$$

where $z(x,t) = \sum_{m=1}^{\infty} (a_m \cos((m\pi)^2 t) + b_m \sin((m\pi)^2 t)) \sin(m\pi x)$ is the solution of the system

$$\begin{cases} z_{tt}(x,t) + z_{xxxx}(x,t) = 0, & \Omega\times]0,+\infty[\\ z(x,0) = z_0(x),\ z_t(x,0) = z_1(x), & \Omega \\ z(0,t) = z(1,t) = z_{xx}(0,t) = z_{xx}(1,t) = 0, &]0,+\infty[\end{cases}$$

We deduce from Theorem 8.4 that the control

$$u(t) = -\frac{(\max\{\|y_t\|_{L^2(\Omega)}, 1\}) \displaystyle\int_0^1 |y_t(x,t)|^2 dx}{\|y_t\|^2_{L^2(\Omega)} + \|y_{xx}\|^2_{L^2(\Omega)}} \tag{8.47}$$

exponentially stabilizes the system (8.46).

8.3 Regional Stabilization of Second Order Semilinear Systems

In this section we study regional stabilization of the system defined on an open, regular and bounded domain $\Omega \subset \mathbb{R}^n$, $n \geq 0$ by

$$\begin{cases} y_{tt}(t) = Ay(t) + u(t)By_t(t), \ t \geq 0 \\ y(0) = y_0, \ y_t(0) = y_1, \end{cases} \tag{8.48}$$

where A is an unbounded dissipative operator on a functional space $H = H(\Omega)$ endowed with the inner product $\langle .,. \rangle$ and the corresponding norm $\| \ \|_H$, $B : H \to H$ is a bounded operator and possibly nonlinear, u denotes the control function. Let $V = D(A^{\frac{1}{2}}) = V(\Omega)$ endowed with the norm $\|v\|_V = \|A^{\frac{1}{2}}v\|_H$ and $X = V \times H$ be the state space.

8.3.1 Regional Stabilization of Bilinear Systems

In this section we study regional stabilization of system (8.48) that can be written in the form

$$\begin{cases} \dfrac{d}{dt}\tilde{y}(t) = \tilde{A}\tilde{y}(t) + \tilde{B}\tilde{y}(t) \\ \tilde{y}(0) = \tilde{y}_0 \end{cases} \tag{8.49}$$

where $\tilde{y}(t) = (y, y_t)$ and $\tilde{A}$ given by

$$\tilde{A}\tilde{y}(t) = \begin{pmatrix} 0 & I \\ A & 0 \end{pmatrix} \tilde{y}(t)$$

and the operator $\tilde{B}$ is given by

$$\tilde{B}\tilde{y}(t) = \begin{pmatrix} 0 \\ u(t)By_t(t) \end{pmatrix}$$

System (8.49) has a unique weak solution in $C([0,T],X)$ (see [1]), given by the variation of constants formula

$$\tilde{y}(t) = S(t)\tilde{y}_0 + \int_0^t S(t-s)\tilde{B}\tilde{y}(s)ds. \tag{8.50}$$

where $S(t)$ is the semigroup generated by the operator $\tilde{A}$.

Let ω be an open subregion of Ω; $V(\omega)$ and $H(\omega)$ are respectively functional spaces of the restriction of the functions of V and H. We define by

$$\begin{aligned} \chi_\omega^1 : V(\Omega) &\longrightarrow V(\omega) \\ z_1 &\longmapsto z_1|_\omega \end{aligned}$$

and

$$\begin{aligned} \chi_\omega^2 : H(\Omega) &\longrightarrow H(\omega) \\ z_2 &\longmapsto z_2|_\omega \end{aligned}$$

the restrictions operators to ω while χ_ω^{1*} and χ_ω^{2*} denote respectively the adjoint operators of χ_ω^1 and χ_ω^2. Let us denote $i_\omega = \chi_\omega^{1*}\chi_\omega^1$ and $j_\omega = \chi_\omega^{2*}\chi_\omega^2$.

In what follows, we assume that

$$\langle i_\omega B y_t, y_t\rangle\langle B y_t, y_t\rangle \geq 0, \quad \forall y_t \in H. \tag{8.51}$$

and

$$\langle j_\omega A y, y\rangle \leq 0, \quad \forall y \in \mathcal{D}(A). \tag{8.52}$$

On the other hand, we consider the following uncontrolled system

$$\begin{cases} z_{tt}(t) = Az(t), \\ z(0) = y_0,\, z_t(0) = y_1, \end{cases} \tag{8.53}$$

Here we give sufficient conditions for regional strong stabilization of system (8.48) when B is linear.

Theorem 8.5 *Suppose that there exist some positive constants T and α such that the solution z of system (8.53) satisfies*

$$\|\chi_\omega^1 y_0\|^2_{V(\omega)} + \|\chi_\omega^2 y_1\|^2_{H(\omega)} \leq \alpha \int_0^T |\langle i_\omega B z_t(t), z_t(t)\rangle| dt \tag{8.54}$$

then the control

$$u(t) = -\langle i_\omega B y_t(t), y_t(t)\rangle \tag{8.55}$$

stabilizes regionally strongly system (8.48).

Proof From the relation

$$\langle i_\omega B z_t, z_t\rangle = \langle i_\omega B(z_t - y_t), z_t\rangle - \langle i_\omega B y_t, y_t - z_t\rangle + \langle i_\omega B y_t, y_t\rangle,$$

and since B and χ^1_ω are bounded, then there exists $\delta > 0$ such that

$$\begin{aligned} |\langle i_\omega B z_t, z_t\rangle| \le\ & \delta^2 \|B\|_H \|z_t - y_t\|_H \|z_t\|_H \\ & + \delta^2 \|B\|_H \|y_t - z_t\|_H \|y_t\|_H + |\langle i_\omega B y_t, y_t\rangle| \end{aligned} \tag{8.56}$$

Differentiating the energy $E_y(t) = \frac{1}{2}\{\|y_t\|_H^2 + \|y\|_V^2\}$ and since A is dissipative, we get

$$\frac{d}{dt} E_y(t) \le -f(y_t(t))$$

where $f(y_t(t)) = \langle i_\omega B y_t, y_t\rangle \langle B y_t, y_t\rangle$.

Then

$$E_y(t) \le -\int_0^t f(y_s(s))ds + E_y(0)$$

It follows from (8.51), that

$$E_y(t) \le E_y(0) \tag{8.57}$$

Moreover, we have

$$\|y_t\|_H^2 \le \|(y, y_t)\|_X^2 \le E_y(t) \quad \text{and} \quad \|z_t - y_t\|_H^2 \le \|\psi - \phi\|_X^2 \tag{8.58}$$

where $\psi = (z, z_t)$ and $\phi = (y, y_t)$.

Using (8.56), (8.57) and (8.58), we obtain

$$|\langle i_\omega B z_t, z_t\rangle| \le 2\delta^2 \|B\|_H \|\psi - \phi\|_X \sqrt{E_y(0)} + |\langle i_\omega B y_t, y_t\rangle| \tag{8.59}$$

From (8.50) and since A is dissipative, we have

$$\|\psi - \phi\|_X \le \sqrt{E_y(0)} \left(\int_0^t \langle i_\omega B y_s(s), y_s(s)\rangle ds\right)$$

It follows from Schwartz's inequality, that

$$\|\psi - \phi\|_X \le \sqrt{T E_y(0)} \left(\int_0^T |\langle i_\omega B y_s(s), y_s(s)\rangle|^2 ds\right)^{\frac{1}{2}} \tag{8.60}$$

Integrating (8.59) over the interval $[0, T]$, and with (8.60) we obtain

$$\int_0^T |\langle i_\omega B z_t, z_t\rangle| dt \leq (2E_y(0)\|B\|_H + 1)\sqrt{T} \times \left(\int_0^T |\langle i_\omega B y_s(s), y_s(s)\rangle|^2 ds\right)^{\frac{1}{2}}$$

Condition (8.54) gives

$$\|\chi_\omega^1 y_0\|_{V(\omega)}^2 + \|\chi_\omega^2 y_1\|_{H(\omega)}^2 \leq \beta \left(\int_0^T |\langle i_\omega B y_s(s), y_s(s)\rangle|^2 ds\right)^{\frac{1}{2}}$$

where $\beta = \alpha(2E_y(0)\|B\|_H + 1)\sqrt{T}$.

Now replacing (y_0, y_1) by (y, y_t), we get

$$\|\chi_\omega^1 y(t)\|_{V(\omega)}^2 + \|\chi_\omega^2 y_t(t)\|_{H(\omega)}^2 \leq \beta \left(\int_t^{t+T} |\langle i_\omega B y_s(s), y_s(s)\rangle|^2 ds\right)^{\frac{1}{2}}$$

Then $\|\chi_\omega^1 y(t)\|_{V(\omega)}^2 + \|\chi_\omega^2 y_t(t)\|_{H(\omega)}^2 \longrightarrow 0$, *as* $t \longrightarrow +\infty$, which completes the proof.

The following result provides sufficient conditions for regional exponential stabilization of system (8.48) when B is linear.

Theorem 8.6 *Assume that the solution z of system (8.53) satisfies the condition*

$$\|\chi_\omega^1 y_0\|_{V(\omega)} + \|\chi_\omega^2 y_1\|_{H(\omega)} \leq \alpha \int_0^T |\langle i_\omega B z_t(t), z_t(t)\rangle| dt, \ (for\ some\ T, \alpha > 0) \tag{8.61}$$

then the control

$$u(t) = \begin{cases} -\dfrac{\langle i_\omega B y_t(t), y_t(t)\rangle}{\|(y, y_t)\|_X^2}, & (y, y_t) \neq (0,0) \\ 0, & (y, y_t) = (0,0) \end{cases} \tag{8.62}$$

stabilizes regionally exponentially system (8.48), in other word there exist $M > 0$ and $\lambda > 0$ such that for every $(y_0, y_1) \in X$

$$\|\chi_\omega^1 y(t)\|_{V(\omega)}^2 + \|\chi_\omega^2 y_t(t)\|_{H(\omega)}^2 \leq M e^{-\lambda t} \|(y_0, y_1)\|_X^2.$$

Proof Let us consider the solution z of system (8.53) with the natural energy associated with (8.48)

$$E_y(t) = \frac{1}{2}\{\|y(t)\|_V^2 + \|y_t(t)\|_H^2\}$$

Since A is dissipative, we have

$$\frac{d}{dt} E_y(t) \leq -\frac{\langle i_\omega B y_t, y_t\rangle}{2E_y(t)} \langle B y_t, y_t\rangle \tag{8.63}$$

Integrating this inequality, gives

$$E_y(t) \leq -\int_0^t \frac{\langle i_\omega B y_s(s), y_s(s)\rangle}{2E_y(s)} \langle B y_s(s), y_s(s)\rangle ds + E_y(0)$$

Then

$$E_y(t) \leq E_y(0) \tag{8.64}$$

Furthermore, we have

$$\|y_t\|_H^2 \leq \|(y, y_t)\|_X^2 \leq 2E_y(t) \quad and \quad \|z_t - y_t\|_H^2 \leq \|\psi - \phi\|_X^2 \tag{8.65}$$

where $\psi = (z, z_t)$ and $\phi = (y, y_t)$.

Using (8.56) and combining (8.64) and (8.65), we obtain

$$|\langle i_\omega B z_t, z_t\rangle| \leq 2\delta^2 \|B\|_H \|\psi - \phi\|_H \sqrt{2E_y(0)} + |\langle i_\omega B y_t, y_t\rangle| \tag{8.66}$$

Furthermore, from (8.50), we have

$$\|\psi - \phi\|_X \leq \|B\|_H \left\{ \int_0^t \frac{|\langle i_\omega B y_s(s), y_s(s)\rangle|}{2E_y(s)} \sqrt{2E_s(s)} ds \right\}$$

Schwartz's inequality, gives

$$\|\psi - \phi\|_X \leq \|B\|_H \sqrt{2T E_y(0)} \left\{ \int_0^T \frac{|\langle i_\omega B y_s(s), y_s(s)\rangle|^2}{2E_y(s)} ds \right\}^{\frac{1}{2}} \tag{8.67}$$

Using (8.64), we have

$$|\langle i_\omega B y_t, y_t\rangle| \leq \frac{|\langle i_\omega B y_t, y_t\rangle|}{2E_y(t)} \sqrt{2E_y(t)} \sqrt{2E_y(0)} \tag{8.68}$$

Integrating (8.66) over the interval $[0, T]$ and taking into account (8.67) and (8.68), we obtain

$$\int_0^T |\langle i_\omega B z_t, z_t\rangle| dt \leq (2\|B\|_H + 1)\|B\|_H \sqrt{T 2E_y(0)} \left\{ \int_0^T \frac{|\langle i_\omega B y_s(s), y_s(s)\rangle|^2}{2E_y(s)} ds \right\}^{\frac{1}{2}}$$

It follows from (8.61) that

$$\|\chi_\omega^1 y_0\|_{V(\omega)} + \|\chi_\omega^2 y_1\|_{H(\omega)} \leq \beta \left\{ \int_0^T \frac{|\langle i_\omega B y_s(s), y_s(s)\rangle|^2}{2E_y(s)} ds \right\}^{\frac{1}{2}}$$

where $\beta = \alpha(2\|B\|_H + 1)\|B\|_H \sqrt{2T E_y(0)}$.

Now replacing (y_0, y_1) by (y, y_t), we get

$$\|\chi_\omega^1 y_t(t)\|_{V(\omega)} + \|\chi_\omega^2 y(t)\|_{H(\omega)} \leq \beta\left\{\int_t^{t+T} \frac{|\langle i_\omega B y_s(s), y_s(s)\rangle|^2}{2E_y(s)} ds\right\}^{\frac{1}{2}}.$$

Then

$$\|\chi_\omega^1 y(t)\|_{V(\omega)}^2 + \|\chi_\omega^2 y_t(t)\|_{H(\omega)}^2 \leq \beta^2 \int_t^{t+T} \frac{|\langle i_\omega B y_s(s), y_s(s)\rangle|^2}{2E_y(s)} ds \tag{8.69}$$

Moreover, we have

$$\frac{d}{dt}(\|\chi_\omega^1 y(t)\|_{V(\omega)}^2 + \|\chi_\omega^2 y_t(t)\|_{H(\omega)}^2) \leq -\frac{|\langle i_\omega B y_t, y_t\rangle|^2}{2E_y(t)} \tag{8.70}$$

Integrating inequality (8.70) from nT to $(n+1)T$, and since $E_y(t)$ decreases, we obtain

$$\|\chi_\omega^1 y(nT)\|_{V(\omega)}^2 + \|\chi_\omega^2 y_t(nT)\|_{H(\omega)}^2 - \|\chi_\omega^1 y((n+1)T)\|_{V(\omega)}^2 - \|\chi_\omega^2 y_t((n+1)T)\|_{H(\omega)}^2$$

$$\geq \int_{nT}^{(n+1)T} \frac{|\langle i_\omega B y_s(s), y_s(s)\rangle|^2}{2E_y(s)} ds$$

From (8.69), we get

$$\begin{aligned}&\|\chi_\omega^1 y((n+1)T)\|_{V(\omega)}^2 + \|\chi_\omega^2 y_t((n+1)T)\|_{H(\omega)}^2 \\ &\leq r(\|\chi_\omega^1 y(nT)\|_{V(\omega)}^2 + \|\chi_\omega^2 y_t(nT)\|_{H(\omega)}^2)\end{aligned}$$

where $r = \frac{1}{\beta^2}$.

By recurrence, we show that

$$\|\chi_\omega^1 y(nT)\|_{V(\omega)}^2 + \|\chi_\omega^2 y_t(nT)\|_{H(\omega)}^2 \leq r^n(\|\chi_\omega^1 y_0\|_{V(\omega)}^2 + \|\chi_\omega^2 y_1\|_{H(\omega)}^2)$$

Since $E_y(t)$ decreases and taking n the integer part of $\frac{t}{T}$, it follows that

$$\|\chi_\omega^1 y(t)\|_{V(\omega)}^2 + \|\chi_\omega^2 y_t(t)\|_{H(\omega)}^2 \leq M e^{-\lambda t}\|(y_0, y_1)\|_X^2$$

where $M = (2\|B\|_H + 1)\delta\|B\|_H\sqrt{T}$ and $\lambda = \dfrac{\ln((2\|B\|_H + 1)\|B\|_H\sqrt{T})}{T}$.

In other word system (8.48) is regionally exponentially stabilizable by the control (8.62). □

8.3.2 Regional Stabilization of Semilinear Systems

In this section we study regional strong and exponential stabilization of system (8.48) when B is nonlinear.

Let us consider the system

$$\begin{cases} z_{tt}(t) = Az(t), \\ z(0) = y_0, \quad z_t(0) = y_1, \end{cases} \tag{8.71}$$

The following result gives sufficient conditions for strong stabilization of system (8.48).

Theorem 8.7 *Assume that B is locally Lipschitz and there exist $T > 0$ and $\delta > 0$ such that the solution (z, z_t) of system (8.71) verifies the inequality*

$$\int_0^T |\langle i_\omega B z_t(t), z_t(t)\rangle| dt \geq \delta\{\|\chi_\omega^2 y_t(0)\|^2_{H(\omega)} + \|\chi_\omega^1 y(0)\|^2_{V(\omega)}\}, \tag{8.72}$$

then the control

$$u(t) = -\langle i_\omega B y_t(t), y_t(t)\rangle \tag{8.73}$$

stabilizes regionally strongly system (8.48).

Proof We show first that $h : y_t \mapsto \langle i_\omega B y_t, y_t\rangle B y_t$ is locally Lipschitz.

Since B is locally Lipschitz, then for all $R > 0$, there exists $L > 0$ such that

$$\begin{aligned} &\|By_t - Bz_t\|_H \leq L\|y_t - z_t\|_H, \ \forall y_t, z_t \in H : \\ &0 < \|y_t\|_H \leq \|z_t\|_H \leq R. \end{aligned} \tag{8.74}$$

Let remark that

$$\begin{aligned} \langle i_\omega B z_t, z_t\rangle B z_t - \langle i_\omega B y_t, y_t\rangle B y_t &= \langle i_\omega B z_t, z_t\rangle (B z_t - B y_t) \\ &\quad + (\langle i_\omega B z_t, z_t\rangle - \langle i_\omega B y_t, y_t\rangle) B y_t \\ &= (\langle i_\omega B z_t, z_t - y_t\rangle - \langle i_\omega B z_t - i_\omega B y_t, y_t\rangle) B y_t \\ &\quad + \langle i_\omega B z_t, z_t\rangle (B z_t - B y_t) \end{aligned}$$

Using (8.74), we have

$$\begin{aligned} \|h(z_t) - h(y_t)\|_H &\leq L^2 \|z_t\|^2_H \|z_t - y_t\|_H \\ &\quad + L^2 \|z_t - y_t\|_H (\|z_t\|_H + \|y_t\|_H) \|y_t\|_H \\ &\leq \mathcal{L} \|z_t - y_t\|_H \end{aligned}$$

where $\mathcal{L} = 3LR^2$.

It follows that system (8.48) has a unique weak solution (y, y_t) (see [1]).

Using (8.56) and (8.74), we obtain

$$|\langle i_\omega Bz_t, z_t\rangle| \le L\|z_t - y_t\|_H\|z_t\|_H + L\|y_t - z_t\|_H\|y_t\|_H + |\langle i_\omega By_t, y_t\rangle| \quad (8.75)$$

Now, we differentiate $E_y(t)$

$$\frac{d}{dt}E_y(t) \le -\langle i_\omega By_t, y_t\rangle\langle By_t, y_t\rangle$$

Then

$$E_y(t) \le -\int_0^t \langle i_\omega By_s(s), y_s(s)\rangle\langle By_s(s), y_s(s)\rangle ds + E_y(0)$$

It follows from (8.51), that

$$E_y(t) \le E_y(0) \quad (8.76)$$

Moreover, we have

$$\|y_t\|_H^2 \le \|(y, y_t)\|_X^2 \le 2E_y(t) \quad \text{and} \quad \|z_t - y_t\|_H^2 \le \|\psi - \phi\|_X^2 \quad (8.77)$$

where $\psi = (z, z_t)$ and $\phi = (y, y_t)$.

From (8.75) and combining (8.76) with (8.77), we get

$$|\langle i_\omega Bz_t, z_t\rangle| \le 2L\|\psi - \phi\|_X\sqrt{2E_y(0)} + |\langle i_\omega By_t, y_t\rangle| \quad (8.78)$$

From (8.50), we have

$$\|\psi - \phi\|_X \le \sqrt{2E_y(0)}\int_0^t |\langle i_\omega By_t, y_t\rangle|^2 ds$$

For a fixed $T > 0$, Schwarz inequality, yields

$$\|\psi - \phi\|_X \le \sqrt{2TE_y(0)g(0)} \quad (8.79)$$

where $g(t) = \int_t^{t+T} |\langle i_\omega By_s(s), y_s(s)\rangle|^2 ds$.

Integrating (8.78) on the interval $[0, T]$ and using (8.79), we obtain

$$\int_0^T |\langle i_\omega Bz_t, z_t\rangle| dt \le (2E_y(0) + 1)L\sqrt{Tg(0)}$$

Condition (8.72) allows

$$\|\chi_\omega^2 y_1\|_{H(\omega)}^2 + \|\chi_\omega^1 y_0\|_{V(\omega)}^2 \le \frac{(2E_y(0) + 1)L\sqrt{T}}{\delta}\sqrt{g(0)}$$

Now, replacing (y_0, y_1) by (y, y_t), we obtain

$$\|\chi_\omega^2 y_t\|_{H(\omega)}^2 + \|\chi_\omega^1 y\|_{V(\omega)}^2 \le \beta\sqrt{g(t)}$$

where $\beta = \dfrac{(2E_y(0)+1)L\sqrt{T}}{\delta}$.

It follows that $\|\chi_\omega^1 y(t)\|_{V(\omega)}^2 + \|\chi_\omega^2 y_t(t)\|_{H(\omega)}^2 \longrightarrow 0$ as $t \longrightarrow +\infty$, we deduce that the control (8.73) stabilizes regionally strongly system (8.48). □

The following result gives sufficient conditions for regional exponential stabilization of system (8.48).

Theorem 8.8 *Assume that B is locally Lipschitz and there exist $T > 0$ and $C > 0$ such that the solution (z, z_t) of system (8.71) satisfy the following condition*

$$\int_0^T |\langle i_\omega B z_t(t), z_t(t)\rangle| dt \ge C\{\|\chi_\omega^2 y_1\|_{H(\omega)} + \|\chi_\omega^1 y_0\|_{V(\omega)}\}, \tag{8.80}$$

then the control

$$u(t) = \begin{cases} -\dfrac{\langle i_\omega B y_t(t), y_t(t)\rangle}{\|(y, y_t)\|_X^2}, & (y, y_t) \ne (0,0) \\ 0, & (y, y_t) = (0,0) \end{cases} \tag{8.81}$$

stabilizes exponentially system (8.48) on ω.

Proof Let $\phi = (y, y_t)$ et $\psi = (z, z_t)$. We show first that the operator $f : \phi \mapsto \dfrac{\langle i_\omega B y_t, y_t\rangle}{\|(y, y_t)\|_X^2} B y_t$ is locally Lipschitz.

Since B is locally Lipschitz, then for all $R > 0$, there exists $K > 0$ such that

$$\|By_t - Bz_t\|_H \le K\|y_t - z_t\|_H, \ \forall y_t, z_t \in H : 0 < \|y_t\|_H \le \|z_t\|_H \le R. \tag{8.82}$$

Then

$$f(\psi) - f(\phi) = \frac{\|\phi\|_X^2(\langle i_\omega B z_t, z_t\rangle B z_t - \langle i_\omega B y_t, y_t\rangle B y_t)}{\|\psi\|_X^2\|\phi\|_X^2} - \frac{(\|\psi\|_X^2 - \|\phi\|_X^2)\langle i_\omega B y_t, y_t\rangle B y_t}{\|\psi\|_X^2\|\phi\|_X^2}.$$

It follows that

$$\begin{aligned}\|f(\psi) - f(\phi)\|_X &\le \frac{\|\langle i_\omega B z_t, z_t\rangle B z_t - \langle i_\omega B y_t, y_t\rangle B y_t\|_H}{\|\psi\|_X^2} \\ &\quad + K^2 |\|\psi\|_X^2 - \|\phi\|_X^2|.\frac{\|\phi\|_X}{\|\psi\|_X^2} \\ &\le \frac{\|\langle i_\omega B z_t, z_t\rangle (B z_t - B y_t)}{\|\psi\|_X^2} + \frac{(\langle i_\omega B z_t, z_t - y_t\rangle}{\|\psi\|_X^2} \\ &\quad - \frac{\langle i_\omega B y_t - i_\omega B z_t, y_t\rangle) B y_t\|_H}{\|\psi\|_X^2} + 2K^2\|\psi - \phi\|_X\end{aligned}$$

Furthermore, we have

$$\|z_t - y_t\|_H^2 \leq \|(\psi - \phi)\|_X^2 \tag{8.83}$$

Using (8.83), we get $\|f(\psi) - f(\phi)\|_X \leq L\|\psi - \phi\|_X$ where $L = 5K^2$.

We conclude that system (8.48) has a unique weak solution (y, y_t) (see [1]).

Using (8.82) and (8.56), then there exists $\alpha > 0$ such that

$$\begin{aligned} |\langle i_\omega B z_t, z_t\rangle| \leq & K\alpha^2 \|z_t - y_t\|_H \|z_t\|_H \\ & + K\alpha^2 \|y_t - z_t\|_H \|y_t\|_H + |\langle i_\omega B y_t, y_t\rangle| \end{aligned} \tag{8.84}$$

Moreover, we have

$$\|y_t\|_H^2 \leq \|(y, y_t)\|_X^2 \leq 2E_y(t)$$

Since A is dissipative, then

$$\frac{d}{dt}E_y(t) \leq -\frac{\langle i_\omega B y_t, y_t\rangle \langle B y_t, y_t\rangle}{\|y_t\|_H^2 + \|y\|^2}$$

Thus

$$E_y(t) = -\int_0^t \frac{\langle i_\omega B y_s(s), y_s(s)\rangle \langle B y_s(s), y_s(s)\rangle}{\|y_s(s)\|_H^2 + \|y(s)\|^2} ds + E_y(0)$$

It follows from (8.51), that

$$E_y(t) \leq E_y(0) \tag{8.85}$$

Using (8.83), (8.84) and (8.85), we obtain

$$|\langle i_\omega B z_t, z_t\rangle| \leq 2K\alpha^2 \|\psi - \phi\|_X \sqrt{2E_y(0)} + |\langle i_\omega B y_t, y_t\rangle| \tag{8.86}$$

From (8.50), we deduce that

$$\|\psi - \phi\|_X \leq K \int_0^t |f(s)| \sqrt{2E_y(s)} ds$$

where $f(s) = \dfrac{\langle i_\omega B y_s(s), y_s(s)\rangle}{2E_y(s)}$.

Schwarz inequality, gives

$$\|\psi - \phi\|_X \leq K\sqrt{2T} \left(\int_0^T |f(s)|^2 E_y(s) ds \right)^{\frac{1}{2}} \tag{8.87}$$

Using (8.85), we get

$$|\langle i_\omega B y_t, y_t\rangle| \leq |f(t)| \sqrt{2E_y(t)} \sqrt{2E_y(0)} \tag{8.88}$$

Integrating (8.86) on the interval $[0, T]$ and using (8.87) and (8.88), we obtain

$$\int_0^T |\langle i_\omega B z_s(s), z_s(s)\rangle| ds \le 2(2K^2\alpha^2+1)\sqrt{TE_y(0)}\left(\int_0^T |f(s)|^2 E_y(s)ds\right)^{\frac{1}{2}}$$

It follows from (8.80), that

$$\|\chi_\omega^2 y_t(0)\|_{H(\omega)} + \|\chi_\omega^1 y(0)\|_{V(\omega)} \le \frac{2(2K\alpha^2+1)}{C}\sqrt{TE_y(0)}\left(\int_0^T |f(s)|^2 E_y(s)ds\right)^{\frac{1}{2}}$$

Replacing (y_0, y_1) by (y, y_t), we get

$$\|\chi_\omega^2 y_t(t)\|_{H(\omega)} + \|\chi_\omega^1 y(t)\|_{V(\omega)} \le \frac{2(2K\alpha^2+1)}{C}\sqrt{TE_y(t)}\left(\int_t^{t+T} |f(s)|^2 E_y(s)ds\right)^{\frac{1}{2}}$$

It follows that

$$\|\chi_\omega^1 y(t)\|^2_{V(\omega)} + \|\chi_\omega^2 y_t(t)\|^2_{H(\omega)} \le \gamma \int_t^{t+T} |f(s)|^2 E_y(s)ds \tag{8.89}$$

where $\gamma = \dfrac{2(2K\alpha^2+1)}{C}\sqrt{TE_y(0)}$.

Integrating from kT to $(k+1)T$ the inequality

$$\frac{d}{dt}(\|\chi_\omega^1 y(t)\|^2_{V(\omega)} + \|\chi_\omega^2 y_t(t)\|^2_{H(\omega)}) \le -\frac{|\langle i_\omega B y_t, y_t\rangle|^2}{2E_y(t)}$$

and since $(\|\chi_\omega^1 y(t)\|^2_{V(\omega)} + \|\chi_\omega^2 y_t(t)\|^2_{H(\omega)})$ decreases, we obtain

$$\begin{aligned}&\|\chi_\omega^1 y(kT)\|^2_{V(\omega)} + \|\chi_\omega^2 y_t(kT)\|^2_{H(\omega)}\\ &-\|\chi_\omega^1 y((k+1)T)\|^2_{V(\omega)} - \|\chi_\omega^2 y_t((k+1)T)\|^2_{H(\omega)}\\ &\ge \int_{kT}^{(k+1)T} |f(s)|^2 E_y(s)ds\end{aligned}$$

From (8.89), we get

$$\begin{aligned}&\|\chi_\omega^1 y((k+1)T)\|^2_{V(\omega)} + \|\chi_\omega^2 y_t((k+1)T)\|^2_{H(\omega)}\\ &\le r(\|\chi_\omega^1 y(kT)\|^2_{V(\omega)} + \|\chi_\omega^2 y_t(kT)\|^2_{H(\omega)})\end{aligned}$$

where $r = \dfrac{1}{\gamma}$.

This implies that

$$\|\chi^1_\omega y((k+1)T)\|^2_{V(\omega)} + \|\chi^2_\omega y_t((k+1)T)\|^2_{H(\omega)} \leq r^k(\|\chi^1_\omega y_0\|^2_{V(\omega)} + \|\chi^2_\omega y_1\|^2_{H(\omega)}) \tag{8.90}$$

Applying (8.90) to the integer part $k = E(\frac{t}{T})$ of $\frac{t}{T}$, we deduce that

$$\|\chi^1_\omega y(t)\|^2_{V(\omega)} + \|\chi^2_\omega y_t(t)\|^2_{H(\omega)} \leq Me^{-\lambda t}\|(y_0, y_1)\|^2_X$$

where $M = \frac{2\alpha\sqrt{T}}{C}(2K+1)$ and $\lambda = \dfrac{\ln(\frac{2\sqrt{T}}{C}(2K+1))}{T}$.

Then control (8.81) stabilizes exponentially system (8.48) on ω. □

8.4 Simulations Results

In this section we illustrate the previous results by simulations. The developed results show that the stabilizing control is given by (8.14) for exponential stabilization and by (8.6) for strong stabilization. The computation of such control may be achieved using the following algorithm:
Step 1: Initial data : threshold $\epsilon > 0$ *and initial condition* y_0;
Step 2: Apply the control $u(t_i) = \langle By_t(t_i), y_t(t_i)\rangle$;
Step 3: Solving the system (8.1) *using Lax-Wendroff method given in* [7], *gives* $y(t_{i+1})$;
Step 4: If $\|y(t_{i+1})\| < \epsilon$ *stop, otherwise;*
Step 5: $i = i + 1$ *and go to step 2.*

Example 8.5 On $\Omega =]0, 1[$, consider the following system

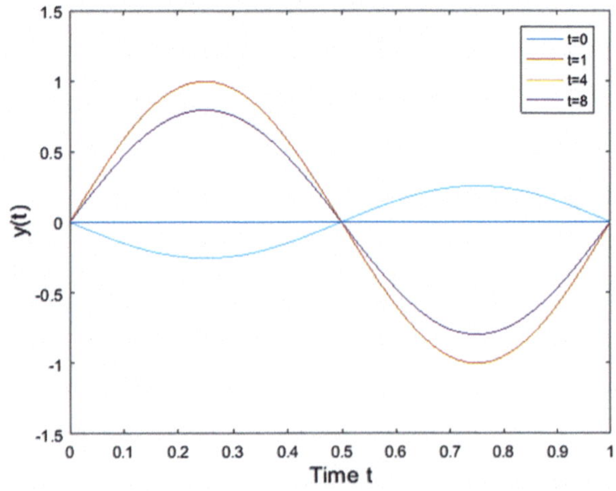

Fig. 8.1 Initial position

$$\begin{cases} y_{tt}(x,t) - y_{xx}(x,t) = u(t)y_t(x,t), & \Omega\times]0,\infty[\\ y(x,0) = \sin(2\pi x),\ \ y_t(x,0) = 0, & \Omega \\ y(0,t) = y(1,t) = 0, &]0,\infty[\end{cases} \tag{8.91}$$

where the control

$$u(t) = -\int_0^1 |y_t(x,t)|^2 dx.$$

System (8.91) has a unique solution on the state space $H_0^1(\Omega) \times L^2(\Omega)$ ([6]).

Figure 8.1 shows that the state is strongly stabilized on $\Omega =]0, 1[$ with stabilization error equals to 1.03×10^{-4}.

Figure 8.2 shows that the energy of the system decreases to zero (Fig. 8.3).

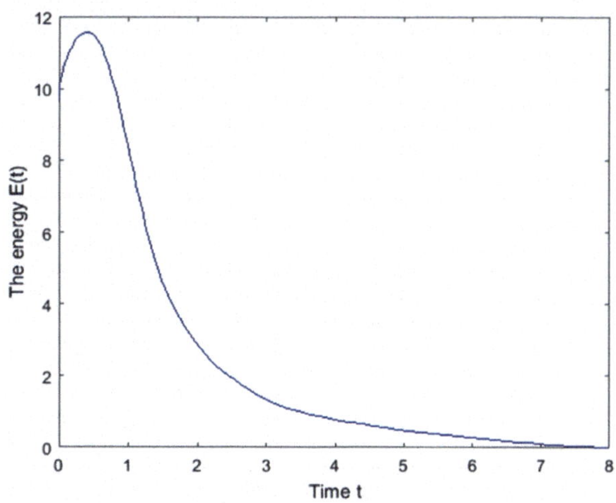

Fig. 8.2 The energy of the system

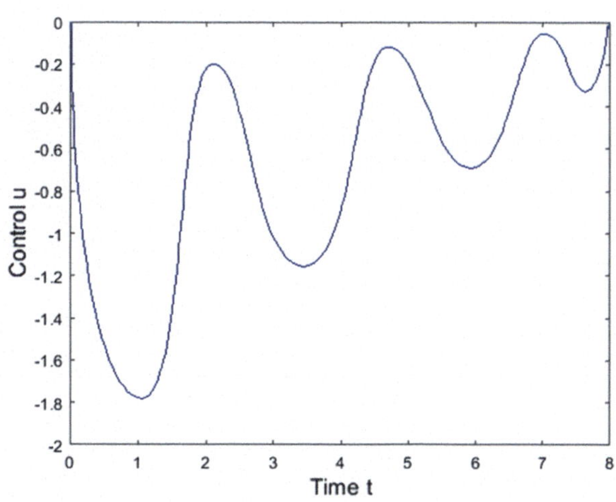

Fig. 8.3 Evolution of the control

References

1. Pazy, A.: Semigroups of Linear Operators and Applications to Partial Differential Equations. Springer, New York (1983)
2. Ball, J.M., Slemrod, M.: Feedback stabilization of distributed semilinear control systems. Appl. Math. Optim. **5**, 169–179 (1979)
3. Bardos, C., Lebeau, G., Rauch, J.: Sharp sufficient conditions for the observation, control and stabilization of waves from the boundary. SIAM J. Control Optim. **30**, 1024–1065 (1992)
4. Lebeau, G.: Contrôle de l'équation de Schrödinger. J. Math. Pures Appl. **71**, 267–291 (1992)
5. Lasiecka, I., Tataru, D.: Uniform boundary stabilisation of semilinear wave equation with non-linear boundary damping. J. Differ. Integr. Equ. **6**, 507–533 (1993)
6. Komornik, V., Loreti, P.: Fourier Series in Control Theory. Springer Monographs in Mathematics (2005)
7. Lax, P., Wendroff, B.: Systems of conservation laws. Commun. Pure Appl. Math. **13**, 217–237 (1960)

Chapter 9
Gradient Stabilization of Infinite Dimensional Linear Systems

The notion of gradient stabilization of infinite dimensional system is very important since it modelizes many real applications. For instance the problem of thermal insulation where the purpose is to keep a constant temperature of the system with regards to the outside environment assumed to be with fluctuating temperature. Thus one has to regulate the system temperature in order to vanish the exchange thermal flux. This is the case inside a car where one has to change the level of the internal air conditioning with respect to the external temperature. As we can not always have external measurements, we use a sensor to measure the flux, which is a transducer producing a signal that is proportional to the local heat flux.

The goal of this chapter is to study the gradient stabilization of infinite dimensional linear systems. First we give definitions and properties of the gradient stability. Then we characterize controls that stabilize the gradient of such a system and the one that stabilizes the gradient and minimizes a functional cost. The obtained results lead to numerical approach successfully illustrated by simulations.

9.1 Gradient Stability—Definitions—Examples

This section is devoted to some preliminaries concerning definition and characterization of gradient stability for infinite dimensional linear systems.

Let Ω be an open regular subset of $\mathbb{R}^m$ and let us consider the state-space system

$$\begin{cases} \dfrac{\partial y}{\partial t} = Ay \\ y(0) = y_0 \in X \end{cases} \tag{9.1}$$

E. H. Zerrik and O. Castillo, *Stabilization of Infinite Dimensional Systems*,
Studies in Systems, Decision and Control 355,
https://doi.org/10.1007/978-3-030-68600-0_9

where $A : \mathcal{D}(A) \subset X \to X$ is a linear operator generating a strongly continuous semigroup $S(t), t \geq 0$, on the state space X which is continuously embedded in $H^1(\Omega)$. The space X is endowed with its a complex inner product $<,>$ and the corresponding norm $\|.\|$.

We define the gradient operator by:

$$\begin{aligned} \nabla \; : X &\to (L^2(\Omega))^m \\ y &\mapsto \left(\frac{\partial y}{\partial x_1}, \frac{\partial y}{\partial x_2}, \ldots, \frac{\partial y}{\partial x_m} \right) \end{aligned} \tag{9.2}$$

$(L^2(\Omega))^m$ is endowed with its usual complex inner product $<,>_m$ and the corresponding norm $\|.\|_m$ where:

$$\begin{aligned} \langle .,. \rangle \; (L^2(\Omega))^m \times (L^2(\Omega))^m &\to \mathbb{C} \\ (f, g) &\longmapsto \sum_{i=1}^{m} \int_\Omega f_i(x)\overline{g_i}(x)dx \end{aligned}$$

with $f = (f_1, f_2, \ldots, f_m)$ and $g = (g_1, g_2, \ldots, g_m)$ where $f_i, g_i \in L^2(\Omega)$ $i = 0, 1, \ldots, m$. The mild solution of (9.1) is given by $y(t) = S(t)y_0$. Let ∇^* the adjoint operator of ∇, and we define the operator $G = \nabla^*\nabla$, which is a bounded operator applying X into itself.

Let $\mathcal{L}(X, (L^2(\Omega))^m)$ be the space of bounded linear operators mapping X into $(L^2(\Omega))^m$ endowed with the uniform norm of operators $|||.|||$.

Definition 9.1 The system (9.1) is said to be

1. Gradient weakly stable (g.w.s) if $\forall y_0 \in X$, the corresponding solution $y(t)$ of (9.1) satisfies

$$< \nabla y(t), z >_m \longrightarrow 0 \text{ as } t \longrightarrow \infty, \quad \forall z \in (L^2(\Omega))^m \tag{9.3}$$

2. Gradient strongly stable (g.s.s) if for any initial condition $y_0 \in X$ the corresponding solution $y(t)$ of (9.1) satisfies

$$||\nabla y(t)||_m \longrightarrow 0 \text{ as } t \longrightarrow \infty \tag{9.4}$$

3. Gradient exponentially stable (g.e.s) if there exist $M, \alpha > 0$ such that

$$\|\nabla y(t)\|_m \leq M e^{-\alpha t} \|y_0\|, \quad t \geq 0, \quad y_0 \in X \tag{9.5}$$

Remark 9.1 From the above definitions we have:

1. g.e.s $\Rightarrow$ g.s.s $\Rightarrow$ g.w.s.
2. If the system (9.1) is stable then it is also gradient stable.
3. We can find systems gradient stable but not stable. This is illustrated in the following example.

Example 9.1 Let $\Omega =]0, 1[$, on $H^1(\Omega)$ we consider the system

$$\begin{cases} \frac{\partial y}{\partial t}(t) &= Ay(t) \qquad \Omega\times]0, +\infty[\\ \frac{\partial y}{\partial t}(0, t) &= \frac{\partial y}{\partial t}(1, t) = 0 \;]0, +\infty[\\ y(., 0) &= y_0 \in H^1(\Omega) \end{cases} \tag{9.6}$$

where $Ay = \Delta y + y$ and $\Delta = \frac{d^2}{dx^2}$ is the Laplace operator.

The eigenpairs (λ_i, ϕ_i), $i \in \mathbb{N}$ of A are given by:

$$\begin{cases} \lambda_i &= 1 - (i\pi)^2 \qquad i \geq 0 \\ \phi_i(x) &= \sqrt{\frac{2}{1+(i\pi)^2}} \cos(i\pi x) \; i \geq 0 \end{cases}$$

A generates a strongly continuous semigroup $S(t)$ given by

$$S(t)y_0 = \sum_{i\geq 0} e^{\lambda_i t} \langle y_0, \phi_i \rangle \phi_i$$

Since $\lambda_0 > 0$ then (9.6) isn't stable but

$$\begin{aligned} \|\nabla S(t)y_0\|_m^2 &\leq e^{2\lambda_1 t} \sum_{i>0} \langle y_0, \phi_i \rangle^2 \|\nabla \phi_i\|_m^2 \\ &\leq e^{2\lambda_1 t} \|y_0\|^2 \end{aligned}$$

Therefore the system (9.6) is g.e.s.

9.2 Characterizations of Gradient Stability

The following result links gradient stability of the system (9.1) to the spectrum properties of the operator A. Let us consider the sets

$$\sigma^1(A) = \{\lambda \in \sigma(A),\, Re(\lambda) \geq 0,\, N(A - \lambda I) \not\subset N(G)\}$$

and

$$\sigma^2(A) = \{\lambda \in \sigma(A),\, Re(\lambda) < 0,\, N(A - \lambda I) \not\subset N(G)\}$$

where $\sigma(A)$ and $N(A)$ are the points spectrum and the kernel of the operator A.

Proposition 9.1 *(1) If the system (9.1) is g.w.s then* $\sigma^1(A) = \emptyset$.

(2) Assume that the state space X *has an orthonormal basis* $(\phi_n)_n$ *of eigenfunctions of* A, *if* $\sigma^1(A) = \emptyset$ *and, for some* $\alpha > 0$, $Re(\lambda) \leq -\alpha$ *for all* $\lambda \in \sigma^2(A)$, *then the system (9.1) is g.e.s.*

Proof (1) Assume that there exists $\lambda \in \sigma(A)$ such that $Re(\lambda) \geq 0$ and there exists $\phi \in X$ such that $A\phi = \lambda\phi$ for $y_0 = \phi$, the solution of (9.1) is $S(t)\phi = e^{\lambda t}\phi$, so

$$|\langle \nabla S(t)\phi, \nabla\phi\rangle_m| \geq e^{Re(\lambda)t}\|\nabla\phi\|_m^2 \geq \|\nabla\phi\|_m^2 > 0$$

hence the system (9.1) is not g.w.s.

(2) For $y_0 \in X$ we have

$$\nabla S(t)y_0 = \sum_{n\geq 0} e^{\lambda_n t}\sum_{k=1}^{r_n}\langle y_0, \phi_{n,k}\rangle \nabla\phi_{n,k}$$

where r_n is the multiplicity of the eigenvalue λ_n.
$\sigma^1(A) = \emptyset$, gives

$$\|\nabla S(t)y_0\|_m \leq Me^{-\alpha t}\|y_0\|, \text{ for some } M > 0.$$

So we have the g.e.s of the system (9.1). □

As example we consider (9.6). We have $\sigma^1(A) = \emptyset$ and $\forall\lambda \in \sigma^2(A)$, $Re(\lambda) \leq 1 - \pi^2$, then the system (9.6) is g.e.s.

For the gradient exponential stability, we need the following lemma.

Lemma 9.1 *Assume that there exists a function $M(t) \in L^2(0, +\infty; \mathbb{R}^+)$ such that*

$$|||\nabla S(t+s)||| \leq M(t)|||\nabla S(s)||| \quad \forall\, t, s \geq 0. \tag{9.7}$$

then the operators $(\nabla S(t))_{t\geq 0}$ are uniformly bounded.

Proof Let us show that $\sup_{t\geq 0}|||\nabla S(t)||| < +\infty$. Otherwise there exists a sequence $(t_1 + \tau_k), t_1 > 0$ and $\tau_k \to \infty$ such that $|||\nabla S(t_1 + \tau_k)|||$ is increasing without bound.

Now we have $\displaystyle\int_0^\infty \|\nabla S(s+\tau_k)y\|_m^2 ds = \int_{\tau_k}^\infty \|\nabla S(s)y\|_m^2 ds$ and the right-hand side goes to zero when $k \to \infty$.

By Fatou's Lemma $\liminf \|\nabla S(s+\tau_k)y\|_m = 0$ when $k \to \infty$, almost everywhere $0 \leq s < \infty$.

Hence for some $s_0 < t_1$ we can find a subsequence $\{\tau_{k_n}\}$ such that $\lim_n \|\nabla S(s_0 + \tau_{k_n})y\|_m = 0$. But with (9.7) we have $\|\nabla S(t_1 + \tau_{k_n})y\|_m \leq M(t_1 - s_0)\|\nabla S(s_0 + \tau_{k_n})y\|_m \to 0$ when $n \to +\infty$, which is a contradiction. The conclusion follows from the uniform boundedness principle. □

Proposition 9.2 *If (9.7) is satisfied and*

$$|||\nabla S(nt)||| \leq |||\nabla S(t)|||^n \;\forall\, t \geq 0,\; \forall n \in \mathbb{N}^* \tag{9.8}$$

then the system (9.1) is g.e.s if and only if

$$\int_0^{\infty} ||\nabla S(t)y||_m^2 dt < \infty, \quad \forall y \in X \tag{9.9}$$

Proof We have

$$\begin{aligned}
t\|\nabla S(t)y\|_m^2 &= \int_0^t \|\nabla S(t)y\|_m^2 ds \\
&\leq \int_0^t \|\nabla S(s+t-s)y\|_m^2 ds \\
&\leq \int_0^t M^2(s)\|\nabla S(t-s)y\|_m^2 ds \qquad \text{from (9.7)} \\
&\leq N\|y\|^2 \qquad \text{from lemma (9.1)}
\end{aligned}$$

where $N > 0$, then $\ln |||\nabla S(t)||| < 0$, $\forall t \geq t_0$ for some $t_0 > 0$, hence

$$w_0 = \inf_{t>0} \frac{\ln |||\nabla S(t)|||}{t} < 0$$

Now we show that $w_0 = \lim\limits_{t\to+\infty} \dfrac{\ln |||\nabla S(t)|||}{t}$.

Let $t_1 > 0$ and $N' = \sup\limits_{t\in[0,t_1]} \|S(t)\|$, there exists $n \in IN$ such that $nt_1 \leq t < (n+1)t_1$ for each $t \geq t_1$, then

$$\frac{\ln |||\nabla S(t)|||}{t} \leq \frac{\ln |||\nabla S(nt_1)|||}{t} + \frac{\ln \|S(t-nt_1)\|}{t}$$

with (9.8) we have

$$\frac{\ln |||\nabla S(t)|||}{t} \leq \frac{nt_1}{t}\frac{\ln |||\nabla S(t_1)|||}{t_1} + \frac{\ln N'}{t}$$

therefore

$$\limsup_{t\to\infty} \frac{\ln |||\nabla S(t)|||}{t} \leq \inf_{t>0} \frac{\ln |||\nabla S(t)|||}{t} \leq \liminf_{t\to\infty} \frac{\ln |||\nabla S(t)|||}{t}$$

then $w_0 = \lim\limits_{t\to+\infty} \dfrac{\ln |||\nabla S(t)|||}{t}$.

Hence for all $\omega \in]0, -\omega_0[$, there exists M' such that $||\nabla S(t)y||_m \leq M'e^{-\omega t}\|y\|$, $\forall y \in X$, $t \geq 0$. So the system is g.e.s.

The converse is immediate. □

Example 9.2 The system (9.6) satisfies the conditions (9.7) and (9.8). Indeed, let $t > 0$, and $y \in H^1(\Omega)$. We have

$$\nabla S(t)y = \sum_{i>0} e^{\lambda_i t}\langle y, \phi_i\rangle \nabla\phi_i$$

which implies

$$\begin{aligned}\|\nabla S(t)y\|_m^2 &\leq e^{2\lambda_1 t}\sum_{i>0}\langle y,\phi_i\rangle^2\|\nabla\phi_i\|_m^2\\ &\leq e^{2\lambda_1 t}\|y\|^2\end{aligned}$$

we can show that $\||\nabla S(t)\|| = e^{2\lambda_1 t}$. We have $\int_0^{+\infty} \|\nabla S(t)y\|^2 dt < +\infty$. Therefore the system (9.6) is g.e.s.

Corollary 9.1 *Under conditions (9.7) and (9.8) and assume, in addition, that there exists a self-adjoint positive operator $P \in \mathcal{L}(X)$ such that*

$$< Ay, Py > + < Py, Ay > + < Ry, y >= 0, \;\; y \in \mathcal{D}(A) \tag{9.10}$$

where $R \in \mathcal{L}(X)$ is a self-adjoint operator satisfying

$$< Ry, y > \geq c\; ||\nabla y||_m^2, \; \textit{for some} \;\; c > 0 \tag{9.11}$$

then system (9.1) is g.e.s.

Proof We define the function $V(y) = < Py, y >$, $\forall y \in X$. For $y_0 \in \mathcal{D}(A)$, we have $y(t) = S(t)y_0$ and

$$\begin{aligned}\frac{d}{dt}V(y(t)) &= < PAS(t)y_0, S(t)y_0 > + < PS(t)y_0, AS(t)y_0 >\\ &= - < RS(t)y_0, S(t)y_0 >\end{aligned}$$

Thus

$$\int_0^{+\infty} < RS(s)y_0, S(s)y_0 > ds \leq V(y_0)$$

By (9.11), we obtain $\int_0^{+\infty} ||\nabla S(s)y_0||_m^2 ds < \infty$. Since $\mathcal{D}(A)$ is dense in X we can extend this inequality to all $y_0 \in X$, and the Proposition 9.2 gives the conclusion.■

For the gradient strong stability we have the following result.

Proposition 9.3 *Assume that the equation*

$$< Ay, Py > + < Py, Ay > + < Ry, y >= 0, \;\; y \in \mathcal{D}(A) \tag{9.12}$$

has a self-adjoint positive solution $P \in \mathcal{L}(X)$, where $R \in \mathcal{L}(X)$ is a self-adjoint operator satisfying (9.11). Moreover if the following condition holds

$$Re < GAy, y > \leq 0, \;\; y \in \mathcal{D}(A) \tag{9.13}$$

then (9.1) is g.s.s.

Proof Let us consider the function $V(y) = < Py, y >, \ \forall y \in X, \ \ t \geq 0$,
For $y_0 \in \mathcal{D}(A)$, we have $y(t) = S(t)y_0$ and

$$\begin{aligned}\frac{d}{dt}V(y(t)) &= < PAS(t)y_0, S(t)y_0 > + < PS(t)y_0, AS(t)y_0 > \\ &= - < RS(t)y_0, S(t)y_0 >\end{aligned}$$

we obtain $\int_0^{+\infty} < RS(s)y_0, S(s)y_0 > ds \leq V(y_0)$. By (9.11), $\int_0^{+\infty} ||\nabla S(s)y_0||_m^2$ $ds < \infty$ and from (9.13), we have $\frac{\partial}{\partial t}||\nabla S(t)y_0||_m^2 \leq 0$.

Then $t||\nabla S(t)y_0||_m^2 = \int_0^t ||\nabla S(t)y_0||_m^2 ds \leq \int_0^t ||\nabla S(s)y_0||_m^2 ds$. We deduce

$$||\nabla S(t)y_0||_m^2 \leq \frac{\alpha(y_0)}{t}, t > 0, \, y_0 \in \mathcal{D}(A) \text{ for some } \alpha(y). \tag{9.14}$$

From the density of $\mathcal{D}(A)$ in X, and the continuity of $\alpha(.)$, then (9.14) is satisfied for all $y_0 \in X$. This means that the gradient of system (9.1) is strongly stable. □

9.3 Gradient Stabilization for Infinite Dimensional System

Let us consider the system

$$\begin{cases} \dfrac{\partial y(t)}{\partial t} = Ay(t) + Bu(t) \\ y(.,0) = y_0 \in X \end{cases} \tag{9.15}$$

with the same assumptions on A, and B is a bounded linear operator mapping U, the space of controls (assumed to be Hilbert space), into X.

Definition 9.2 The system (9.15) is said to be gradient weakly (respectively strongly, exponentially) stabilizable if there exists a bounded operator $K \in \mathcal{L}(X, U)$ such that the system

$$\begin{cases} \dfrac{\partial y(t)}{\partial t} = (A + BK)y(t) \\ y(.,0) = y_0 \in X \end{cases} \tag{9.16}$$

is g.w.s (respectively g.s.s, g.e.s).

Remark 9.2 1. If a system is stabilizable, then it is also gradient stabilizable.
2. Gradient stabilization is cheaper than state stabilization.
   Indeed if we consider the cost functional

$$q(u) = \int_0^{+\infty} \|u(t)\|^2 dt$$

and the spaces

$$\mathcal{U}_{ad} = \{u \in L^2(0, +\infty; U); \text{ u stabilizes the gradient}\}$$

and

$$\mathcal{U}^1_{ad} = \{u \in L^2(0, +\infty; U); \text{ u stabilizes the state}\}$$

then we have

$$\mathcal{U}_{ad} \supset \mathcal{U}^1_{ad}$$

and therefore

$$\min_{u \in \mathcal{U}_{ad}} q(u) \leq \min_{u \in \mathcal{U}^1_{ad}} q(u)$$

3. The gradient stabilization may be seen as a special case of output stabilization with output operator ∇.

In the following we give the feedback control which stabilizes the gradient of the system (9.15), by two approaches. The first is an extension of state space decomposition and the second one is based on algebraic Riccati equation.

9.3.1 State Space Decomposition

Let $\delta < 0$ be a fixed real and consider the subsets $\sigma_u(A)$ and $\sigma_s(A)$ of the spectrum $\sigma(A)$ of A defined by

$$\sigma_u(A) = \{\lambda : \lambda \in \sigma(A), Re(\lambda) \geq \delta\}$$

and

$$\sigma_s(A) = \{\lambda : \lambda \in \sigma(A), Re(\lambda) < \delta\}$$

Assume that $\sigma_u(A)$ is bounded and is separated from the set $\sigma_s(A)$ in such a way that a rectifiable simple closed curve can be drawn so as to enclose an open set containing $\sigma_s(A)$ in its interior and $\sigma_u(A)$ in its exterior. This is the case, for example, where A is self-adjoint with compact resolvent, there are at most finitely many nonnegative eigenvalues of A, and each with finite dimensional eigenspace. Then the state space X can be decomposed according to:

$$X = X_u + X_s \tag{9.17}$$

with $X_u = PX$, $X_s = (I - P)X$, and $P \in \mathcal{L}(X)$ is the projection operator given by

$$P = \frac{1}{2\pi i} \int_C (\lambda I - A)^{-1} d\lambda$$

where C is a closed curve surrounding $\sigma_s(A)$.

The system (9.15) may be decomposed into the two subsystems

$$\begin{cases} \frac{\partial y_u(t)}{\partial t} &= A_u y_u(t) + P B v(t) \\ y_{0u} &= P y_0 \\ y_u &= P y \end{cases} \tag{9.18}$$

and

$$\begin{cases} \frac{\partial y_s(t)}{\partial t} &= A_s y_s(t) + (I - P) B v(t) \\ y_{0s} &= (I - P) y_0 \\ y_s &= (I - P) y \end{cases} \tag{9.19}$$

where A_s and A_u are the restrictions of A to X_s and X_u, and are such that $\sigma_s(A) = \sigma(A_s)$, $\sigma_u(A) = \sigma(A_u)$ and A_u is a bounded operator on X_u. The solutions of (9.18) and (9.19) are given by

$$y_u(t) = S_u(t) y_{0u} + \int_0^t S_u(t - \tau) P B v(\tau) d\tau \tag{9.20}$$

and

$$y_s(t) = S_s(t) y_{0s} + \int_0^t S_s(t - \tau)(I - P) B v(\tau) d\tau \tag{9.21}$$

where $S_u(t)$ and $S_s(t)$ denote the restrictions of $S(t)$ to X_u and X_s, which are strongly continuous semigroups generated by A_u and A_s.

For the system state, it is known (see [1]) that if the operator A_s satisfies the spectrum growth assumption

$$\lim_{t \to +\infty} \frac{\ln \|S_s(t)\|}{t} = supRe(\sigma(A_s)) \tag{9.22}$$

then stabilizing (9.15) comes back to stabilize (9.18).

The following proposition gives an extension of this result to the gradient case.

Proposition 9.4 *Let the state space satisfy the decomposition (9.17) and A_s satisfy the following inequality*

$$\lim_{t \to +\infty} \frac{\ln |||\nabla S_s(t)|||}{t} \leq supRe(\sigma(A_s)). \tag{9.23}$$

(1) If the system (9.18) is gradient exponentially (respectively strongly) stabilizable by a feedback control $\bar{u} = K_u G y_u$, with $K_u \in \mathcal{L}(X, U)$, then the system (9.15)

is gradient exponentially (respectively strongly) stabilizable using the control $v = (\bar{u}, 0)$.

(2) *If the system (9.18) is gradient exponentially (resp strongly) stabilizable by the feedback control:*

$$v = K_u y_u, \text{ with } K_u \in \mathcal{L}(X, U)$$

then the system (9.15) is gradient exponentially (respectively strongly) stabilizable using the feedback operator $K = (K_u, 0)$.

Proof We give the proof for the exponential case. In view of the above decomposition, we have $supRe(\sigma(A_s)) \leq \delta$. Hence if A_s satisfies (9.23) then for some M_1 and $\beta \in]0, -\delta[$, we have:

$$|||\nabla S_s(t)||| \leq M_1 e^{-\beta t}, \ t \geq 0. \tag{9.24}$$

It follows that the system (9.19) is gradient exponentially stabilizable taking v(t) = 0.

Let K_u be such that $y_u(t) = e^{F_u t} y_{0u}$, with $F_u = A_u + PBK_u G \in \mathcal{L}(y_u)$ and there exists $\alpha > 0$, $M_2 > 0$ such that $\|\nabla y_u(t)\|_m < M_2 e^{-\alpha t} \|y_{0u}\|$. Then with the feedback control $v = K_u G y_u$ we have $\|v(t)\| \leq M_3 \|K_u\| e^{-\alpha t} \|y_{0u}\|$, with $M_3 > 0$. From (9.21) and (9.24) we have

$$\begin{aligned}
\|\nabla y_s(t)\|_m &\leq M_1 e^{-\beta t} \|y_{0s}\| + M_3 \|y_{0u}\| \int_0^t e^{-\beta(t-s)} e^{-\alpha s} ds \\
&\leq M_1 e^{-\beta t} \|y_{0s}\| + M_4 \|y_{0u}\| \frac{e^{-\beta t} - e^{-\alpha t}}{\alpha - \beta}
\end{aligned}$$

with $M_4 > 0$. Thus the system (9.15) excited by $v(t) = Ky(t)$ satisfies

$$\|\nabla y(t)\|_m \leq (M_1 e^{-\beta t} + M_4 \frac{e^{-\beta t} - e^{-\alpha t}}{\alpha - \beta} + M_2 e^{-\alpha t}) \|y_0\|$$

which shows that the system (9.15) is gradient exponentially stabilizable.

(2) The case of strong stabilization follows from similar above techniques. □

Corollary 9.2 *Let A satisfy the spectrum decomposition assumption (9.17) and suppose that (9.23) is satisfied. If in addition*

1. X_u *is a finite dimensional space*
2. *The system (9.18) is controllable on* X_u

then the system (9.15) is gradient exponentially stabilizable.

Proof The system (9.18) is of finite dimension and is controllable on the space X_u then it is stabilizable on the same space, hence it is gradient stabilizable, the conclusion is obtained with the Proposition 9.4. □

9.3.2 Riccati Method

Let us consider the system (9.15) with the same assumptions on A and B. We denote by $S_K(t), t \geq 0$ the strongly continuous semigroup generated by $A + BK$, where K is the feedback operator $K \in \mathcal{L}(X, U)$.

Let $R \in \mathcal{L}(X)$ be a self-adjoint operator such that (9.11) is satisfied and suppose that the steady state Riccati equation

$$< Ay, Py > + < Py, Ay > - < B^*Py, B^*Py > + < Ry, y >= 0, \quad y \in \mathcal{D}(A) \tag{9.25}$$

has a self-adjoint positive solution $P \in \mathcal{L}(X)$, and let $K = -B^*P$.

Proposition 9.5 *1. If $S_K(t)$ satisfies the conditions (9.7) and (9.8), then the system (9.15) is gradient exponentially stabilizable by the control $u^*(t) = Ky(t)$.*

2. If $< G(A + BK)y, y > \leq 0, \quad y \in \mathcal{D}(A)$ then the system (9.15) is gradient strongly stabilizable.

3. Suppose that the system (9.15) is gradient exponentially stabilizable. If in addition the feedback operator K satisfies

$$< Gy, y >\geq cRe < (A + BK)y, y >, y \in \mathcal{D}(A), \text{ for some } c > 0 \tag{9.26}$$

then the state of the system (9.16) remains bounded.

Proof The first and second points are deduced from the second section.

For the thirst point, let $y_0 \in \mathcal{D}(A)$, we have

$$Re(\langle (A + BK)y(t), y(t)\rangle) = \frac{1}{2}\frac{\partial}{\partial t}\|y(t)\|^2$$

and from (9.26) we obtain $\displaystyle\int_0^t \|\nabla y(s)\|_m^2 ds \geq \frac{c}{2}(\|y(t)\|^2 - \|y_0\|^2)$.

Since the system is gradient exponentially stabilizable then $\displaystyle\int_0^{+\infty} \|\nabla y(t)\|_m^2 dt < +\infty$, so there exists $M > 0$ such that $\|y(t)\| \leq M$, for all $t \geq 0$ and by the density of $\mathcal{D}(A)$ in X we have the conclusion. □

9.4 Gradient Stabilization Control Problem

Here we explore the control that stabilizes the gradient of the system (9.15) as a solution of the minimization problem

$$\begin{cases} min\, q(u) \\ u \in \mathcal{U}_{ad} \end{cases} \tag{9.27}$$

where

$$q(u) = \int_0^{+\infty} \langle Ry(t), y(t)\rangle dt + \int_0^{+\infty} \|u(t)\|^2 dt \tag{9.28}$$

with

$$\mathcal{U}_{ad} = \{u \in L^2(0, +\infty; U); q(u) < +\infty\}$$

and R is a linear bounded operator mapping X into itself and satisfying (9.11).

We recall the classical result known for state stabilization, if $\mathcal{U}_{ad} \neq \emptyset$ for each initial state y_0, then there exists a unique control u^* that minimizes (9.28) and given by $u^*(t) = -B^*Py(t)$ where P is a positive operator, solution of the steady state Riccati equation (9.25).

If in addition the operator R is coercive then the state of system (9.15) is exponentially stabilizable (see [2]).

In the following we give an extension of the above result to the gradient case. We suppose that $\mathcal{U}_{ad} \neq \emptyset$ for each initial state $y_0 \in X$, and R satisfies (9.11).

Proposition 9.6 *The control given by $u^*(t) = -B^*Py^*(t)$ minimizes $q(u)$ where P assumed to be a self-adjoint, positive operator, and satisfies the steady state Riccati equation (9.25), if in addition the semigroup $S_K(t)$ satisfies the conditions (9.7) and (9.8) then the same control stabilizes the gradient of system (9.15)*

Proof The proof follows from [2], and the Proposition 9.5. □

9.5 Numerical Algorithm and Simulations

In this section we present an algorithm which allows the calculation of the solution of problem (9.27) stabilizing the gradient of the system (9.15).By the previous result this control may be obtained by solving the algebraic Riccati equation (9.25).

Let $X_n = span\{e_i, \ i = 1, 2, \dots, n\}$ where $\{e_i, \ i \geq 1\}$ is a hilbertian basis of X. X_n is a subspace of X endowed with the restriction of the inner product of X. The projection operator $\Pi_n : X \to X_n$ is defined by

$$\Pi_n(y) = \sum_{i=1}^{n} \langle e_i, y\rangle e_i \qquad \forall y \in X \tag{9.29}$$

The projection of (9.25) on space X_n, is given formally by:

$$P_n A_n + A_n P_n - P_n B_n B_n^* P_n + R_n = 0 \tag{9.30}$$

where A_n, P_n and R_n are respectively the projections of A, P and R on X_n, and B_n the projection of B which is mapping U the space of controls into X_n.

We have $\lim_{n\to\infty} \|P_n\Pi_n y - Py\| = 0$, that is $P_n\Pi_n$ converges to P strongly in X, (see [3]).

We can write the projection of (9.15) like

$$\frac{\partial y_n(t)}{\partial t} = A_n y_n(t) - B_n B_n^* P_n y_n(t) \;\; ; y_n(0). \tag{9.31}$$

The solution of this system is given explicitly by:

$$y_n(t) = e^{(A_n - B_n B_n^* P_n)t} y_n(0) \tag{9.32}$$

To calculate the matrix exponential we use the Padé approximation with scaling and squaring (see [4]).

If we denote $\tilde{y}_n^i(t) = \langle y_n(t), e_i \rangle$, we have

$$\nabla y_n(t) = \sum_{i=1}^{n} \tilde{y}_n^i(t) \nabla e_i \tag{9.33}$$

Let consider a time sequence $t_i = i\delta$, $i \in N$ where $\delta > 0$ small enough.

With these notations, the gradient stabilization control may be obtained following the algorithm steps.

Remark 9.3 The crucial steps of the algorithm are the resolution of Riccati equation and system (9.31) using formula (9.32) where one have to be careful with respect to problem data.

Example 9.3 Let $\Omega =]0, 1[$, on the space $X = \{y \in H^1(\Omega)$ such that $\frac{dy}{dx}(0) = \frac{dy}{dx}(1) = 0\}$ we consider the following system

$$\begin{cases} \dfrac{\partial y(t)}{\partial t} & = Ay(t) + \chi_D u(t) \;\; \Omega \times]0, +\infty[\\ \dfrac{\partial y(0,t)}{\partial x} & = \dfrac{\partial y(1,t)}{\partial x} = 0 \quad \forall t \geq 0 \\ y(x, 0) & = \frac{1}{6} x^2 (3 - 2x) \quad \Omega \end{cases} \tag{9.34}$$

where $Ay = 0.02\Delta y + 0.5y$, $u(t) \in X \; \forall t \geq 0$, χ_D is the restriction operator on $D =]0.2, 0.9[$, and we consider the problem (9.27) with $R = \nabla^* \nabla$.

A generates a strongly continuous semigroup $S(t)$ given by:

$$S(t)y = \sum_{i \geq 0}^{+\infty} e^{\lambda_i t} \langle y, \phi_i \rangle \phi_i.$$

where $\lambda_i = -0.01(i\pi)^2 + 0.5$ and $\phi_i(x) = \alpha_i \cos(i\pi x)$ with $\alpha_i = \sqrt{\frac{2}{1+(i\pi)^2}}$.

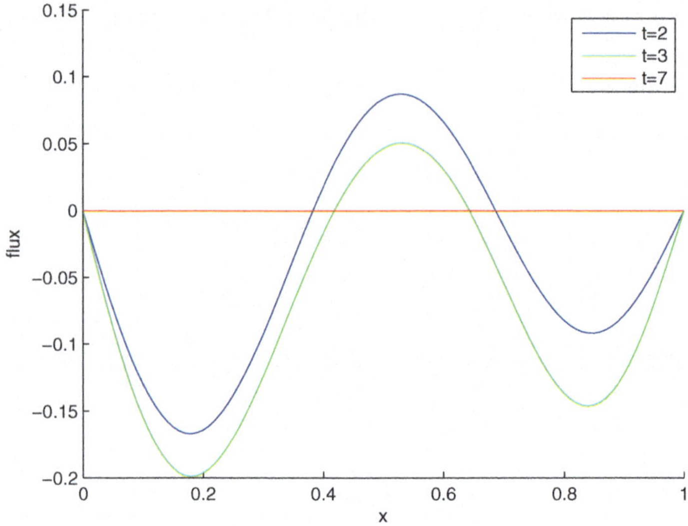

Fig. 9.1 The gradient evolution for the Neumann boundary condition case

The state and the gradient of system (9.34) are unstable since $\lambda_0 > 0$ and $\lambda_1 > 0$. Let consider the subspace

$$X_n = Span\{\alpha_{i-1} \cos((i-1)\pi x), 1 \leq i \leq n, x \in \Omega\}$$

Applying the algorithm taking the truncation at $n = 5$, we obtain Fig. 9.1 which illustrates the evolution of the system gradient and shows how the gradient evolves close to zero when the time t increases.

The gradient is stabilized with error equals 9.9836×10^{-7} and cost equals 2.6982×10^{-4}. This shows the efficiency of the developed algorithm.

In Table 9.1 we give the cost of gradient stabilization of system (9.34) for different supports control D. The Table 9.1 shows that there is relation between area of control support and the cost of gradient stabilization, more precisely more this area decreases more cost increases.

Algorithm 1. Let $\varepsilon > 0$ the threshold, a time sequence $t_i = i\delta,\ i \in \mathbb{N}$ where $\delta > 0$ small enough, n the dimension of the projection space.
2. Solve (9.30) using Schur-type methods (see [5])

Table 9.1 Support control-cost stabilization

D	]0, 0.1[	]0, 0.3[	]0, 0.5[	]0, 0.7[	]0, 0.9[	]0, 1[
Cost	9.0097	1.921	0.9408	0.817	0.2868	0.1636

Fig. 9.2 The gradient evolution for Dirichlet boundary condition

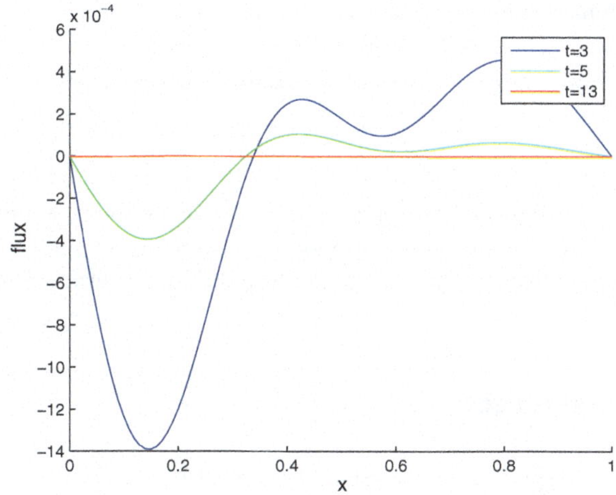

3. Solve system (9.31) by formula (9.32) using the Padé approximation with scaling and squaring (see [4]).
4. Calculate $\nabla y_n(t_i)$ by the formula (9.33).
5. If $\| \nabla y_n(t_i) \| < \epsilon$ stop, else
6. i=i+1 and go to 3

Example 9.4 Let $\Omega =]0, 1[$, on $X = H_0^1(\Omega)$ we consider the system (9.34) with Dirichlet boundary conditions:

$$\begin{cases} \dfrac{\partial y(t)}{\partial t} = Ay(t) + \chi_D u(t) & \Omega \times]0, +\infty[\\ y(0, t) = y(1, t) = 0 & \forall t \geq 0 \\ y(x, 0) = (1 - x)^2 x^2 & x \in \Omega \end{cases} \tag{9.35}$$

where $Ay = 0.01\Delta y + 0.5y$, $u(t) \in X \; \forall t \geq 0$, $D =]0, 0.3[$, and we consider the problem (9.27) with $R = \nabla^* \nabla$.

The eigenpairs of A are given by $(\lambda_i, \alpha_i \sin(i\pi x))$, $i \geq 1$, with $\lambda_i = -0.01(i\pi)^2 + 0.5$ and $\alpha_i = \sqrt{\frac{2}{1+(i\pi)^2}}$.

The state and the gradient of system (9.35) are unstable since $\lambda_1 > 0$.

We consider the subspace $X_n = \{\alpha_i \sin(i\pi x), i = 1, \ldots, n\}$ with $\alpha_i = \sqrt{\dfrac{2}{1 + (i\pi)^2}}$.

Applying the previous algorithm with truncation (n = 5), the Fig. 9.2 shows the gradient evolution at times $t = 3, 5$, and $t = 13$.

The gradient is stabilized with error equals 9.7961×10^{-7} and cost equals 3.2689×10^{-4}.

Table 9.2 Support control-cost stabilization

D	]0, 0.1[	]0, 0.3[	]0, 0.5[	]0, 0.7[	]0, 0.9[	]0, 1[
Cost	0.0247	0.0039	0.0016	4.7255×10^{-4}	2.6982×10^{-4}	2.6081×10^{-4}

In Table 9.2 we present the cost of gradient stabilization of system (9.34) for different zone control support D Also in this example, we remark that more the area of control support increases more the cost of gradient stabilization decreases.

References

1. Triggiani, R.: On the stabilizability problem in Banach space. J. Math. Anal. Appl. **52**, 383–403 (1975)
2. Curtain, R.F., Zwart, H.J.: An Introduction to Infinite Dimensional Linear Systems Theory. Springer, New York (1995)
3. Banks, H., Kunisch, B.: The linear regulator problem for parabolic systems. SIAM J. Control Optim. **22**, 684–696 (1984)
4. Higham, N.J.: The scaling and squaring method for the matrix exponential. SIAM Rev. **51**(4), 747–767 (2009)
5. Arnold, W.F., Laub, A.J.: Generalized eigenproblem: algorithms and software for algebraic Riccati equations. Proc. IEEE **72**, 1746–1754 (1984)

Chapter 10
Regional Gradient Stabilization of Infinite Dimensional Linear Systems

This chapter deals with regional gradient stabilization of infinite dimensional linear system evolving on a spacial domain Ω. Then we study different degrees of the gradient stabilization on $\omega \subset \Omega$ and moreover we characterize the stabilizing control which minimizes a quadratic functional cost. The developed approaches are illustrated by examples and numerical simulations.

10.1 Regional Gradient Stability

10.1.1 Definitions

Let Ω be an open regular subset of $\mathbb{R}^n$ and $Q = \Omega\times]0, \infty[$, we consider a system governed by the following state equation

$$\begin{cases} \dfrac{\partial y}{\partial t} = Ay & \text{on } Q \\ \\ y(0) = y_0 \in X \end{cases} \tag{10.1}$$

where $A : \mathcal{D}(A) \subset X \longrightarrow X$ a linear operator that generates a C_0-semigroup $S(t)$, $t \geq 0$, on a Hilbert state space X that is continuously injected in $H^1(\Omega)$. X is endowed with the induced inner product of $H^1(\Omega)$ noted $\langle ., . \rangle$ and the corresponding norm $\|.\|$.

E. H. Zerrik and O. Castillo, *Stabilization of Infinite Dimensional Systems*,
Studies in Systems, Decision and Control 355,
https://doi.org/10.1007/978-3-030-68600-0_10

We define the gradient operator:

$$\begin{aligned} \nabla : X &\rightarrow (L^2(\Omega))^n \\ y &\mapsto (\frac{\partial y}{\partial x_1}, \frac{\partial y}{\partial x_2}, \cdots, \frac{\partial y}{\partial x_n}) \end{aligned} \tag{10.2}$$

$(L^2(\Omega))^n$ provided with the usual inner product

$$<,>= \sum_{i=1}^{n} \int_{\Omega} f_i(x)\overline{g_i}(x)dx$$

and the associated norm $\|.\|$; with $f = (f_1, f_2, \cdots, f_n)$ and $g = (g_1, g_2, \cdots, g_n)$; $f_i, g_i \in L^2(\Omega)$.

Let ω be a subregion of Ω, we define the operator ∇_ω by

$$\nabla_\omega = (\chi_\omega \frac{\partial .}{\partial x_1}, \chi_\omega \frac{\partial .}{\partial x_2}, \cdots, \chi_\omega \frac{\partial .}{\partial x_n})$$

Let ∇_ω^* be it's adjoint operator, and consider the operator $G_\omega = \nabla_\omega^* \nabla_\omega$.

Definition 10.1 The system (10.1) is said regionally
1. weakly gradient stable (R.G-stable) on ω if, for all $y_0 \in X$, we have

$$< \nabla_\omega S(t)y_0, z > \longrightarrow 0 \text{ when } t \longrightarrow \infty, \quad z \in (L^2(\omega))^n \tag{10.3}$$

2. strongly gradient stable on ω if, for all $y_0 \in X$, we have

$$||\nabla_\omega S(t)y_0|| \longrightarrow 0 \text{ when } t \longrightarrow \infty \tag{10.4}$$

3. exponentially gradient stable on ω if, for all $y_0 \in X$, there exist M and $\alpha > 0$ such that

$$\|\nabla_\omega S(t)y_0\| \le Me^{-\alpha t}\|y_0\|, \quad t \ge 0, \ \forall y_0 \in X \tag{10.5}$$

Remark 10.1 1. In the case where $\omega = \Omega$, the system (10.1) is said weakly (respectively strongly, exponentially) G-stable.
2. If the system (10.1) is stable then it is G-stable.
3. If the system (10.1) is G-stable on ω, then it is G-stable on any subregion $\omega' \subset \omega$.
4. The notion of regional stability of the gradient is more general that the regional state stability. Indeed there exist unstable systems that are gradient stable on some subregion ω.

Example 10.1 On $\Omega =]0, 1[\times]0, 1[$, we consider the system

$$\begin{cases} \frac{\partial y}{\partial t}(x_1, x_2, t) = Ay(x_1, x_2, t) & \Omega\times]0, +\infty[\\ \frac{\partial y}{\partial \nu}(\zeta, \xi, t) = 0 & \partial\Omega\times]0, +\infty[\\ y(., 0) = y_0 \in H^1(\Omega) \end{cases} \tag{10.6}$$

where $Ay = \Delta y + 6\pi^2\langle y, \varphi_{20}\rangle\varphi_{20}$. The operator A has an orthogonal basis (φ_{ij}) of eigenfunctions and (λ_{ij}) the associated eigenvalues. We have

$$\varphi_{ij}(x_1, x_2) = 2a_{ij}\cos(i\pi x_1)\cos(j\pi x_2)$$

where $a_{ij} = (1 - \lambda_{ij})^{-\frac{1}{2}}$. The eigenvalues are given by

$$\lambda_{ij} = \begin{cases} -(i^2 + j^2)\pi^2 & (i, j) \neq (2, 0) \\ 2\pi^2 & (i, j) = (2, 0) \end{cases}$$

A generates a C_0-semigroup $(S(t))_{t\geq 0}$ on $H^1(\Omega)$ given by

$$S(t)y = \sum_{(i,j)}^{\infty} e^{\lambda_{ij}t}\langle y, \varphi_{ij}\rangle\varphi_{ij}$$

The state and the gradient of system (10.6) are unstable on $\omega = \{\frac{1}{2}\} \times [0, 1]$. But $\|\nabla_\omega\varphi_{20}\| = 0$ then $\nabla_\omega y(t) \to 0$ exponentially when $t \to \infty$ and consequently the system (10.6) is exponentially G-stable on ω.

Example 10.2 On $\Omega =]0, 1[$ we consider the system described by the equation

$$\begin{cases} \frac{\partial y}{\partial t}(t) = Ay(t) & \Omega\times]0, +\infty[\\ \frac{\partial y}{\partial t}(0, t) = \frac{\partial y}{\partial t}(1, t) = 0 &]0, +\infty[\\ y(., 0) = y_0 \in H^1(\Omega) \end{cases} \tag{10.7}$$

The operator $Ay = \Delta y + y$ generates a C_0-semigroup given by

$$S(t)y_0 = \sum_{i\geq 0} e^{\lambda_i t}\langle y_0, \phi_i\rangle\phi_i$$

where

$$\phi_i(x) = \sqrt{\frac{2}{1 + (i\pi)^2}}\cos(i\pi x) \quad , \quad \lambda_i = 1 - (i\pi)^2$$

Since $\lambda_0 > 0$, the system (10.7) is unstable. However it is exponentially G -stable as

$$\begin{aligned}\|\nabla S(t)y_0\|^2 &\leq e^{2\lambda_1 t}\sum_{i>0}\langle y_0,\phi_i\rangle^2\|\nabla\phi_i\|^2\\ &\leq e^{2\lambda_1 t}\|y_0\|^2\end{aligned}$$

Consequently (10.7) is exponentially G-stable on Ω.

10.1.2 Characterizations

We have seen that the exponential stability is obtained under the condition (2.21), the following result gives an extension to the gradient stability.

Proposition 10.1 *Suppose that there exists a function $M_\omega \in L^2(0,+\infty;\mathbb{R}^+)$ such that, for all $y \in X$, we have*

$$\|\nabla_\omega S(t+s)y\| \leq M_\omega(t)\|\nabla_\omega S(s)y\| \quad , \quad t,s \geq 0 \tag{10.8}$$

and

$$\|\nabla_\omega S(mt)y\| \leq \|\nabla_\omega S(t)\|^m \;\forall t \geq 0 \quad , \quad m \in \mathbb{N}^* \tag{10.9}$$

Then the system (10.1) is exponentially G-stable on ω if and only if

$$\int_0^\infty ||\nabla_\omega S(t)y||^2 dt < \infty \quad , \quad y \in X \tag{10.10}$$

Proof 1. We show that the sequence of operators $(\nabla_\omega S(t))_{t\geq 0}$ is uniformly bounded. Indeed, we have $\sup_{t\geq 0}\|\nabla_\omega S(t)\| < +\infty$, otherwise there exists a sequence $(t_1+\tau_k)_k$, $t_1 > 0$ and $\tau_k \to \infty$ when $k \to +\infty$, such that $\|\nabla_\omega S(t_1+\tau_k)\| \to +\infty$ when $k \to \infty$. So

$$\int_0^\infty \|\nabla_\omega S(s+\tau_k)y\|^2 ds = \int_{\tau_k}^\infty \|\nabla S(s)y\|^2 ds$$

As the right term tends to 0 when $k \to \infty$. Fatou lemma imply that lim inf $\|\nabla_\omega S(s+\tau_k)y\| = 0$ when $k \to \infty$, almost for all $s > 0$.

Consequently for $s_0 < t_1$ there exists a subsequence $\{\tau_{k_m}\}$ such that $\lim_{m\to\infty}\|\nabla_\omega S(s_0+\tau_{k_m})y\| = 0$. But with (10.8) and when $m \to +\infty$, we have

$$\|\nabla_\omega S(t_1+\tau_{k_m})y\| \leq M(t_1-s_0)\|\nabla S(s_0+\tau_{k_m})y\| \to 0$$

which is impossible. Using Banach-Steinhaus theorem we deduce that $(\nabla_\omega S(t))_{t\geq 0}$ is uniformly bounded.

2. We show that

$$\sigma_0' = \inf_{t>0}\frac{1}{t}\ln|||\nabla_\omega S(t)|||_{\mathcal{L}(X,(L^2(\Omega))^n)} = \lim_{t\to+\infty}\frac{1}{t}\ln|||\nabla_\omega S(t)|||_{\mathcal{L}(X,(L^2(\Omega))^n)}$$

Indeed, we have

$$\begin{aligned} t\|\nabla_\omega S(t)y\|^2 &= \int_0^t \|\nabla_\omega S(t)y\|^2 ds \\ &\leq \int_0^t \|\nabla_\omega S(s+t-s)y\|^2 ds \end{aligned}$$

By the condition (10.8) and as the operator $\nabla_\omega S(t)$ is uniformly bounded, there exists a constant $N > 0$ such that

$$\begin{aligned} t\|\nabla_\omega S(t)y\|^2 &\leq \int_0^t M^2(s)\|\nabla_\omega S(t-s)y\|^2 ds \\ &\leq N\|y\|^2 \end{aligned}$$

So there exists $t_0 > 0$ such that, for all $t \geq t_0$,

$$\ln |||\nabla_\omega S(t)|||_{\mathcal{L}(X,(L^2(\Omega))^n)} < 0$$

We deduce that $\sigma_0' < 0$.

It remains to show that $\sigma_0' = \lim_{t\to+\infty} \frac{1}{t} \ln |||\nabla_\omega S(t)|||_{\mathcal{L}(X,(L^2(\Omega))^n)}$.

Let $t_1 > 0$ and $N' = \sup_{t\in[0,t_1]} |||S(t)|||_{\mathcal{L}(X)}$, there exists $p \in \mathbb{N}$ such that $pt_1 \leq t < (p+1)t_1$, for all $t \geq t_1$. Then

$$\frac{1}{t} \ln |||\nabla_\omega S(t)||| \leq \frac{1}{t} \ln |||\nabla_\omega S(pt_1)||| + \frac{1}{t} \ln |||S(t-pt_1)|||$$

and with (10.9) we obtain

$$\frac{1}{t} \ln |||\nabla_\omega S(t)||| \leq (\frac{pt_1}{t}) \frac{1}{t_1} \ln |||\nabla_\omega S(t_1)||| + \frac{\ln N'}{t}$$

which gives

$$\begin{aligned} \limsup_{t\to\infty} \frac{1}{t} \ln |||\nabla_\omega S(t)||| &\leq \inf_{t>0} \frac{1}{t} \ln |||\nabla_\omega S(t)||| \\ &\leq \liminf_{t\to\infty} \frac{1}{t} \ln |||\nabla_\omega S(t)||| \end{aligned}$$

Finally, this gives $\sigma_0' = \lim_{t\to+\infty} \frac{1}{t} \ln |||\nabla_\omega S(t)|||_{\mathcal{L}(X,(L^2(\Omega))^n)}$. So for all $\sigma \in]0, -\sigma_0'[$, there exists M' such that

$$||\nabla_\omega S(t)y||_n \leq M' e^{-\sigma t}\|y\| \quad , \ y \in X, \ t \geq 0$$

Consequently the system is exponentially G-stable on ω. The converse is immediate. □

Corollary 10.1 *Let us assume that the conditions (10.8) and (10.9) hold and there exists $t_0 > 0$, such that $|||\nabla_\omega S(t_0)||| < 1$, then the system (10.1) is exponentially G-stable on ω.*

Proof According to the second step of the Proposition 10.1, we have

$$\sigma_0' = \lim_{t\to+\infty} \frac{1}{t} \ln |||\nabla_\omega S(t)||$$

this imply that $\sigma_0' < 0$. So (10.1) is exponentially G-stable on ω. □

The following result gives a sufficient condition for strong stability of the gradient on a subregion ω.

Proposition 10.2 *Let us assume that there exists an operator $P \in \mathcal{L}(X)$ positive and self adjoint auto-adjoint solution of the equation*

$$< Ay, Py > + < Py, Ay > + < Ry, y >= 0, \quad y \in \mathcal{D}(A) \tag{10.11}$$

where $R \in \mathcal{L}(X)$ is a self adjoint operator such that

$$\langle Ry, y\rangle \geq \alpha\|\nabla_\omega y\|^2 \; \forall y \in X, \textit{ for some real } \alpha > 0 \tag{10.12}$$

If, moreover, the condition

$$Re(< G_\omega Ay, y >) \leq 0, \quad y \in \mathcal{D}(A) \tag{10.13}$$

holds, then (10.1) is strongly G-stable on ω.

Proof We consider the function $V(y) = < Py, y >$, for all $y \in X$. For $y_0 \in \mathcal{D}(A)$, the relation

$$\begin{aligned}\frac{d}{dt}V(S(t)y_0) &= < PAS(t)y_0, S(t)y_0 > + < PS(t)y_0, AS(t)y_0 > \\ &= - < RS(t)y_0, S(t)y_0 >\end{aligned}$$

gives

$$\int_0^{+\infty} < RS(s)y_0, S(s)y_0 > ds \leq V(y_0)$$

By (10.12), we have $\int_0^{+\infty} ||\nabla_\omega S(s)y_0||^2 ds < \infty$. Using (10.13), we obtain $\frac{\partial}{\partial t}||\nabla_\omega S(t)y_0||^2 \leq 0$.

Thus

$$t||\nabla_{\omega}S(t)y_0||^2 = \int_0^t ||\nabla_{\omega}S(t)y_0||^2 ds \leq \int_0^t ||\nabla_{\omega}S(s)y_0||^2 ds$$

We deduce that for all $y_0 \in \mathcal{D}(A)$, we have

$$||\nabla_{\omega}S(t)y_0||^2 \leq \frac{f(y_0,t)}{t} \quad , \ t > 0 \tag{10.14}$$

with $f(y_0,t) = \int_0^t ||\nabla_{\omega}S(s)y_0||^2 ds$ which is continuous with respect to t.

It follows $\lim_{t\to+\infty} \|\nabla_{\omega}S(t)y_0\| = 0$.

Since $\mathcal{D}(A)$ is dense in X, and the function $y_0 \longrightarrow f(y_0,t)$ is continuous, the inequality (10.14) holds for all $y_0 \in X$. So (10.1) is strongly G-stable on ω. □

Corollary 10.2 *Let us assume there exists an operator $P \in \mathcal{L}(X)$ self adjoint solution of the equation*

$$< Ay, Py > + < Py, Ay > + < G_{\omega}y, y >= 0, \ \forall \ y \in \mathcal{D}(A)$$

If moreover

$$Re(< G_{\omega}Ay, y >) \leq 0 \ \forall \ y \in \mathcal{D}(A)$$

then the system (10.1) is strongly G-stable on ω.

The result deduced since G_{ω} verify the conditions of the above proposition with $(\alpha = 1)$.

The weak stability of system (10.1) is linked to the spectral properties of the dynamic of the system. Indeed, in finite dimension the system (10.1) is weakly stable if and only if all the poles of the system dynamic are located in the open left half plan. In infinite dimension we have the following result.

Proposition 10.3 *1. If the system (10.1) is weakly G-stable on ω then*

$$\forall\lambda \in \sigma(A), \ Re(\lambda) \geq 0 \Rightarrow \|\nabla_{\omega}\phi\| = 0 \ \ , \ \forall\phi \in Ker(A - \lambda I)$$

2. Let us suppose that X is an Hilbert space which has an orthogonal basis $(\phi_m)_m$, of eigenfunctions of the operator A.

If in addition $\exists\alpha > 0$ such that, $\forall\lambda \in \sigma(A)$, there exists $\phi \in Ker(A - \lambda I)$ such that

$$\|\nabla_{\omega}\phi\| \neq 0 \Rightarrow Re(\lambda) \leq -\alpha$$

then the system (10.1) is exponentially G-stable on ω.

Proof 1. We assume there exists $\lambda \in \sigma(A)$ such that $Re(\lambda) \geq 0$ and $\phi \in X$ satisfying

$$A\phi = \lambda\phi \quad \text{and} \quad \|\nabla_\omega \phi\| \neq 0$$

The corresponding solution of the system (10.1) to the initial state $y_0 = \phi$ is given by $y(t) = e^{\lambda t}\phi$, so:

$$\langle \nabla_\omega S(t)\phi, \nabla_\omega \phi \rangle = e^{Re(\lambda)t} \|\nabla_\omega \phi\|^2 \geq \|\nabla_\omega \phi\|^2 > 0$$

consequently the system (10.1) is not weakly G-stable on ω.

2. Let $y_0 \in X$, we have, $\nabla_\omega S(t) y_0 = \sum_{m \geq 0} e^{\lambda_n t} \sum_{k=1}^{r_m} \langle y_0, \phi_{m,k} \rangle \nabla_\omega \phi_{m,k}$ where r_m is the multiplicity of the eigenvalue λ_m, then

$$\|\nabla_\omega S(t) y_0\| \leq e^{-\alpha t} \|y_0\|$$

We conclude that the system (10.1) is exponentially G-stable on ω. □

As illustration of the second point of the above result, we consider the system (10.6) and $\omega = \{\frac{1}{2}\} \times [0, 1]$. We have, for all $i, j \in \mathbb{N}$, $\nabla_\omega \varphi_{ij} \neq 0 \Rightarrow Re(\lambda_{ij}) \leq -\pi^2$ and so the system (10.6) is exponentially G-stable on ω.

If the semigroup is compact, then the weak stability imply the exponential one. This is expressed by the following result.

Proposition 10.4 *We suppose that the semigroup $S(t)$ verify the conditions (10.8) and (10.9) and there exists t_0 such that $S(t_0)$ be compact. If the system (10.1) is weakly G-stable on ω then it is exponentially G-stable on ω.*

Proof We show that $\lim_{t \to \infty} \| \nabla_\omega S(t) \| = 0$.

Since $S(t_0)$ is compact, then $S(t)$ is compact for all $t > t_0$, and using ∇_ω is continuous then $\nabla_\omega S(t)$ is compact for all $t > t_0$.

If the system (10.1) is weakly G-stable then $\| \nabla_\omega S(t) \| \to 0$ when $t \to \infty$.

By Banach Steinhaus theorem there exists $M > 0$ such that $\| \nabla_\omega S(t) \| < M$.

Let $\varepsilon > 0$, as $S(t_0)$ is compact then there exist $z_1, z_2, \ldots, z_N$ in X such that

$$\{S(t_0)y, \| y \| \leq 1\} \subset \bigcup_{i=1,N} \mathcal{B}(z_i, \frac{\varepsilon}{M})$$

where $\mathcal{B}(z_i, \frac{\varepsilon}{M})$ is the ball of center 0 and radius $\dfrac{\varepsilon}{M}$.

Let t_ε such that $\| \nabla_\omega S(t) z_i \| \leq \varepsilon$ for all $t > t_\varepsilon$ and $i = 1, \ldots, N$.

For $t > t_\varepsilon + t_0$ and $\| y \| \leq 1$, we have

$$\begin{aligned}\| \nabla_\omega S(t)y \| &= \| \nabla_\omega S(t-t_0)S(t_0)y - \nabla_\omega S(t-t_0)z_i + \nabla_\omega S(t-t_0)z_i \| \\ &\leq \| \nabla_\omega S(t-t_0)(S(t_0)y - z_i) \| + \varepsilon \\ &\leq M\frac{\varepsilon}{M} + \varepsilon = 2\varepsilon\end{aligned}$$

So there exist $t_1 > 0$ such that $\| \nabla_\omega S(t_1) \| < 1$ and by the corollary (10.1) the system (10.1) is exponentially G-stable on ω. □

10.2 Gradient Stabilization

In the previous section we establish sufficient conditions and necessary and sufficient other ones for an infinite dimensional system be exponentially (resp. strongly, weakly) gradient stable on ω. The goal of this section is to characterize a control that stabilizes the gradient of the system (10.1) and minimizes a stabilisation cost.

10.2.1 Stabilizing Control

We consider the system described by the equation

$$\begin{cases} \dfrac{\partial y(t)}{\partial t} = Ay(t) + Bu(t) \text{ on } Q \\ y(.,0) = y_0 \in X \end{cases} \tag{10.15}$$

where A generates a C_0-semigroup $S(t)$, and $B \in \mathcal{L}(V, X)$ a linear continuous operator; V is the space of controls supposed to be Hilbert. The considered controls are of feedback type $u(t) = Ky(t)$ with $K \in \mathcal{L}(X, V)$.

The system (10.15) is written in the form

$$\begin{cases} \dfrac{\partial y(t)}{\partial t} = (A + BK)y(t) \text{ on } Q \\ y(.,0) = y_0 \in X \end{cases} \tag{10.16}$$

In this case the operator $A + BK$ generate a C_0-semigroup $T(t)$. We consider the following definitions.

Definition 10.2 1. The system (10.15) is said exponentially G-stabilizable on ω if there exists an operator $K \in \mathcal{L}(X, V)$, such that the system (10.16) be exponentially G-stable on ω. 2. If there exists an operator $K \in \mathcal{L}(X, V)$, such that the system (10.16) be strongly (resp. weakly) G-stable on ω, the system (10.15) is said strongly (resp. weakly) G-stabilizable on ω.

Remark 10.2 1. The regional stabilisation of the gradient is a general notion that extends the stabilisation of gradient.
2. The cost of the regional stabilisation of the gradient lower than the one of the stabilisation of the gradient.
3. The cost of the gradient stabilisation is lower than the one of the state stabilisation. Indeed, if we consider the cost function

$$q(u) = \int_0^{+\infty} \|u(t)\|^2 dt$$

and the sets

$$V_{ad} = \{u \in L^2(0, +\infty; V);\ \text{u stabilise the gradient on } \omega\}$$

$$V_{ad}^1 = \{u \in L^2(0, +\infty; V);\ \text{u stabilise the gradient on } \Omega\}$$

$$V_{ad}^2 = \{u \in L^2(0, +\infty; V);\ \text{u stabilise the state on } \omega\}$$

$$V_{ad}^3 = \{u \in L^2(0, +\infty; V);\ \text{u stabilise the state on } \Omega\}$$

Then we have $V_{ad}^3 \subset V_{ad}^2 \subset V_{ad}^1 \subset V_{ad}$. Consequently

$$\min_{u \in V_{ad}} q(u) \leq \min_{u \in V_{ad}^1} q(u) \text{ and } \min_{u \in V_{ad}^1} q(u) \leq \min_{u \in V_{ad}^2} q(u) \leq \min_{u \in V_{ad}^3} q(u)$$

In general, a characterization of a stabilizing control of the gradient on ω or on Ω may be obtained by two approaches, the first one leads to resolution of Riccati equation and the second one is based on the decomposition of the state space.

10.2.1.1 Riccati Approach

Let us recall that algebraic Riccati equation allows a control that stabilises exponentially the state of infinite dimensional system on ω. Here we give extension to the gradient stabilization .

We suppose that algebraic Riccati equation (10.11) has a solution $P \in \mathcal{L}(X)$ positive and self adjoint, if R verify the condition (10.12) then we have the following result.

Proposition 10.5 *If the semigroup $T(t)$ verify the conditions (10.8) and (10.9) then 1. The system (10.15) is exponentially G-stabilizable on ω. 2. If there exists $d > 0$ such that*

$$< G_\omega y, y > \geq dRe(< (A + BK)y, y >) \ , \ y \in \mathcal{D}(A) \tag{10.17}$$

then the state of the system (10.16) remains bounded on Ω.

Proof 1. Since P is solution of (10.11) and R verify (10.12) then

$$\int_0^{+\infty} \|\nabla_\omega T(t)y\|^2 dt < +\infty, \quad y \in X$$

and we apply the Theorem 10.1 to conclude.
2. The condition (10.17) gives

$$d(\|y(t)\|^2 - \|y_0\|^2) \le \int_0^t \|\nabla_\omega T(s)y_0\|^2 ds$$

As system (10.15) is exponentially G-stabilizable on ω, then the application

$$y_0 \longrightarrow \int_0^t \|T(s)y_0\|^2 ds$$

is uniformly bounded, this proves the state remains bounded on the whole evolution domain. □

In the regional case the next result concerns the behaviour of the gradient on the residual part $\Omega \setminus \omega$ of Ω.

Proposition 10.6 *If the semigroup $T(t)$ verify the conditions (10.8) and (10.9) then if there exists $d > 0$ such that:*

$$< G_\omega y, y > \ge d\mathcal{R}e < G_\Omega(A + BK)y, y > \quad , \ y \in \mathcal{D}(A) \tag{10.18}$$

then the gradient of the system (10.16) remains bounded on Ω.

Proof The semigroup $(T(t))$ verify (10.8) and (10.9), the previous proposition shows that the system (10.15) is exponentially G-stabilizable on the subregion ω.

The condition (10.18) leads to, for all $y_0 \in \mathcal{D}(A)$, we have

$$\|\nabla y(t)\|^2 \le \frac{1}{d} \int_0^t \|\nabla_\omega T(t)y_0\|^2 dt$$

Since ∇_ω and ∇ are continuous then the last inequality holds for all $y_0 \in X$. As, moreover, $\int_0^{+\infty} \|\nabla_\omega T(t)y_0\|^2 dt < +\infty$, we deduce the result. □

Remark 10.3 1. The condition (10.17) imply that the gradient of (10.16) remains bounded on Ω (since the state of (10.16) is bounded).
2. The condition (10.18) gives the same result as the condition (10.17), but without taking into account the behaviour of the state.
3. The condition (10.18) may be substituted by the following condition

$$< G_{\omega}y, y > \geq d'\frac{d}{dt} < (G_{\Omega\backslash\omega}y, y > \quad , \ y \in \mathcal{D}(A) \text{ with } d' > 0 \tag{10.19}$$

in view to limit the gradient of system (10.16) on the whole domain Ω.

10.2.2 State Space Decomposition

The aim of this paragraph is to extend the decomposition method used for the state stabilization to the gradient.

Let $\delta > 0$ and consider the subsets:

$$\sigma_u(A) = \{\lambda : Re(\lambda) \geq -\delta\} \quad \text{and} \quad \sigma_s(A) = \{\lambda : Re(\lambda) < -\delta\}$$

We suppose that $\sigma_u(A)$ is bounded and may be separated from $\sigma_s(A)$ by a simple closed curve, then the state space X may be decomposed in sum of two orthogonal subspaces $X = X_u \oplus X_s$ with $X_u = \Pi X, \ X_s = (Id - \Pi)X$, and $\Pi \in \mathcal{L}(X)$ is the projection operator given by

$$\Pi = \frac{1}{2\pi i}\int_C (\lambda Id - A)^{-1} d\lambda \tag{10.20}$$

where C is a curve surrounding $\sigma(A)$ [1].

Thus the system (10.15) may be decomposed into the following subsystems:

$$\begin{cases} \dfrac{dy_u(t)}{dt} &= A_u y_u(t) + \Pi B v(t) \\ y_{0u} &= \Pi y_0 \\ y_u &= \Pi y \end{cases} \tag{10.21}$$

and

$$\begin{cases} \dfrac{dy_s(t)}{dt} &= A_s y_s(t) + (Id - \Pi) B v(t) \\ y_{0s} &= (Id - \Pi) y_0 \\ y_s &= (Id - \Pi) y \end{cases} \tag{10.22}$$

where A_s and A_u are respectively the restrictions of A on X_s and X_u. $S_u(t)$ and $S_s(t)$ are the restrictions of $S(t)$ on X_u and X_s, which are respectively the C_0- semigroups generated by A_u and A_s.

We have the following result.

Proposition 10.7 *If the semigroup $S_s(t)$ verify the inequality*

$$\lim_{t \to +\infty} \frac{1}{t} \ln \|\nabla_{\omega} S_s(t)\| \leq supRe(\sigma(A_s)) \tag{10.23}$$

and that the system (10.21) is exponentially (resp. strongly) G-stabilizable on ω by the control $v = K_u \nabla_\omega y_u$, with $K_u \in \mathcal{L}((L^2(\omega))^n, V)$, then the control $\widetilde{v} = (v, 0)$, exponentially (resp. strongly) G-stabilizes the system (10.15) on ω.

Proof We show strong G-stabilization on ω. Indeed the previous decomposition gives sup $Re(\sigma(A_s)) \leq -\delta$, there exist $M_1 > 0$ and $\beta \in]0, \delta[$, such that

$$\|\nabla_\omega S_s(t)\| \leq M_1 e^{-\beta t}, \ t \geq 0 \tag{10.24}$$

Without control ($v = 0$), the system (10.22) is exponentially G-stabilisable on ω. If the system (10.21) is strongly G-stabilizable on ω, this means that

$$\lim_{t \to +\infty} \|\nabla_\omega y_u(t)\| = 0 \text{ on } \omega \tag{10.25}$$

So for all $\epsilon > 0$ there exists $t_0 > 0$ such that if $t > t_0$, we have $\|v(t)\| < \dfrac{\epsilon\beta}{2}$. The solution of (10.22) is given by:

$$y_s(t) = S_s(t) y_{0s} + \int_0^t S_s(t - \tau)(Id - \Pi) B v(\tau) d\tau \tag{10.26}$$

Using (10.24), there exists $M_1 > 0$ such that

$$\|\nabla_\omega y_s(t)\|_n \leq M_1 e^{-\beta t} \|y_{0s}\| + M_1 |||(Id - \Pi)B||| \int_0^t e^{-\beta(t-\tau)} \|v(\tau)\| d\tau$$

It remains to show that $\lim_{t \to +\infty} \|\nabla_\omega y_s(t)\| = 0$.

We have

$$\begin{aligned}
\int_0^t e^{-\beta(t-\tau)} \|v(\tau)\| d\tau &= \int_0^{t_0} e^{-\beta(t-\tau)} \|v(\tau)\| d\tau + \int_{t_0}^t e^{-\beta(t-\tau)} \|v(\tau)\| d\tau \\
&\leq \gamma e^{-\beta(t-t_0)} + \frac{\epsilon}{2}
\end{aligned}$$

where $\gamma = \dfrac{1}{\sqrt{2\beta}} \left(\displaystyle\int_0^{t_0} \|v(t)\|^2 dt \right)^{\frac{1}{2}}$.

As

$$\lim_{t \longrightarrow +\infty} \int_0^t e^{-\beta(t-\tau)} \|v(\tau)\| d\tau = 0$$

then $\lim_{t \to +\infty} \|\nabla_\omega y_s(t)\| = 0$. Finally if we consider the system (10.15) controlled by $\widetilde{v}(t)$ we obtain

$$\|\nabla_\omega y(t)\| \leq \|\nabla_\omega y_u(t)\| + \|\nabla_\omega y_s(t)\|$$

and so the system (10.15) is strongly G-stabilisable on ω. □

10.2.3 Lower Cost Gradient Stabilization Problem

In this paragraph we characterize a control that stabilises the gradient of the system (10.15), solution of the problem

$$\begin{cases} \min q(u) \\ u \in V_{ad} \end{cases} \tag{10.27}$$

where

$$q(u) = \int_0^{+\infty} \langle Ry(t), y(t) \rangle dt + \int_0^{+\infty} \|u(t)\|^2 dt \tag{10.28}$$

with

$$V_{ad} = \{u \in L^2(0, +\infty; V); q(u) < +\infty\}$$

and $R : X \to X$ is a bounded linear operator which verify (10.12). We suppose that (10.11) has a solution P, a linear positive and self adjoint operator, and $\forall y_0 \in X,\ V_{ad} \neq \emptyset$, then we have the following result.

Proposition 10.8 *If the semigroup $T(t)$ satisfy the conditions (10.8) and (10.9), then the control*

$$u^*(t) = -B^* P y^*(t) \tag{10.29}$$

is solution of the problem (10.27) and exponentially stabilises the gradient of system (10.15) on ω.

Proof 1. The control (10.29) stabilizes the gradient of system (10.15) on ω. Indeed, since $T(t)$ verify (10.8) and (10.9), the system (10.15) is exponentially G-stabilisable on ω.
2. The control (10.29) is solution of problem (10.27).
Let $y_0 \in \mathcal{D}(A)$, since $q(u) < \infty$ then [2], the Riccati differential equation

$$\begin{aligned} \frac{d}{dt}\langle y, P(t)y \rangle &= \langle P(t)Ay, y \rangle + \langle y, P(t)Ay \rangle \\ &\quad - \langle B^* P(t)y, B^* P(t)y \rangle + \langle Ry, y \rangle \end{aligned} \tag{10.30}$$

admits a solution $P(t)$ with $P(0) = 0$. The sequence $P(m)$, $m > 0$, is uniformly bounded and for all $y_0 \in X$, $\lim_{m \longrightarrow +\infty} P(m)y_0 = P_\omega y_0$, where P_ω is a solution of equation (10.30).

Moreover, $0 \leq P(m) \leq P(m+1)$ and we conclude that [2] there exists a linear positive and self adjoint operator $P_\omega \in \mathcal{L}(X)$ such that for all $y_0 \in X$, $\lim_{m \longrightarrow +\infty} P(m)y_0 = P_\omega y_0$ and $P(m) \leq P_\omega$ for $m \geq 0$.

For all $t \geq 0$, let $m \in \mathbb{N}$ such that $m \leq t \leq m+1$, and according to [2], we have $P(m) \leq P(t) \leq P(m+1)$. Hence

$$\lim_{t \longrightarrow +\infty} P(t)y_0 = P_\omega y_0$$

Since $P_\omega(t)$ satisfy (10.30) for all $y \in \mathcal{D}(A)$, and passing to limit in equation (10.30) we obtain

$$\langle P_\omega Ay, y\rangle + \langle y, P_\omega Ay\rangle - \langle B^* P_\omega y, B^* P_\omega y\rangle + \langle Ry, y\rangle = 0$$

It remains to show that the control (10.29) is solution of problem (10.27). Let $u \in L^2(0, \infty, V)$, with the Theorem 6.1.13 in [2], we have

$$q(y_0, u, m) = \int_0^m \langle Ry(t), y(t)\rangle dt + \int_0^m \| u(t) \|^2 \, dt \leq q(y_0, u)$$

Thus

$$\begin{aligned} \inf_{u \in L^2([0,\infty[,V)} q(y_0, v) &\geq \inf_{u \in L^2([0,\infty[,V)} q(y_0, u, m) \\ &= \inf_{u \in L^2([0,m],V)} q(y_0, u, m) \\ &= \langle y_0, P(m) y_0\rangle \end{aligned}$$

When $m \to \infty$, we have

$$\inf_{u \in L^2([0,\infty[,V)} q(y_0, u) \geq \langle y_0, P(m) y_0\rangle \tag{10.31}$$

Since P_ω is positive, we have

$$\begin{aligned} q(y_0, u, m) \leq\ & \langle y_0, P_\omega y_0\rangle \\ &+ \int_0^m \langle [u(s) + B^* P_\omega y(s)], [u(s) + B^* P_\omega y(s)]\rangle ds \end{aligned}$$

Let $\tilde{y}(s) = T(s)y_0$ the solution of (10.15) controlled by $\tilde{u}(s) = B^* P_\omega T(s) y_0$ and with (10.31) we have $q(y_0, m, \tilde{v}) \leq \langle y_0, P_\omega y_0\rangle$.

We passe to limit when $m \to \infty$ we have

$$q(y_0, \tilde{u}) = \lim_{m \longrightarrow +\infty} q(y_0, m, \tilde{u}) \leq \langle y_0, P_\omega y_0\rangle \tag{10.32}$$

By (10.31) and (10.32), we have

$$\langle y_0, P_\omega y_0\rangle \leq \inf_{u \in L^2([0,\infty[,V)} q(y_0, u) \leq q(y_0, \tilde{u}) \leq \langle y_0, P(m) y_0\rangle$$

We deduce that

$$\inf_{u \in L^2([0,\infty[,V)} q(y_0, u) = \langle y_0, P(m) y_0 \rangle$$

Consequently the control given by (10.29) is solution of problem (10.27). □

10.2.4 Numerical Approach and Simulations

Here we present a numerical approach that allows the computation of the control u^* solution solution of the problem (10.27) and stabilizes the gradient of system (10.15) on ω.

We have seen that this control may be calculated by the resolution of algebraic Riccati equation (10.11).

Let $X_N = Vect\{e_i,\ i = 1, 2, \cdots, N\}$ where $\{e_i,\ i \geq 1\}$ a basis of the Hilbert space X. X_N is a subspace of X endowed with the induced inner product of X.

We consider the projection operator $\Pi_N : X \to X_N$ defined by:

$$\Pi_N(y) = \sum_{i=1}^{N} \langle e_i, y \rangle e_i \quad , \ y \in X \tag{10.33}$$

The projection of (10.11) on X_N, is formally given by

$$P_N A_N + A_N P_N - P_N B_N B_N^* P_N + R_N = 0 \tag{10.34}$$

where A_N, P_N and R_N are respectively the projections of the operators A, P and R on H_N, and B_N is the projection of $B : V \to X_N$.

So we have $\lim\limits_{N\to\infty} \|P_N \Pi_N y - P y\| = 0$, and hence $P_N \Pi_N$ strongly converges P when $N \to +\infty$ (see [3]).

The projection of system (10.15) is given by:

$$\frac{\partial y_N(t)}{\partial t} = A_{_N}\ y_N(t) - B_N B_N^* P_N\ y_N(t) \tag{10.35}$$

with

$$y_N(t) = e^{(A_N - B_N B_N^* P_N)t}\ y_N(0) \tag{10.36}$$

To calculate the exponential of a matrix, we use the Padé approximations (see [4]).

If we note $\tilde{y}_N^i(t) = \langle y_N(t), e_i \rangle$, we have

$$\nabla y_N(t) = \sum_{i=1}^{N} \tilde{y}_N^i(t) \nabla e_i \tag{10.37}$$

We consider a time regular subdivision $t_i = ih,\ i \in \mathbb{N}$, and $h > 0$ (small enough) the subdivision step.

The control that stabilizes the system (10.15) and solution of problem (10.27), may be obtained by the following algorithm;

Algorithm

Step 1 : Initial data:
tolerance $\varepsilon > 0$, target subregion ω, support of control D
and N the dimension of the projection space.
Step 2 : Solving equation (10.34) with the Schur method
(see [5])
Step 3 : Solving equation (10.35) using the formula (10.36), gives $y_N(t_i)$
Step 4 : Computation of $\nabla_\omega y_N(t_i)$ by formula (10.37).
Step 5 : If $\| \nabla_\omega y_N(t_i) \| < \varepsilon$ stop, otherwise $i := i + 1$ and go to step 3.

Example 10.3 On $\Omega =]0, 1[$, we consider the system

$$\begin{cases} \dfrac{\partial y(t)}{\partial t} = Ay(t) + \chi_D u(t) & \Omega \times]0, +\infty[\\ \dfrac{\partial y(0,t)}{\partial x} = \dfrac{\partial y(1,t)}{\partial x} = 0 & t \geq 0 \\ y(x,0) = x^2(\dfrac{1}{2} - \dfrac{1}{3}x) & \Omega \end{cases} \tag{10.38}$$

where $Ay = 0.02\Delta y + 0.5y$. The state space is

$$X = \{y \in H^1(\Omega) \text{ such that } \frac{dy}{dx}(0) = \frac{dy}{dx}(1) = 0\}$$

is of Hilbert and $u(t) \in X$, for all $t \geq 0$.

We take $D =]0.2, 0.9[$ and we consider the problem (10.27) with $R = \nabla^*\nabla$.

The operator A generates a C_0-semigroup given by

$$S(t)y = \sum_{i \geq 0}^{+\infty} e^{\lambda_i t} \langle y, \phi_i \rangle \phi_i$$

where $\lambda_i = -0.01(i\pi)^2 + 0.5$ and $\phi_i(x) = \alpha_i \cos(i\pi x)$ with $\alpha_i = \sqrt{\frac{2}{1+(i\pi)^2}}$.

Since $\lambda_0, \lambda_1 > 0$ the state and the gradient of system (10.38) are unstable.

We consider the subspace

$$X_N = Span\{\alpha_{i-1} \cos((i-1)\pi x),\ 1 \leq i \leq N, x \in \Omega\}$$

We apply the above algorithm with truncation order $N = 5$. we obtain the following results.

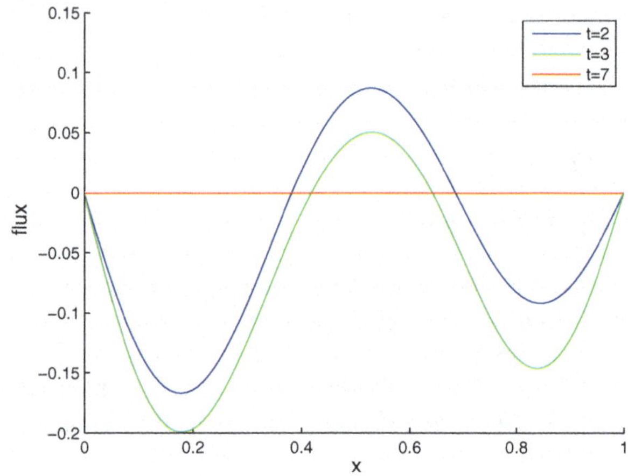

Fig. 10.1 Evolution of the gradient

Global case. The next figure represents the gradient evolution of system (10.38) (Fig. 10.1).

We remark that the goes close to 0 when time t increases. The stabilisation error is about 9.9836×10^{-7} and the cost equals to 2.6982×10^{-4}.

Studying the variation of cost stabilization of the gradient with respect to the control zone area we obtain the following array.

D	]0,0.1[	]0,0.3[	]0,0.5[	]0,0.7[	]0,0.9[	]0,1[
Cost	9.0097	1.921	0.9408	0.817	0.2868	0.1636

This array shows that there exists a link between the zone area of the control and the cost of the gradient stabilization; more precisely more the area increases more the cost decreases.

Regional case. We apply the above algorithm with $\omega =]0.3, 0.8[$ and $D =]0, 0.3[$. We obtain the following figure (Fig. 10.2).

The gradient is stabilised on $\omega =]0.3, 0.8[$ with error equals 3.7933×10^{-9} and a cost equals 14.99×10^{-2}.

Also in this case, for $\omega =]0, 0.3[$ fixed, we have the following array.

D	]0.3,0.4[	]0.3,0.5[	]0.3,0.6[	]0.3,0.7[	]0.3,0.8[	]0.3,1[
Cost	1.1802	1.1801	1.18	1.1798	1.1794	1.1792

Once more we remark that the cost of the gradient stabilization and the zone area of the control action D are inversely proportional.

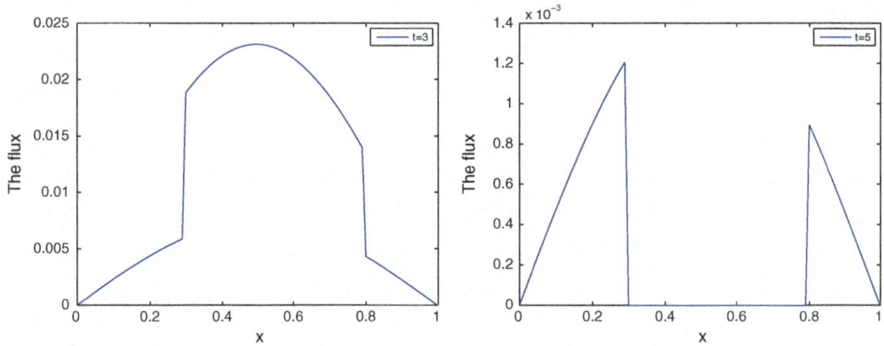

Fig. 10.2 Evolution of the gradient

Now if we fix the zone of the control action and we compute the error and cost for the stabilization of the gradient for different subregions ω. The array bellow gives $E = \text{Error} \times 10^9$ and $J = \text{cost} \times 10^2$.

ω	]0,0.3[	]0,0.4[	]0,0.6[	]0,0.7[	]0,0.9[	]0,1[
E	6.6039	6.7428	7.6575	8.5953	9.1983	9.3011
J	117.83	135.17	155.13	159.52	162.03	165.18

We remark that if the area of the subregion ω increases the cost and the error of the stabilisation of the gradient increases.

Example 10.4 On $\Omega =]0, 1[$, we consider the system

$$\begin{cases} \dfrac{\partial y(t)}{\partial t} = Ay(t) + \chi_{_D} u(t) \ \Omega\times]0, +\infty[\\ y(x, 0) = (1 - x)^2 x^2 \qquad x \in \Omega \end{cases} \tag{10.39}$$

where $Ay = 0.01\Delta y + 0.5y$. The state space is $X = H_0^1(\Omega)$ and $u(t) \in X$, for all $t \geq 0$.

We take $D =]0, 0.3[$ and we consider the problem (10.27) with $R = \nabla_\omega^* \nabla_\omega$.

The eigenvalues and the eigenfunctions of A are given, for $i \geq 1$, by

$$\varphi_i = \alpha_i \sin(i\pi x) \ \ and \ \ \lambda_i = -0.01(i\pi)^2 + 0.5$$

As $\lambda_1 > 0$, then the state and the gradient of system (10.39) are unstable.

We consider the projection subspace $X_N = \{\alpha_i \sin(i\pi x), i = 1, \cdots, N\}$ with $\alpha_i = \sqrt{\frac{2}{1+(i\pi)^2}}$.

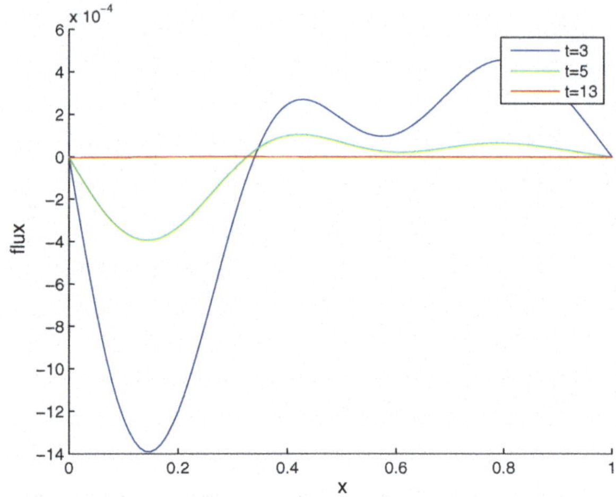

Fig. 10.3 Evolution of the gradient

Global case. If we apply the previous algorithm with a truncation order $N = 5$, then we obtain the figure (Fig. 10.3).
which shows the gradient evolves close to 0 when t increases.

The stabilisation error is equal to 9.7961×10^{-7} and cost equals 3.2689×10^{-4}.

The following array gives the cost of the gradient stabilisation of system (10.39) depending on different localisations of the control zone D.

D	]0, 0.1[	]0, 0.3[	]0, 0.5[	]0, 0.7[	]0, 0.9[	]0, 1[
$Cost$	0.0247	0.0039	0.0016	$4.7255\ 10^{-4}$	$2.6982\ 10^{-4}$	$2.6081\ 10^{-4}$

We remark that if the area of control zone increases then the stabilization cost decreases.

Regional case. Now we consider $\omega =]0, 0.45[$, and $D =]0.0.8[$, we obtain the figure (Fig. 10.4)

The cost and error of gradient stabilization are respectively 4.53×10^{-2} and 9.9659×10^{-6}.

The next array gives errors and costs of gradient stabilization for different subregions ω.

ω	]0,0.17[	]0,0.33[	]0,0.36[	]0,0.45[	]0,0.78[
Error $\times 10^6$	9.9530	9.9534	9.9609	9.9659	9.9729
Cost	$3.68\ 10^{-2}$	$4.26\ 10^{-2}$	$4.3\ 10^{-2}$	$4.53\ 10^{-2}$	$6.4\ 10^{-2}$

It is clear that the error and the cost increase when the area of the target subregion increases.

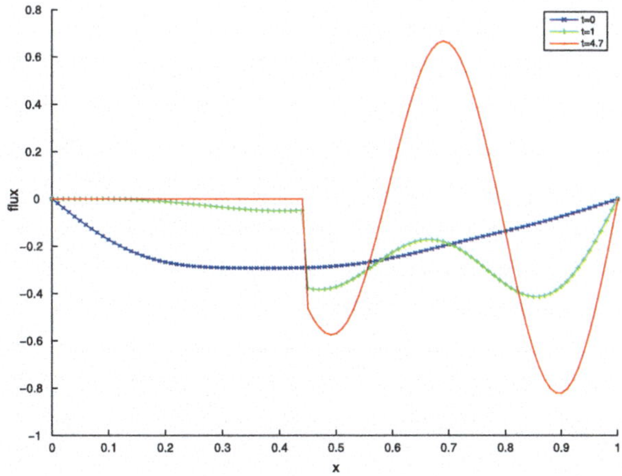

Fig. 10.4 Evolution du gradient

We fix $\omega =]0, 0.45[$ and we study the link between cost of the gradient stabilisation and the area of the control support, we obtain the following array.

D	]0,0.79[	]0,0.75[	]0,0.7[	]0,0.65[	]0,0.6[	]0,55[	]0,0.5[
Cost $\times 10^2$	4.57	4.74	5	5.37	5.87	6.38	6.76

Which shows that are inversely proportional as in previous example.

References

1. Kato, T.: Perturbation Theory for Linear Operators, 2nd edn. Springer, Berlin (1980)
2. Curtain, R.F., Zwart, H.J.: An Introduction to Infinite Dimensional Linear Systems Theory. Springer, New York (1995)
3. Banks, H., Kunisch, B.: The linear regulator problem for parabolic systems. SIAM J. Control Optim. **22**, 684–696 (1984)
4. Higham, N.J.: The scaling and squaring method for the matrix exponential. SIAM Rev. **51**(4), 747–767 (2009)
5. Arnold, W.F., Laub, A.J.: Generalized eigenproblem: algorithms and software for algebraic Riccati equations. Proc. IEEE **72**, 1746–1754 (1984)

Chapter 11
Gradient Stabilization of Infinite Dimensional Bilinear Systems

In this chapter we study gradient stabilization of infinite dimensional bilinear systems. Then under sufficient conditions, we establish exponential, strong and weak gradient stabilization of bilinear systems on the whole geometrical domain $\Omega \subset \mathbb{R}^n$ and also on a subregion ω of Ω.

11.1 Gradient Stabilization of Bilinear Systems

In this section, we deal with the gradient stabilization of a bilinear system defined on a regular spatial domain $\Omega \subset \mathbb{R}^n$ by

$$\begin{cases} \dfrac{dy(t)}{dt} = Ay(t) + u(t)By(t), \quad [0, +\infty[\\ y(0) = y_0 \end{cases} \tag{11.1}$$

where the operator A generates a semigroup of contractions $S(t)$ on $H_0^1(\Omega)$ endowed with the inner product, denoted by $\langle ., . \rangle$ and $\|.\|$ the associated norm, $u \in L^2(0, +\infty)$ is a scalar valued control and B is a linear and bounded operator mapping $H_0^1(\Omega)$ into itself.

We define the gradient operator by

$$\begin{array}{rcl} \nabla : H_0^1(\Omega) & \longrightarrow & (L^2(\Omega))^n \\ y & \longmapsto & \left(\dfrac{\partial y(x)}{\partial x_1}, \dfrac{\partial y(x)}{\partial x_2}, \ldots, \dfrac{\partial y(x)}{\partial x_n}\right) \end{array}$$

$(L^2(\Omega))^n$ is endowed with its usual inner product $\langle ., . \rangle_n$, and $\|.\|_n$ is the corresponding norm.

E. H. Zerrik and O. Castillo, *Stabilization of Infinite Dimensional Systems*, Studies in Systems, Decision and Control 355,
https://doi.org/10.1007/978-3-030-68600-0_11

Let us recall that the gradient of system (11.1) is exponentially (respectively strongly, weakly) stabilizable, if for any initial condition $y_0 \in H_0^1(\Omega)$ the corresponding solution $y(t)$ of (11.1) is unique and $\nabla y(t)$ converges to zero exponentially (respectively strongly, weakly) as $t \longrightarrow +\infty$.

The purpose of this Chapter is to study the gradient stabilization of system (11.1) where operators A and B verify the conditions:

$$\langle \nabla By, \nabla y\rangle_n \langle By, y\rangle \geq 0, \quad \forall y \in H_0^1(\Omega). \tag{11.2}$$

and

$$\langle \nabla Ay, \nabla y\rangle_n \leq 0, \quad \forall y \in \mathcal{D}(A). \tag{11.3}$$

So using a nonlinear control, we establish exponential stabilization under a condition that we call the gradient's observability inequality. Furthermore, under the same assumption, we obtain a strong stabilization using a feedback control, and the weak stabilization is obtained under a weaker condition.

11.1.1 Gradient Exponential Stabilization

In this section, we study the gradient exponential stabilization of system (11.1) using the following control

$$u(t) = \begin{cases} -\dfrac{\langle \nabla By(t), \nabla y(t)\rangle_n}{\|y(t)\|^2}, & y(t) \neq 0 \\ 0, & y(t) = 0 \end{cases} \tag{11.4}$$

Let us denote $h(y) = \dfrac{\langle \nabla By, \nabla y\rangle_n}{\|y\|^2}$ for all $y \neq 0$, $h(0) = 0$.

If we formally differentiate $\|\nabla y(t)\|_n^2$ along the trajectories of (11.1), we obtain

$$\frac{d}{dt}\|\nabla y(t)\|_n^2 \leq 2\langle \nabla Ay(t), \nabla y(t)\rangle_n + u(t)\langle \nabla By(t), \nabla y(t)\rangle_n, \; \forall t \geq 0.$$

Using the assumption (3), we get

$$\frac{d}{dt}\|\nabla y(t)\|_n^2 \leq u(t)\langle \nabla By(t), \nabla y(t)\rangle_n, \; \forall t \geq 0.$$

Then the control (11.4) yields the dissipating energy inequality

$$\frac{d}{dt}\|\nabla y(t)\|_n^2 \leq -\frac{|\langle \nabla By(t), \nabla y(t)\rangle_n|^2}{\|y(t)\|^2}, \; \forall t \geq 0.$$

The following result gives sufficient conditions for the gradient exponential stabilization of the system (11.1).

Theorem 11.1 *Assume that A generates a semigroup of contractions* $(S(t))_{t\geq 0}$ *and there exist* $T,\ \mu > 0$ *such that*

$$\int_0^T |\langle \nabla BS(s)z, \nabla S(s)z\rangle_n| ds \geq \mu \|\nabla z\|_n^2, \forall z \in H_0^1(\Omega) \tag{11.5}$$

then control (11.4) exponentially stabilizes the gradient of system (11.1).

Proof We know from [1] that for each $y_0 \in H_0^1(\Omega)$, system (11.1) has a unique mild solution $y \in C([0, t_{\max}[, H_0^1(\Omega))$ defined on a maximal interval $[0, t_{\max}[$ and given by the variation of constants formula

$$y(t) = S(t)y_0 + \int_0^t u(s)S(t-s)By(s)ds \tag{11.6}$$

Since $S(t)$ is a semigroup of contractions (A is dissipative), then

$$\|y(t)\|^2 \leq \|y_0\|^2 - 2\int_0^t h(y(s))\langle By(s), y(s)\rangle ds$$

From (11.2), it follows that

$$\|y(t)\| \leq \|y_0\|. \tag{11.7}$$

Differentiating $\|\nabla y(t)\|_n^2$ and using (11.3), we have

$$\frac{d}{dt}\|\nabla y(t)\|_n^2 \leq -2|h(y(t))|^2\|y(t)\|^2.$$

Integrating this inequality from mT to $(m+1)T$, $(m \in \mathbb{N})$, we obtain

$$\|\nabla y(mT)\|_n^2 - \|\nabla y((m+1)T)\|_n^2 \geq 2\int_{mT}^{(m+1)T} |h(y(s))|^2\|y(s)\|^2 ds \tag{11.8}$$

Moreover, for all $y_0 \in H_0^1(\Omega)$ and $t \geq 0$, we have

$$\begin{aligned}\langle \nabla BS(t)y_0, \nabla S(t)y_0\rangle_n &= \langle \nabla By(t), \nabla y(t)\rangle_n \\ &\quad - \langle \nabla By(t), \nabla(y(t) - S(t)y_0)\rangle_n \\ &\quad + \langle \nabla B(S(t)y_0 - y(t)), \nabla S(t)y_0\rangle_n\end{aligned}$$

Using (11.7) and since ∇ is bounded, there exists $\delta > 0$ such that

$$|\langle \nabla BS(t)y_0, \nabla S(t)y_0\rangle_n| \leq 2\delta^2\|B\|\|y(t) - S(t)y_0\|\|y_0\| \\ + |\langle \nabla By(t), \nabla y(t)\rangle_n| \tag{11.9}$$

It follows from (11.6), that

$$\|y(t) - S(t)y_0\| \leq \|B\| \int_0^t |h(y(s))|\|y(s)\|ds$$

Schwarz's inequality yields

$$\|y(t) - S(t)y_0\| \leq \|B\|\left\{T \int_0^T |h(y(s))|^2\|y(s)\|^2 ds\right\}^{\frac{1}{2}} \tag{11.10}$$

Using (11.7), we get

$$|\langle \nabla By(t), \nabla y(t)\rangle_n| \leq |h(y(t))|\|y(t)\|\|y_0\| \tag{11.11}$$

Integrating (11.9) over the interval $[0, T]$ and taking into account (11.10) and (11.11), we obtain

$$\int_0^T |\langle \nabla BS(s)y_0, \nabla S(s)y_0\rangle_n|ds \leq 2\delta^2\|B\|^2 T^{\frac{3}{2}}\|y_0\| \\ \times\left\{\int_0^T |h(y(s))|^2\|y(s)\|^2 ds\right\}^{\frac{1}{2}} \\ +\sqrt{T}\|y_0\|\left\{\int_0^T |h(y(s))|^2\|y(s)\|^2 ds\right\}^{\frac{1}{2}} \tag{11.12}$$

Replacing y_0 by $y(t)$ in (11.12), we deduce that

$$\int_0^T |\langle \nabla BS(s)y(t), \nabla S(s)y(t)\rangle_n|ds \leq 2\delta^2\|B\|^2 T^{\frac{3}{2}}\|y(t)\|\sqrt{g(t)} \\ +\|y(t)\|\sqrt{Tg(t)}$$

where $g(t) = \displaystyle\int_t^{t+T} |h(y(s))|^2\|y(s)\|^2 ds$.

From Poincarés inequality there exists a constant $C > 0$ such that

$$\int_0^T |\langle \nabla BS(s)y(t), \nabla S(s)y(t)\rangle_n|ds \leq (2\delta^2\|B\|^2 T + 1) \\ \times\sqrt{T}C\|\nabla y(t)\|_n\sqrt{g(t)}$$

Condition (11.5), gives

$$\mu\|\nabla y(t)\|_n \leq (2\delta^2\|B\|^2T+1)\sqrt{T}C\sqrt{g(t)}$$

Then,

$$\rho\|\nabla y(t)\|_n^2 \leq g(t) \tag{11.13}$$

where $\rho = \dfrac{\mu^2}{(2\delta^2\|B\|^2T+1)^2TC^2}$.

Combining (11.13) and (11.8), we deduce that

$$(1+2\rho)\|\nabla y((m+1)T)\|_n^2 \leq \|\nabla y(mT)\|_n^2$$

Thus,

$$\|\nabla y((m+1)T)\|_n \leq \beta\|\nabla y(mT)\|_n$$

where $\beta = \dfrac{1}{\sqrt{1+2\rho}}$.

By recurrence, we show that

$$\|\nabla y(mT)\|_n \leq \beta^m\|\nabla y_0\|_n$$

Taking $m = E(\frac{t}{T})$ the integer part of $\frac{t}{T}$, we conclude that

$$\|\nabla y(t)\|_n \leq Me^{-\alpha t}\|y_0\|, \ \forall t \geq 0$$

where $M = \sqrt{\delta^2(1+2(\frac{\mu}{\rho})^2}$ and $\alpha = \dfrac{\ln(1+2(\frac{\mu}{\rho})^2)}{2T}$, which completes the proof. ■

Remark 11.1 1. Under the gradient's observability inequality (11.5), exponential stabilization of the gradient may also be obtained using the following switching control

$$u(t) = -r\ \mathrm{sign}(\langle\nabla By(t), \nabla y(t)\rangle_n), \quad \text{for suitable} \quad r > 0.$$

2. If the system is stabilizable then the gradient is stabilizable. The converse is true for strong and exponential stabilization on $H_0^1(\Omega)$ (Poincaré's inequality).
3. The gradient stabilization can be obtained under a lower cost than the state stabilization: let us consider the cost $q(u)$ and

$$\mathcal{U}_{ad}^1 = \{u \in L^\infty(0,+\infty)|q(u) < +\infty \ and \ \nabla y(t) \longrightarrow 0 \ as \ t \longrightarrow +\infty \ on \ L^2(\Omega)\}$$

and

$$\mathcal{U}_{ad}^2 = \{u \in L^\infty(0,+\infty)|q(u) < +\infty \ and \ y(t) \longrightarrow 0 \ as \ t \longrightarrow +\infty \ on \ L^2(\Omega)\}$$

we have $\mathcal{U}_{ad}^2 \subset \mathcal{U}_{ad}^1$, so

$$\min_{u\in\mathcal{U}_{ad}^1} q(u) \leq \min_{u\in\mathcal{U}_{ad}^2} q(u).$$

Example 11.1 . On $\Omega =]0, 1[$, the system, defined by

$$\begin{cases} \dfrac{\partial^2 y}{\partial t^2}(x,t) = \Delta y(x,t) + u(t)\dfrac{\partial y}{\partial t}(x,t) \; on \; \Omega\times]0,+\infty[\\ y(x,0) = y_0, \; \dfrac{\partial y}{\partial t}(x,0) = y_1 \quad on \; \Omega \\ y(0,t) = y(1,t) = 0 \quad on \; [0,+\infty[\end{cases} \tag{11.14}$$

has a unique solution $(y, \partial_t y)$ in the state space $X = H^2(\Omega) \cap H_0^1(\Omega) \times H_0^1(\Omega)$ endowed with the inner product $\langle (y_1, y_2), (z_1, z_2)\rangle = \langle \nabla y_1, \nabla z_1\rangle_{H_0^1(\Omega)} + \langle y_2, z_2\rangle_{H_0^1(\Omega)}$.

Let $\tilde{A} = -\Delta$, with domain $D(\tilde{A}) = \{y \in H_0^1(0,1)/ -\Delta y \in H_0^1(0,1)\}$.

The eigenvalues of the operator $\tilde{A}$ are $\lambda_m = (m\pi)^2$, corresponding to eigenfunctions $\phi_m(x) = \sqrt{\frac{2}{1+(m\pi)^2}}\sin(m\pi x), \; \forall m \in \mathbb{N}^*$.

System (11.14) has the form of (11.1) if we set

$$A = \begin{pmatrix} 0 & I \\ -\tilde{A} & 0 \end{pmatrix} \; and \; B = \begin{pmatrix} 0 & 0 \\ 0 & I \end{pmatrix}$$

Let $y = \begin{pmatrix} y_1 \\ y_2 \end{pmatrix} \in D(\tilde{A}) \times D(\tilde{A}^{\frac{1}{2}})$, condition (11.3) is verified, indeed

$$\langle \nabla Ay, \nabla y\rangle = \langle \Delta y_2, \Delta y_1\rangle_{L^2(\Omega)} - \langle \Delta y_1, \Delta y_2\rangle_{L^2(\Omega)} = 0.$$

Let $z = \begin{pmatrix} z_1 \\ z_2 \end{pmatrix} \in H$, we have $z = \sum_{m=1}^{\infty} \begin{pmatrix} a_m \\ \lambda_m^{\frac{1}{2}} b_m \end{pmatrix} \phi_m$, where $(a_m, b_m) \in \mathbb{R}^2, \; \forall m \in \mathbb{N}^*$.

The assumption (11.2) is satisfied, indeed

$$\langle \nabla Bz, \nabla z\rangle_{H_0^1(\Omega)\times L^2(\Omega)} \langle Bz, z\rangle = \|\nabla z_2\|^2_{L^2(0,1)} \|z_2\|^2_{H_0^1(0,1)} \geq 0.$$

The operator A generates a semigroup given by

$$S(s)z = \sum_{m=1}^{+\infty} \begin{pmatrix} a_m \cos(m\pi s) + b_m \sin(m\pi s) \\ b_m m\pi \cos(m\pi s) - a_m m\pi \sin(m\pi s) \end{pmatrix} \phi_m, \quad \forall s \geq 0.$$

Then

$$\begin{aligned}\langle \nabla BS(s)z, \nabla S(s)z\rangle_{H_0^1(\Omega)\times L^2(\Omega)} &= \sum_{m=1}^{+\infty} \frac{\lambda_m^2}{1+\lambda_m} \\ &\times \big(a_m \sin(m\pi s) - b_m \cos(m\pi s)\big)^2 \\ &= \sum_{m=1}^{+\infty} \frac{\lambda_m^2}{1+\lambda_m} \big(a_m^2 \sin^2(m\pi s) \\ &-b_m a_m \sin(2m\pi s) + b_m^2 \cos^2(m\pi s)\big).\end{aligned}$$

Inequality (11.5) is verified for $T = 2$, indeed

$$\begin{aligned}\int_0^2 |\langle \nabla BS(s)z, \nabla S(s)z\rangle_{H_0^1(\Omega)\times L^2(\Omega)}|ds &= \sum_{m=1}^{+\infty} \frac{\lambda_m^2}{1+\lambda_m}(a_m^2 + b_m^2) \\ &\geq \|\nabla z\|^2_{L^2(\Omega)}\end{aligned}$$

We deduce that, the gradient of system (11.14) is exponentially stabilizable by the control

$$u(t) = \begin{cases} -\dfrac{\|\nabla \partial_t y(t)\|^2_{L^2(\Omega)}}{\|\partial_t y(t)\|^2_{H_0^1(\Omega)} + \|\nabla y(t)\|^2_{H_0^1(\Omega)}}, & if \quad (y(t), \partial_t y(t)) \neq (0,0) \\ 0, & if \quad (y(t), \partial_t y(t)) = (0,0) \end{cases}$$

11.1.2 Gradient Strong Stabilization

Here, we study the gradient strong stabilization of system (11.1), and we give an illustrating example.

Theorem 11.2 *Let A generate a semigroup of contractions $(S(t))_{t\geq 0}$ and assume that conditions* (11.2), (11.3) *and* (11.5) *hold, then the control*

$$u(t) = -\langle \nabla By(t), \nabla y(t)\rangle_n \tag{11.15}$$

strongly stabilizes the gradient of system (11.1).

Proof Differentiating $\|\nabla y(t)\|_n^2$ and using (11.3), we obtain

$$\frac{d}{dt}\|\nabla y(t)\|_n^2 \leq -2|\langle \nabla By(t), \nabla y(t)\rangle_n|^2, \ \forall t \geq 0.$$

Integrating this inequality over $[0, t]$, we get

$$\int_0^t |\langle \nabla By(s), \nabla y(s)\rangle_n|^2 ds \leq \frac{1}{2}\|\nabla y_0\|_n^2, \ \forall t \geq 0.$$

Hence

$$\int_0^{+\infty} |\langle \nabla By(s), \nabla y(s)\rangle_n|^2 ds < +\infty. \tag{11.16}$$

Let $\Psi(t) = \int_0^t S(t-s)u(s)By(s)ds$ and using (11.6) we have

$$\langle \nabla BS(t)y_0, \nabla S(t)y_0\rangle_n = \langle \nabla By(t), \nabla y(t)\rangle_n - \langle \nabla B\Psi(t), \nabla S(t)y_0\rangle_n$$
$$-\langle \nabla By(t), \nabla \Psi(t)\rangle_n$$

In the previous result, we found out that (11.2) leads to (11.7), which gives

$$\begin{aligned} |\langle \nabla BS(t)y_0, \nabla S(t)y_0\rangle_n| &\le 2\delta^2 \|B\| \|\Psi(t)\| \|y_0\| \\ &+ |\langle \nabla By(t), \nabla y(t)\rangle_n|, \forall t \in [0, T]. \end{aligned} \tag{11.17}$$

Moreover, we have

$$\|\Psi(t)\| \le \|B\| \|y_0\| \sqrt{T} \Big(\int_0^T |\langle \nabla By(s), \nabla y(s)\rangle_n|^2 ds \Big)^{\frac{1}{2}} \tag{11.18}$$

Integrating (11.17) over the interval $[0, T]$ and taking into account (11.18), we obtain

$$\int_0^T |\langle \nabla BS(s)y_0, \nabla S(s)y_0\rangle_n| ds \le (2\delta^2 T \|B\|^2 \|y_0\|^2 + 1)$$
$$\times \sqrt{T} \Big(\int_0^T |\langle \nabla By(s), \nabla y(s)\rangle_n|^2 ds \Big)^{\frac{1}{2}}$$

Replacing y_0 by $y(t)$, we get

$$\int_0^T |\langle \nabla BS(s)y(t), \nabla S(s)y(t)\rangle_n| ds \le (2\delta^2 T \|B\|^2 \|y_0\|^2 + 1)\sqrt{T}$$
$$\times \Big(\int_t^{t+T} |\langle \nabla By(s), \nabla y(s)\rangle_n|^2 ds \Big)^{\frac{1}{2}}$$

From (11.16), we have

$$\int_t^{t+T} |\langle \nabla By(s), \nabla y(s)\rangle_n|^2 ds \longrightarrow 0, \ as \ t \longrightarrow +\infty.$$

It follows that

$$\int_0^T |\langle \nabla BS(s)y(t), \nabla S(s)y(t)\rangle_n| ds \longrightarrow 0, \ as\ t \longrightarrow +\infty. \tag{11.19}$$

From (11.5) and (11.19), we deduce that $\|\nabla y(t)\|_n \longrightarrow 0,\ as\ t \longrightarrow +\infty$, which completes the proof. ■

Remark 11.2 In fact, the gradient strong stabilization may be ensured by the controls of the form

$$u(t) = -\frac{\langle \nabla By(t), \nabla y(t)\rangle_n}{g(y(t))}$$

where $g : H_0^1(\Omega) \longrightarrow (0, +\infty)$ is a positive function satisfying the dissipating energy inequality.

Example 11.2 . On $\Omega =]0, 1[$, the system, defined by

$$\begin{cases} \dfrac{\partial^2 y}{\partial t^2}(x,t) = -\dfrac{\partial^4 y}{\partial x^4}(x,t) + u(t)\dfrac{\partial y}{\partial t}(x,t) & on\ \Omega \times]0, +\infty[\\ y(x,0) = y_0,\ \dfrac{\partial y}{\partial t}(x,0) = y_1 & on\ \Omega \\ y(\xi,t) = \dfrac{\partial^2 y}{\partial x^2}(\xi,t) = 0,\ \xi = 0, 1 & on\ ]0, +\infty[\end{cases} \tag{11.20}$$

has only one solution $(y, \partial_t y)$ in the state space $X = (H^3(0,1) \cap H_0^1(0,1)) \times H_0^1(0,1)$ endowed with the inner product $\langle (y_1, y_2), (z_1, z_2)\rangle = \langle \Delta y_1, \Delta z_1\rangle_{H_0^1(0,1)} + \langle y_2, z_2\rangle_{H_0^1(0,1)}$. Let $F = \dfrac{\partial^4 y}{\partial x^4}$ with domain $\mathcal{D}(F) = \{y \in H_0^1(0,1) / \dfrac{\partial^4 y}{\partial x^4} \in H_0^1(0,1)$, $y(\xi,t) = \dfrac{\partial^2 y}{\partial x^2}(\xi,t) = 0,\ \xi = 0, 1\}$.

The eigenvalues of the operator Δ^2 are $\lambda_j = (j\pi)^4$, corresponding to eigenfunctions $\varphi_j(x) = \sqrt{\dfrac{2}{1 + (j\pi)^2}} \sin(j\pi x),\ \forall j \in \mathbb{N}^*$.

We set

$$A = \begin{pmatrix} 0 & I \\ -F & 0 \end{pmatrix}\ and\ B = \begin{pmatrix} 0 & 0 \\ 0 & I \end{pmatrix}.$$

Let $y = \begin{pmatrix} y_1 \\ y_2 \end{pmatrix} \in D(F) \times D(F^{\frac{1}{2}})$, condition (11.3) is satisfied, indeed

$$\langle \nabla Ay, \nabla y\rangle = \langle \partial_x^3 y_2, \partial_x^3 y_1\rangle_{L^2(\Omega)} - \langle \partial_x^3 y_1, \partial_x^3 y_2\rangle_{L^2(\Omega)} = 0.$$

Now, let $z = \begin{pmatrix} z_1 \\ z_2 \end{pmatrix} \in H$, we have $z = \sum_{j=1}^{\infty} \begin{pmatrix} c_j \\ \lambda_j^{\frac{1}{2}} d_j \end{pmatrix} \varphi_j$, where $(c_j, d_j) \in \mathbb{R}^2\ \forall j \geq 1$.

The assumption (11.2) is verified, indeed

$$\langle \nabla Bz, \nabla z\rangle_{(H^2(0,1)\cap H_0^1(\Omega))\times L^2(\Omega)}\langle Bz, z\rangle = \|\nabla z_2\|^2_{L^2(0,1)}\|z_2\|^2_{H_0^1(0,1)} \geq 0.$$

The operator A generates a semigroup given by

$$S(s)z = \sum_{j=1}^{\infty} \begin{pmatrix} c_j\cos(\lambda_j^{\frac{1}{2}}s) + d_j\sin(\lambda_j^{\frac{1}{2}}s) \\ d_j\lambda_j^{\frac{1}{2}}\cos(\lambda_j^{\frac{1}{2}}s) - c_j\lambda_j^{\frac{1}{2}}\sin(\lambda_j^{\frac{1}{2}}s) \end{pmatrix}\varphi_j, \quad \forall s \geq 0.$$

Then

$$\begin{aligned}\langle \nabla BS(s)z, \nabla S(s)z\rangle_{(H^2(0,1)\cap H_0^1(0,1))\times L^2(0,1)} &= \sum_{j=1}^{+\infty}\frac{(j\pi)^6}{1+(j\pi)^4} \\ &\times\big(c_j^2\sin^2(j\pi s) \\ &-d_jc_j\sin(2j\pi s) \\ &+d_j^2\cos^2(j\pi s)\big).\end{aligned}$$

For $T = 2$, condition (11.5) holds, indeed

$$\begin{aligned}&\int_0^2 |\langle \nabla BS(s)z, \nabla S(s)z\rangle_{(H^2(0,1)\cap H_0^1(0,1))\times L^2(0,1)}|ds \\ &= \sum_{j=1}^{+\infty}\frac{(j\pi)^6}{1+(j\pi)^4}(c_j^2+d_j^2) \\ &\geq \|\nabla z\|^2_{L^2(\Omega)}\end{aligned}$$

So the control

$$u(t) = -\|\frac{\partial^2 y(x,t)}{\partial x\partial t}\|^2_{L^2(\Omega)}$$

strongly stabilizes the gradient of system (11.20). ■

11.1.3 Gradient Weak Stabilization

This section is devoted to the gradient weak stabilization of system (11.1) .

Theorem 11.3 *Suppose that A generates a semigroup of contractions $(S(t))_{t\geq 0}$, conditions* (11.2), (11.3) *hold and B is compact such that*

$$\langle \nabla BS(t)z, \nabla S(t)z\rangle_n = 0,\ t \geq 0 \Longrightarrow \nabla z = 0 \tag{11.21}$$

holds, then the control

$$u(t) = -\langle \nabla By(t), \nabla y(t)\rangle_n \tag{11.22}$$

weakly stabilizes the gradient of system (11.1).

Proof System (11.1) has a unique mild solution given by

$$y(t) = S(t)y_0 + \int_0^t S(t-s)u(s)By(s)ds$$

Let (t_n) be a sequence of real numbers such that $t_n \longrightarrow +\infty$ as $n \longrightarrow +\infty$.

From (11.7), $y(t_n)$ is bounded in the Hilbert space $H_0^1(\Omega)$, then there exists a subsequence $(t_{\gamma(n)})$ of (t_n) such that

$$y(t_{\gamma(n)}) \rightharpoonup z \in H_0^1(\Omega) \quad as \quad n \longrightarrow +\infty \tag{11.23}$$

The same arguments as in the proof of Theorem 11.2, gives

$$\begin{aligned} &\int_0^T |\langle \nabla BS(s)y(t_{\gamma(n)}), \nabla S(s)y(t_{\gamma(n)})\rangle_n|ds \\ &\leq C\Big(\int_{t_{\gamma(n)}}^{t_{\gamma(n)+T}} |\langle \nabla By(s), \nabla y(s)\rangle_n|^2 ds\Big)^{\frac{1}{2}} \end{aligned} \tag{11.24}$$

where $C = (2\delta^2 T\|B\|^2\|y_0\|^2 + 1)\sqrt{T}$.

Let $n \longrightarrow +\infty$ in (11.24), we obtain

$$\lim_{n\longrightarrow+\infty} \int_0^T \langle \nabla BS(s)y(t_{\gamma(n)}), \nabla S(s)y(t_{\gamma(n)})\rangle_n ds = 0$$

Since B is compact and $S(s)$ is continuous $\forall s \geq 0$, we have

$$\lim_{n\to+\infty} \langle \nabla BS(s)y(t_{\gamma(n)}), \nabla S(s)y(t_{\gamma(n)})\rangle_n = \langle \nabla BS(s)z, \nabla S(s)z\rangle_n \tag{11.25}$$

Moreover, by (11.7) and since the operators ∇ and B are bounded, there exists $\alpha > 0$ such that

$$|\langle \nabla BS(s)y(t_{\gamma(n)}), \nabla S(s)y(t_{\gamma(n)})\rangle_n| \leq \alpha\|y_0\|^2 \tag{11.26}$$

Using (11.25) and (11.26), we deduce from dominated convergence theorem (Bartle, 2014), that

$$\int_0^T |\langle \nabla BS(s)z, \nabla S(s)z\rangle_n|ds = 0$$

which leads to

$$\langle \nabla BS(s)z, \nabla S(s)z\rangle_n = 0, \quad \forall s \in [0, T]$$

Using (11.21), it follows that $\nabla z = 0$.

Since ∇ is continuous, (11.23) leads to $\nabla y(t_{\gamma(n)}) \rightharpoonup 0 \;\; as \;\; n \longrightarrow +\infty$.

Which implies that $\forall \varphi \in (L^2(\Omega))^n$, $\langle \nabla y(t_n), \varphi\rangle_n \longrightarrow 0$ as $n \longrightarrow +\infty$, and hence

$$\nabla y(t) \rightharpoonup 0 \;\; as \;\; t \longrightarrow +\infty.$$

We conclude that control (11.22) weakly stabilizes the gradient of system (11.1). ■

Example 11.3 On $\Omega =]0, 1[$, we consider the heat equation defined by

$$\begin{cases} \dfrac{\partial y(x,t)}{\partial t} = \dfrac{\partial^2 y(x,t)}{\partial x^2} + y(x,t) + u(t)By(x,t), & \Omega\times]0, +\infty[\\ y(x,0) = y_0, & \Omega \\ y(.,t) = 0, & \partial\Omega\times]0, +\infty[\end{cases} \tag{11.27}$$

where $By = \sum_{j=1}^{+\infty} \frac{1}{j^2}\langle y, \varphi_j\rangle\varphi_j, \forall j \in \mathbb{N}^*$.

System (11.27) has only one solution in the state space $H_0^1(\Omega)$.

The eigenvalues of operator $Ay = \Delta y + y$ are $\lambda_j = -(j\pi)^2 + 1$ corresponding to eigenfunctions $\varphi_j(x) = \sqrt{\dfrac{2}{1+(j\pi)^2}}\sin(j\pi x)$, for $j \in \mathbb{N}$.

Conditions (11.2) and (11.3) are verified, indeed

$$\langle \nabla Ay, \nabla y\rangle = \sum_{j=1}^{+\infty} \frac{(j\pi)^2 - (j\pi)^4}{1+(j\pi)^2} |\langle y, \varphi_j\rangle|^2 \le 0$$

and

$$\langle \nabla By, \nabla y\rangle_{L^2(\Omega)}\langle By, y\rangle = \sum_{j=1}^{+\infty} \frac{(j\pi)^2}{j^4(1+(j\pi)^2)} |\langle y, \varphi_j\rangle|^2 \ge 0$$

The operator A generates the semigroup $S(t)y = \sum_{j=1}^{+\infty} e^{\lambda_j t}\langle y, \varphi_j\rangle\varphi_j, \forall j \in \mathbb{N}^*$.

We have

$$\langle \nabla BS(t)y, \nabla S(t)y\rangle_{L^2(\Omega)} = \sum_{j=1}^{+\infty} \frac{\pi^2}{1+(j\pi)^2} e^{2\lambda_j t} |\langle y, \varphi_j\rangle|^2 = 0, \;\; \forall t \ge 0$$

$$\Longrightarrow \nabla y = 0.$$

then (11.21) holds, so we deduce that the control (11.22) weakly stabilizes the gradient of system (11.27).

11.2 Regional Gradient Stabilization of Bilinear Systems

We study in this section the regional gradient stabilization of infinite dimensional bilinear systems and we give different characterizations of control that stabilizes the gradient on a subregion.

Let Ω be a regular open domain of $\mathcal{R}^n$, $Q = \Omega\times]0, \infty[$ and ω a non null measure subregion of Ω. We consider a bilinear system described by the equation

$$\begin{cases} \dot{y}(t) = Ay(t) + u(t)By(t) \quad \text{on } Q \\ y(0) = y_0 \in X = H^1(\Omega) \end{cases} \tag{11.28}$$

where A is a linear operator of domain $\mathcal{D}(A) \subset H^1(\Omega)$ generates a C_0-semigroup $(S(t))_{t\geq 0}$ on $X = H^1(\Omega)$ endowed with its usual inner product $< . >$ and the associated norm $\| \,.\, \|$, $B \in \mathcal{L}(X)$ is a bounded linear operator, and $u(t)$ is a scalar control. We define the operator

$$\begin{aligned} \nabla_\omega \ &: X \longrightarrow (L^2(\omega))^n \\ y &\longmapsto (\chi_\omega \frac{\partial y(x)}{\partial x_1}, \chi_\omega \frac{\partial y(x)}{\partial x_2}, \cdots, \chi_\omega \frac{\partial y(x)}{\partial x_n}) \end{aligned} \tag{11.29}$$

The space $(L^2(\Omega))^n$ is endowed with its usual inner product $< .,. >$ and the associated norm $\| \,.\, \|$, and $(L^2(\omega))^n$ is endowed with the restriction $< .,. >_n$.

We note $G_\omega = \nabla_\omega^* \nabla_\omega$, where ∇_ω^* is the adjoint operator of ∇_ω.

Definition 11.1 The gradient of system (11.28) is said to be

1. weakly stabilizable on ω, if for each initial condition $y_0 \in X$, the corresponding solution $y(t)$ of (11.28) is global and verify
$$< \nabla_\omega y(t), z >_n \longrightarrow 0 \text{ when } t \longrightarrow \infty, \quad z \in (L^2(\Omega))^n.$$
2. strongly stabilizable on ω, if for each initial condition $y_0 \in X$, the corresponding solution $y(t)$ of system (11.28) is global, and $||\nabla_\omega y(t)||_n \longrightarrow 0$ when $t \longrightarrow \infty$.
3. exponentially stabilizable on ω, if for each initial condition $y_0 \in X$, the corresponding solution $y(t)$ of system (11.28) is global and there exist two reals $M, \alpha > 0$ such that:
$$\|\nabla_\omega y(t)\|_n \leq M e^{-\alpha t} \|y_0\| \ , \ t \geq 0$$

In the sequel a system of stabilizable gradient on ω is will be said G-stabilizable sur ω.

If $\omega = \Omega$ the system (11.28) will be said G-stabilizable.

Remark 11.3 If the system (11.28) is stabilizable, then it is G-stabilizable on all $\omega \subset \Omega$. The converse is not true in general as illustrated by the following counter example.

Example 11.4 On $\Omega =]0, 1[^2$, we consider the system

$$\begin{cases} \frac{\partial y}{\partial t}(x,t) = \Delta y(x,t) + 10\pi^2 y(x,t) + u(t)By(x,t) \text{ on } Q \\ \frac{\partial y}{\partial \nu}(\xi,t) = 0 \qquad \partial\Omega\times]0,+\infty[\\ y(0) \quad = y_0 \in H^1(\Omega) \end{cases} \tag{11.30}$$

where Δ is the Laplacien operator. The operator $A = \Delta + 10\pi^2 Id$ has an orthogonal basis of eigenfunctions given by

$$\varphi_{ij}(x,z) = 2a_{ij}\cos(i\pi x)\cos(j\pi z)$$

where $a_{ij} = (1-\lambda_{ij})^{-\frac{1}{2}}$. The associated eigenvalues are given by

$$\lambda_{ij} = \begin{cases} -(i^2+j^2)\pi^2 + 10\pi^2 & \text{if } (i,j) \neq (3,0) \\ \pi^2 & \text{otherwise} \end{cases}$$

A generates a C_0-semigroup given by

$$S(t)y = e^{\pi t}\langle y, \varphi_{30}\rangle\varphi_{30} + \sum_{(i,j)\neq(3,0)} e^{\lambda_{ij}t}\langle y, \varphi_{ij}\rangle\varphi_{ij}$$

The operator $By = \sum_{(i,j)\neq(3,0)} \lambda_{ij}^{-4}\langle y, \varphi_{ij}\rangle\varphi_{ij}$. Moreover, $(\lambda_{ij}^{-4}, \varphi_{ij})$, for $(i,j) \neq (3,0)$, are the eigenfunctions and the eigenvalues of the operator B that satisfy

$$\begin{aligned} \sum_{(i,j)\in\mathcal{N}^2} \|B\varphi_{ij}\|^2 &= \sum_{(i,j)\neq(3,0)} \|B\varphi_{ij}\|^2 \\ &= \sum_{(i,j)\neq(3,0)} |\lambda_{i,j}|^{-2} < +\infty \end{aligned}$$

which shows that B is a compact operator. We have

$$\langle BS(t)y, S(t)y\rangle = \sum_{(i,j)\neq(3,0)} \lambda_{ij}^{-3} e^{2\lambda_{ij}t}\langle y, \varphi_{ij}\rangle^2$$

Hence

$$\langle BT(t)y, T(t)y\rangle = 0 \ , \ t \geq 0 \Rightarrow \langle y, \varphi_{ij}\rangle = 0 \ \ , \ \ (i,j) \neq (3,0) \tag{11.31}$$

As $\nabla_\omega \varphi_{30} = 0$ according to (11.31) we have $\nabla_\omega y = 0$. Consequently the system (11.30) is weakly G-stabilizable on ω.

In the other hand, we remark that, for $y_0 = \varphi_{30}$, the solution of (11.30) is written $y(t) = e^{\pi^2 t}\varphi_{30}$ and so

$$\|y(t)\| = e^{\pi^2 t}\|\varphi_{30}\| \nrightarrow 0 \ \text{ when } t \to +\infty$$

Finally the system (11.30) is not weakly stabilizable on ω.

11.2.1 Regional Stabilizing Control

In this paragraph we characterize the control that assure the weak gradient stabilization of the bilinear system (11.28) on a subregion of the system evolution domain

The solution of system (11.28) is written

$$y(t) = S(t)y_0 + \int_0^t u(s)S(t-s)By(s)ds \tag{11.32}$$

Deriving formally $\|\nabla_\omega y(t)\|^2$ along the trajectories of system (11.28), we obtain

$$\frac{d}{dt}\|\nabla_\omega y(t)\|^2 = \mathcal{R}e(\langle G_\omega Ay(t), y(t)\rangle) + \mathcal{R}e(u(t)\langle G_\omega By(t), y(t)\rangle)$$

If in addition the operator $G_\omega A$ is dissipative ($\mathcal{R}e\langle G_\omega Ay, y\rangle \leq 0$, for each $y \in \mathcal{D}(A)$) then

$$\frac{d}{dt}\|\nabla_\omega y(t)\|^2 \leq \mathcal{R}e(u(t)\langle G_\omega By(t), y(t)\rangle)$$

The choice of control $u(t) = -\langle y(t), G_\omega By(t)\rangle$ allows the energy dissipation inequality.

More general we are interested to the study of quadratic controls of type $u(t) = \langle y(t), Ky(t)\rangle$ where $K \in \mathcal{L}(X)$ is a linear bounded operator.

The following result gives sufficient conditions assuring gradient weak stabilization of system (11.28) by

$$u(t) = -\langle KBy(t), y(t)\rangle \tag{11.33}$$

Proposition 11.1 *If the conditions*

1. *$(S(t))$ is a contraction semigroup,*
2. *B is compact,*
3. *$Re(\langle DBy, y\rangle\langle y, By\rangle) \geq 0$, for all $y \in H^1(\Omega)$,*

4. $\langle DBS(t)y, S(t)y\rangle\langle S(t)y, BS(t)y\rangle = 0$, *for all* $t \geq 0 \Rightarrow \nabla_\omega y = 0$, *are satisfied, then the control (11.33) weakly stabilize the gradient of system (11.28) on* ω.

Proof The existence of the solution is assured by the Proposition 3.3 and the continuity of the application $y_0 \mapsto y(t)$.

Let the function $F(y) = -\langle y, KBy\rangle By$. The range of each ball of X by B is a ball of X, this imply that F is localy Lipshitz, and as B is compact, there exists $y_0 \in X$ such that $y(t) \rightharpoonup y_0$ when $t \to \infty$ and $\langle F(y(t)), y(t)\rangle = 0$, for all $t > 0$ (see [2]).

Hence $\langle KBT(t)y_0, T(t)y_0\rangle\langle T(t)y_0, BT(t)y_0\rangle = 0$, for all $t > 0$ and with the hypothesis 4, we obtain $\nabla_\omega y_0 = 0$.

Let $z \in (L^2(\Omega))^n$, as ∇_ω is a continuous operator then

$$\langle \nabla_\omega y(t), z\rangle \to \langle \nabla_\omega y_0, z\rangle \text{ quand } t \to \infty$$

We deduce that system (11.28) is weakly G-stabilizable on ω. □

Corollary 11.1 *If the hypothesis 1 and 2 of the previous proposition and the following condition*

$$\langle BS(t)y, S(t)y\rangle = 0\ ,\ t \geq 0\ \Rightarrow \nabla_\omega y = 0 \tag{11.34}$$

hold, then the control $u(t) = -\langle y(t), By(t)\rangle$ *weakly stabilizes the gradient of system (11.28) on* ω.

Proof For $K = Id$, the conditions 3. and 4. of the previous proposition hold. □

Remark 11.4 1. If the condition (11.34) is replaced by the condition

$$\langle BS(t)y, S(t)y\rangle = 0\ ,\ t \geq 0 \Rightarrow y = 0$$

then the control $u(t) = -\langle By(t), y(t)\rangle$ weakly stabilizes the system (11.28) (see [2]). But we also have, $y = 0 \Rightarrow \nabla y = \nabla_\omega y = 0$. Hence the system (11.28) is weakly G-stabilizable.

11.2.2 State Space Decomposition

In this paragraph we look for controls that stabilize the gradient of system (11.28), using the decomposition method. We consider the operator (10.20) and we suppose there exist two operators B_1 and B_2 in $\mathcal{L}(X)$, such that

$$\begin{cases} \Pi B &= B_1\Pi \\ (Id - \Pi)B &= B_2(Id - \Pi) \end{cases}$$

then the state space $X = X_u \oplus X_s$ and the system (11.28) is decomposed into two subsystems given by

$$\begin{cases} \dot{y}_u(t) = A_u y_u(t) + v(t) B_u y_u(t) \\ y_u \;\; = \Pi y \\ y_{0u} \;\; = \Pi y_0 \in H_u \end{cases} \tag{11.35}$$

and

$$\begin{cases} \dot{y}_s(t) = A_s y_s(t) + v(t) B_s y_s(t) \\ y_s \;\; = (Id - \Pi) y \\ y_{0s} \;\; = (Id - \Pi) y_0 \in H_s \end{cases} \tag{11.36}$$

where $A_u = \Pi A$, $A_s = (Id - \Pi)A$, $B_u = B_1 \Pi$ et $B_s = B_2(Id - \Pi)$. The solutions of (11.35) and (11.36) are given by

$$y_u(t) = S_u(t) y_{0u} + \int_0^t v(\tau) S_u(t-\tau) B_u y_u(\tau) d\tau \tag{11.37}$$

and

$$y_s(t) = S_s(t) y_{0s} + \int_0^t v(\tau) S_s(t-\tau) B_s y_s(\tau) d\tau \tag{11.38}$$

where $S_u(t)$ and $S_s(t)$ are respectively the restrictions of $S(t)$ to H_u and to H_s, of generators A_u and A_s.

Let us recall that A_s verify the hypothesis of spectral decreasing

$$\lim_{t \to +\infty} \frac{\ln \|S_s(t)\|}{t} = supRe(\sigma(A_s)) \tag{11.39}$$

which gives

$$\|S_s(t)\| \leq a\, e^{-bt},\; t \geq 0,\;\; 0 < b < \delta \tag{11.40}$$

We consider the control

$$u(t) = -\langle G_\omega K_u G_\omega y_u(t), y_u(t) \rangle \tag{11.41}$$

where $K_u \in \mathcal{L}(H_u)$ and $G_\omega = \nabla_\omega^* \nabla_\omega$, then we have the following result.

Proposition 11.2 *If the system (11.28) may be decomposed into subsystems (11.35) and (11.36) and if moreover, A_s verify (11.39), then we have the following two points:*
1. If the control (11.41) exponentially stabilizes the gradient of system (11.35) ω, then the control (11.41) exponentially stabilizes the gradient of system (11.28) on ω.
2. If the state of system (11.35) is bounded, the the state of system (11.28) is bounded on the whole domain Ω.

Proof 1. As the system (11.35) is exponentially G-stabilizable on ω, then the solution y_u is global and we have

$$\|\nabla_\omega y_u(t)\| \leq M e^{-ct} \|y_{0u}\|, \text{ for } M, c > 0 \tag{11.42}$$

The existence of a weak solution y_s of system (11.36) is assured on an interval $[0, t_{\max}[$.

To show that the solution $y_s(t)$ is global, it's sufficient to prove that $y_s(t)$ is bounded. By (11.38) and (11.40), we have

$$\|y_s(t)\| \leq a\, e^{-bt}\|y_{0s}\| + a\|B_s\| \int_0^t |v(\tau)|e^{-b(t-\tau)}\|y_s(\tau)\|d\tau$$

which gives

$$e^{bt}\|y_s(t)\| \leq a\,\|y_{0s}\| + a\|B_s\| \int_0^t |v(\tau)|e^{b\tau}\|y_s(\tau)\|d\tau.$$

By Gronwall lemma, we have

$$e^{bt}\|y_s(t)\| \leq a\|y_{0s}\| \exp(\int_0^t a\|B_s\||v(\tau)|d\tau) \tag{11.43}$$

From (11.41) and (11.42), it follows

$$|v(t)| \leq M^2\|\nabla_\omega D\nabla_\omega^*\|e^{-2ct}\|y_0\|^2 \tag{11.44}$$

using (11.43) and (11.44), there exists $L > 0$ such that

$$\|y_s(t)\| \leq aL\|y_{0s}\|e^{-bt} \tag{11.45}$$

Consequently $y_s(t)$ is bounded, so $y(t)$ is a weak unique and global solution of system (11.28) ($y = y_s + y_u$).

According to the inequality (11.45), we have $\|y_s(t)\| \to 0$ exponentially when $t \to +\infty$, it follows $\|\nabla_\omega y_s(t)\| \to 0$ exponentially when $t \to +\infty$.

and using (11.35), (11.42) and the inequality

$$\|\nabla_\omega y(t)\| \leq \|\nabla_\omega y_u(t)\| + \|\nabla_\omega y_s(t)\|$$

we have the conclusion.

2. This point is immediate considering (11.45). □

The spectral decreasing inequality (11.39) may be replaced by the less restrictive condition

$$\lim_{t\to+\infty} \frac{\ln |||\nabla_\omega S_s(t)|||}{t} \leq supRe(\sigma(A_s)) \tag{11.46}$$

We obtain the following result.

Proposition 11.3 *Let us suppose that the system (11.28) may be decomposed into two subsystems (11.35) and (11.36) and that A_s verify (11.46).*

Assume also there exists a real $N > 0$, such that

$$\|\nabla_\omega B_s y\| \leq N\|\nabla_\omega y\| \ , \ y \in X \tag{11.47}$$

If in addition $y_s(t)$ is a weak, unique and global solution of system (11.36), then with the control (11.41), the results of the Proposition 11.2 hold.

Proof If the system(11.35) is exponentially G-stabilizable on ω, then its weak solution $y_u(t)$ is unique and global.

Moreover the weak solution of $y_s(t)$ of system (11.36) is unique and global, and $y = y_u + y_s$, then $y(t)$ is a weak solution, unique and global of system (11.28).

According to the spectrum decomposition of A, we have $\sup \mathcal{R}e(\sigma(A_s)) \leq -\delta$.

Using (11.46), it follows that $\|\nabla_\omega T_s(t)\|_n \leq a' e^{-b't}$, with $t \geq 0$, $0 < b' < \delta$ and $a' > 0$.

By (11.38) and (11.47), we obtain

$$e^{b't}\|\nabla_\omega y_s(t)\|_n \leq a' \|\nabla_\omega y_{0s}\|_n + a'N \int_0^t |v(\tau)| e^{b'\tau} \|\nabla_\omega y_s(\tau)\|_n d\tau$$

With the same techniques that used to prove the Proposition 11.2, we complete the proof. □

Remark 11.5 1. If $(S_s(t))$ is contraction semigroup and

$$Re(\langle y_s, B_s y_s\rangle\langle K_u y_u, y_u\rangle) \geq 0$$

then $y_s(t)$ is a weak solution unique and global of system (11.36).

2. If $B_s = Id$ and (11.47) is verified, if moreover $(S_s(t))$ is a contraction semigroup, and K_u is a positive operator then $y_s(t)$ is a weak solution, unique and global of system (11.36).

Proposition 11.4 *If the system (11.28) may be decomposed in subsystems (11.35) and (11.36), if moreover the conditions (11.46), and (11.47) are verified and that $y_s(t)$ is a weak, unique and global solution of system (11.36) and $(S_u(t))$ is a contraction semigroup.*

If the state y_u is bounded, there exists $M' > 0$ such that $\|y_u\| \leq M'$, if moreover B_u is compact and satisfy the condition:

$$\langle B_u S_u(t) y_u, S_u(t) y_u\rangle = 0 \ , \ t \geq 0 \Rightarrow \nabla_\omega y_u = 0 \tag{11.48}$$

If B_s verify la condition

$$\|K_u\| M'^2 < \frac{b}{aN\|B_s\|} \tag{11.49}$$

then the control

$$v(t) = -\langle B_u y_u(t), y_u(t)\rangle$$

weakly stabilizes the gradient of system (11.28) on ω.

In the inequality (11.49), the constant b is the same as the one in the inequality (11.40).

Proof With the same techniques that the ones used to prove the Proposition 11.3 and with Gronwall lemma we have

$$e^{bt}\|\nabla_\omega y_s(t)\| \leq a\|y_{0s}\|\exp(\int_0^t aN|||B_s||||v(\tau)|d\tau)$$

$$\leq a\|y_{0s}\|\exp(aN|||B_s|||\,|||K_u|||M'^2 t)$$

This gives

$$\|\nabla_\omega y_s(t)\| \leq a\|y_{0s}\|\exp((aN|||B_s|||\,|||K_u|||M'^2 - b)t)$$

With corollary 11.1 and inequality (11.49), we finish the proof. □

11.2.3 Problem of Regional Gradient Stabilization

This section considers the problem of gradient stabilization of (11.28) on a subregion ω minimizing a given functional cost. We give also a decay estimate of $\|\nabla_\omega y(t)\|^2$.

We consider the functional

$$J(u) = \int_0^{+\infty} |\langle P_\omega By(t), y(t)\rangle|^2 dt + \int_0^{+\infty} |u(t)|^2 dt + \int_0^{+\infty} \langle Ry(t), y(t)\rangle dt$$

The problem may be modeled by

$$\begin{cases} \min J(u) \\ u \in V_{ad} = \{u \mid y(t) \text{ is a global solution and } J(u) < +\infty\} \end{cases} \tag{11.50}$$

where P_ω and R are positives and self adjoint operators such that

$$\int_0^1 |\langle P_\omega BT(t)y, T(t)y\rangle| dt \geq \alpha\|\nabla_\omega y\|_n^2\ , \ y \in X \text{ for some } \alpha > 0 \tag{11.51}$$

We suppose there exists $\beta > 0$ such that

$$\langle P_\omega y, z\rangle \leq \beta \|\nabla_\omega y\| \|\nabla_\omega z\| \ , \ (y, z) \in X \times X \tag{11.52}$$

and reals $a, b > 0$, such that, for all $t \geq 0$ and $y \in X$, we have

$$\|\nabla_\omega T(t) y\| \leq a \|\nabla_\omega y\| \tag{11.53}$$

and

$$\|\nabla_\omega B y\| \leq b \|\nabla_\omega y\| \tag{11.54}$$

We consider the algebraic Riccati equation

$$\langle P_\omega A y, y\rangle + \langle y, P_\omega A y\rangle + \langle R y, y\rangle = 0 \ , \ y \in \mathcal{D}(A) \tag{11.55}$$

The main result of this section is based on the following lemma.

Lemma 11.1 *We note*

$$\mu = \int_0^1 \langle P_\omega B y(t), y(t)\rangle dt \quad and \quad \nu = \int_0^1 |u(t)|^2 dt$$

Assume that (11.51), (11.52), (11.53) and (11.54) are verified, then there exist $\eta > 0$, *and* $\lambda > 0$ *such that*

$$\nu < \eta \ \Rightarrow \ \|\nabla_\omega y_0\|_n^2 < \lambda \sqrt{\mu}$$

Proof Without loss of generality we assume that $\nu < 1$. Let $\psi(t) = y(t) - S(t) y_0$.
1. We show the estimation

$$\text{there exists } C_0 > 0 \text{ such that } \|\nabla_\omega \psi(t)\| \leq C_0 \|\nabla_\omega y_0\| \sqrt{\nu} \ , \ t \in]0, 1[\tag{11.56}$$

$\psi(t) = \int_0^t u(s) S(t-s) B S(s) y_0 ds + \int_0^t u(s) S(t-s) B (y(s) - S(s) y_0) ds,$ and with (11.53) and (11.54) we obtain

$$\begin{aligned} \|\nabla_\omega \psi(t)\| &\leq a^2 b \|\nabla_\omega y_0\| \int_0^t |u(s)| ds \\ &\quad + ab \int_0^t |u(s)| \|\nabla_\omega \psi(s)\| ds \\ &\leq a^2 b \|\nabla_\omega y_0\| \sqrt{\nu} + ab \int_0^t |u(s)| \|\nabla_\omega \psi(s)\| ds \end{aligned}$$

Gronwall lemma allows to get

$$\begin{aligned}\|\nabla_\omega \psi(t)\| &\leq a^2 b \|\nabla_\omega y_0\| \sqrt{\nu} \\ &\quad + a^3 b^2 \sqrt{\nu} \|\nabla_\omega y_0\| \int_0^t |u(s)| \exp(\int_s^t ab|u(\tau)|d\tau) ds \\ &\leq a^2 b \|\nabla_\omega y_0\| \sqrt{\nu} \\ &\quad + a^3 b^2 \sqrt{\nu} \|\nabla_\omega y_0\| \exp(ab)\end{aligned}$$

foe all $t \in]0, 1[$, and $\nu < 1$. Thus (11.56) is verified with $C_0 = a^2 b \|\nabla_\omega y_0\| + a^3 b^2 \|\nabla_\omega y_0\| \exp(ab)$.

The triangular inequality gives

$$\begin{aligned}\int_0^1 |\langle P_\omega BS(t) y_0, S(t) y_0\rangle| dt &\leq \sqrt{\mu} + \int_0^1 |\langle P_\omega BS(t)\psi(t), S(t) y_0\rangle| dt \\ &\quad + \int_0^1 |\langle P_\omega BS(t) y_0, \psi(t)\rangle| dt + \int_0^1 |\langle P_\omega B\psi(t), \psi(t)\rangle| dt\end{aligned}$$

By (11.52), (11.53), (11.54) and (11.56) there exist $\tilde{a}, \tilde{b} > 0$ such that

$$\int_0^1 |\langle P_\omega BS(t) y_0, S(t) y_0\rangle| dt \leq \sqrt{\mu} + \tilde{a}\sqrt{\nu} \|\nabla_\omega y_0\|^2 + \tilde{b}\nu \|\nabla_\omega y_0\|^2$$

As $\nu < 1$, with (11.51) we obtain

$$(\alpha - (\tilde{a} + \tilde{b})\sqrt{\nu}) \|\nabla_\omega y_0\|^2 \leq \sqrt{\mu}$$

It suffices to take

$$\eta = \frac{1}{2} \inf(1, \frac{\alpha^2}{(\tilde{a} + \tilde{b})^2})$$

to finish the proof. □

Proposition 11.5 *Assume that there exists P_ω a positive and self adjoint operator, solution the algebraic Riccati equation (11.55), such that (11.51), (11.52), (11.53) and (11.54) are verified. If moreover the solution $y^*(t)$ corresponding to the control*

$$u^*(t) = -\langle B^* P_\omega y^*(t), y^*(t)\rangle \tag{11.57}$$

is global, then $u^*(t)$ is the unique solution of problem (11.50) that strongly stabilises the gradient of system (11.28) on ω.

In addition if there exists $d > 0$ such that

$$|\langle P_\omega By, y\rangle|^2 + \frac{1}{2}\langle Ry, y\rangle \leq d\mathcal{R}e(\langle y, By\rangle\langle P_\omega y, y\rangle - \langle Ay, y\rangle) \ , \ y \in \mathcal{D}(A) \tag{11.58}$$

then the state of system (11.28) remains bounded on Ω.

Proof We consider the function $F(y) = \langle P_\omega y, y\rangle$, for each $y \in X$, we show that $u^* \in V_{ad}$.

For $y_0 \in \mathcal{D}(A)$, with (11.55) we have

$$\frac{\partial F(y^*(t))}{\partial t} = -2|\langle B^* P_\omega y^*(t), y^*(t)\rangle|^2 - \langle Ry^*(t), y^*(t)\rangle \quad (11.59)$$

Integrating (11.59) over $[0, t]$ and since P_ω is positive we obtain:

$$\int_0^t |\langle P_\omega By^*(s), y^*(s)\rangle|^2 ds + \int_0^t \langle Ry^*(s), y^*(s)\rangle ds \le F(y_0)ds, \ t \ge 0 \quad (11.60)$$

As the application $y_0 \longmapsto y^*(.)$ is continuous from X to $C([0, t]; X)$, then $y \longmapsto F(y)$ and $y \longmapsto \langle Ry, y\rangle$ are continuous from H to C and hence (11.60) is verified for all $y_0 \in X$.

Which imply that for all initial condition $y_0 \in X$, $J(u^*) < +\infty$ and consequently $u^* \in V_{ad}$.

2. We show that each control $u \in V_{ad}$ stabilizes the gradient of system (11.28) on ω.

Let ϵ such that $0 < \epsilon < \eta$. As J is finite, Cauchy criteria for integral convergence gives

$$\exists T > 0, \ \forall t > T : \int_t^{t+1} \langle P_\omega By(s), y(s)\rangle ds < \epsilon \text{ and } \int_t^{t+1} |u(s)|^2 ds < \epsilon$$

Applying Lemma 11.1 with $y_0 = y(t)$, we have $\|\nabla_\omega y(t)\| < \lambda\sqrt{\epsilon}$, $t > T$.

Then $\|\nabla_\omega y(t)\| \to 0$, when $t \to +\infty$.

3. The control $u^*(t)$ is the unique solution of (11.50). Indeed, we have

$$F(y(t)) \le M\|\nabla_\omega y(t)\|^2 , \text{ for } M > 0$$

Thus $F(y(t)) \to 0$ when $t \to +\infty$.

For $y_0 \in \mathcal{D}(A)$, we integrate the relation

$$\begin{aligned}\frac{\partial F(y(t))}{\partial t} &= |\langle B^* P_\omega y(t), y(t)\rangle + u(t)|^2 - |\langle B^* P_\omega y(t), y(t)\rangle|^2 \\ &\quad -|u(t)|^2 - \langle Ry(t), y(t)\rangle\end{aligned}$$

and we obtain

$$J(v) = F(y_0) + \int_0^{+\infty} |\langle B^* P_\omega y(t), y(t)\rangle + v(t)|^2 dt$$

This leads to $J(u^*) \leq J(u)$, for all $u \in V_{ad}$. For $y_0 \in X$, there exist a sequence $(y_{0n}) \subset \mathcal{D}(A)$ such that $y_{0n} \to y_0$ when $n \to +\infty$. Let $u \in V_{ad}$ we have

$$J(u) = F(y_{0n}) + \int_0^{+\infty} |\langle B^* P_\omega y_n(t), y_n(t)\rangle + v(t)|^2 dt \tag{11.61}$$

So $J(u) \geq F(y_{0n})$.

As the application F is continuous we deduce that $J(v) \geq F(y_0) = J(u^*)$ which show that v^* is a solution of problem (11.50).

4. The uniqueness of $u^*(t)$ comes from the uniqueness of $y(t)$ the solution of system (11.28). Indeed, according to (11.61), every optimal control is of form

$$u(t) = -\langle B^* P_\omega y(t), y(t)\rangle$$

Let

$$u_1(t) = -\langle B^* P_\omega y_1(t), y_1(t)\rangle \quad \text{and} \quad u_2 = -\langle B^* P_\omega y_2(t), y_2(t)\rangle$$

be two solutions of problem (11.50) where $y_1(t)$ and $y_2(t)$ are the corresponding solutions of system (11.28). As (11.28) has only one solution, then $y_1(t) = y_2(t)$ and hence $u_1(t) = u_2(t)$.

5. We show the state of system (11.28) remains bounded on Ω.

By the inequality (11.58), we have $\frac{\partial}{\partial t} F(y(t))^2 \geq d\frac{d}{dt}\|y(t)\|^2$. And

$$F(y(t)) - F(y_0) \geq d(\|y(t)\|^2 - \|y_0\|^2)\ , \ y_0 \in \mathcal{D}(A) \tag{11.62}$$

As $y_0 \longmapsto y(.)$ is continuous from X to $C([0,t], X)$ and the application F is also continuous, the inequality (11.62) is satisfied for all $y_0 \in X$. Moreover $F(y(t)) \to 0$ when $t \to +\infty$. □

Remark 11.6 We can take in condition (11.52), $P_\omega = G_\omega P G_\omega$ where $P \in \mathcal{L}(X)$ is a continuous and positive operator, solution of the equation (11.55).

Proposition 11.6 *If the conditions (11.51), (11.52), (11.53), and (11.54) are verified with P_ω solution of the algebraic Riccati equation (11.55) and if*

$$\text{there exists } d > 0 \text{ such that } \langle P_\omega y, y\rangle \geq d\|\nabla_\omega y\|^2 \tag{11.63}$$

$$\text{there exists } e > 0 \text{ such that } \langle Ry, y\rangle \geq e\|\nabla_\omega y\|^2 \tag{11.64}$$

and

$$Re(\langle RAy, y\rangle) \leq \mathcal{R}e(\langle P_\omega By, y\rangle\langle RBy, y\rangle)\ , \ y \in \mathcal{D}(A) \tag{11.65}$$

are satisfied and that the solution $y(t)$ is global, then for all initial condition y_0 such that $\nabla_\omega y_0 \neq 0$, we have the estimation

$$\|\nabla_\omega y(t)\|^2 = O(\frac{1}{t^3}),\ when\ t \to +\infty$$

Proof We consider the function

$$W(y(t)) = \langle P_\omega y(t), y(t)\rangle + \int_0^t \langle Ry(s), y(s)\rangle ds\ ,\ t \geq 0$$

and

$$W_m = \frac{1}{2}\langle P_\omega y^*(m), y^*(m)\rangle + \frac{1}{2}\int_0^m \langle Ry^*(s), y^*(s)\rangle ds\ ,\ t \geq 0$$

We remark that W_m is a positive sequence. We integrate between m and $m+1$ the relation

$$\frac{\partial W(y^*(t))}{\partial t} = -2|\langle P_\omega By^*(t), y^*(t)\rangle|^2$$

we obtain

$$W_{m+1} - W_m = -\int_m^{m+1} |\langle P_\omega By^*(t), y^*(t)\rangle|^2 dt = -\mu$$

$$= -\int_m^{m+1} |u^*(t)|^2 =: -\nu$$

Applying the Lemma 11.1 with $y_0 = y^*(t)$ we obtain

$$W_{m+1} - W_m \leq -\eta \text{ where } W_{m+1} - W_m < \frac{-1}{\lambda^2}\|\nabla_\omega y_0\|^4. \tag{11.66}$$

For y_0 such that $\nabla_\omega y_0 \neq 0$, there exists $M > 0$ such that

$$W_0 = \frac{1}{2}\langle P_\omega y_0, y_0\rangle \leq M\|\nabla_\omega y_0\|^2.$$

The condition (11.63) leads to $W_0 > 0$ and by (11.66) it follows

$$W_{m+1} - W_m \leq -f(y_0)W_m^2 \tag{11.67}$$

with $f(y_0) = \min(\frac{1}{\lambda^2 M^2}, \frac{\eta}{W_0^2})$. Thus W_m is a decreasing sequence.

We note $V_m = \frac{1}{W_m}$, then

$$V_{m+1} - V_m = \frac{W_m - W_{m+1}}{W_{m+1}W_m} \geq \frac{W_m - W_{m+1}}{W_m^2}$$

By (11.67) we obtain $V_{m+1} - V_m \geq f(y_0)$ and so $V_{m+1} \geq V_0 + mf(y_0)$. which gives

$$W_m \leq \frac{W_0}{1 + W_0 f(y_0)m}, \quad m \geq 0$$

and we have the estimation

$$W(y^*(t)) \leq \frac{2W(y_0)}{1 + W(y_0)f(y_0)t}, \quad t \geq 0 \tag{11.68}$$

For $y_0 \in X$ such that $\nabla_\omega y_0 \neq 0$ then there exists a sequence $y_{0n} \in \mathcal{D}(A)$ and $y_{0n} \to y_0$ when $n \to +\infty$, $\nabla_\omega y_{0n} \neq 0$ (∇_ω is a continuous operator). With the same above techniques, we show that(11.68) is verified for y_{0n}. But as $y_0 \longmapsto y^*(.)$ is continuous from X to $C([0,t], X)$, the function $y_0 \longmapsto f(y_0)$ is continuous. And the estimation (11.68) holds for all $y_0 \in X$.

The condition (11.65) leads to $\frac{\partial}{\partial t}\langle Ry(t), y(t)\rangle \leq 0$. It follows that

$$W(y^*(t)) \geq d\|\nabla_\omega y^*(t)\|_n^2 + t\langle Ry^*(t), y^*(t)\rangle$$

and by (11.64), we obtain

$$\begin{aligned} W(y^*(t)) &\geq d\|\nabla_\omega y^*(t)\|_n^2 + e\|\nabla_\omega y^*(t)\|_n^2 t \\ &\geq (d + et)\|\nabla_\omega y^*(t)\|_n^2 \end{aligned}$$

We deduce that

$$\|\nabla_\omega y^*(t)\|_n^2 \leq \frac{2W(y_0)}{(d + et)(1 + W(y_0)f(y_0)t)} \tag{11.69}$$

As $f(y_0)$, $W(y_0)$ and e are non zero, we obtain the desired estimation.

If we consider the problem (11.50) with $R = 0$, we have the following result.

Corollary 11.2 *If the conditions (11.51), (11.52), (11.53), and (11.54) are verified with P_ω, solution of the algebric Riccati equation (11.55) and satisfy the condition (11.63), then for all initial state y_0, such that $\nabla_\omega y_0 \neq 0$, we have the estimation*

$$\|\nabla_\omega y^*(t)\|_n = O(t^{-\frac{1}{2}}) \quad \textit{when } t \longrightarrow +\infty$$

Proof We note

$$W(y(t)) = \langle P_\omega y(t), y(t)\rangle \text{ et } W_m = \frac{1}{2}W(y^*(m))$$

With the same techniques as above we show that the estimation (11.68) and with (11.63), we deduce the desired estimation. □

11.3 Numerical Approach and Simulations

11.3.1 Regional Gradient Stabilization Case

We have seen that the control solution of problem (11.50) is given by (11.57) where P_ω is the unique solution of the equation (11.55).

We consider the system

$$\begin{cases} \dot{y}(t) = Ay(t) + u(t)By(t) \ t > 0 \\ \\ y(0) = y_0 \end{cases} \tag{11.70}$$

and the control

$$u(t) = -\langle P_\omega By(t), y(t)\rangle \tag{11.71}$$

Let us denote $t_m = mh$, with $h > 0$ and $m \in \mathcal{N}$. On $[t_{m-1}, t_m]$, we take $u(t) = u(t_{m-1})$ and the system (11.70) may be approached by the system

$$\begin{cases} \dot{y}_m(t) = Ay_m(t) + u(t_{m-1})By_m(t) \\ \\ y(0) = y_0 \end{cases} \tag{11.72}$$

and the control by

$$\begin{cases} y_m(t_{m-1}) = y_{m-1}(t_{m-1}) \\ \\ u(t_m) = -\langle P_\omega By(t_m), y(t_m)\rangle \\ \\ u(0) = -\langle P_\omega By_0, y_0\rangle \end{cases} \tag{11.73}$$

Solving these equation gives $y(t_m)$ and the control over the next interval is $u(t_m) = -\langle P_\omega By(t_m), y(t_m)\rangle$.

The following result establish a relation between the gradient of system (11.70) and the one of system (11.72).

Proposition 11.7 *We assume that the weak solution of system (11.70) is global and bounded and there exist* $\lambda > 0$ *and* $M \leq 1$ *such that* $\|\nabla_\omega T(t)\|_n \leq Me^{-\lambda t}$, *if moreover B verify (11.54) and* $\alpha_2 \leq \dfrac{\lambda}{M|||B|||}$, *then there exists* $c > 0$ *such that*

$$\|\nabla_\omega y_m(t) - \nabla_\omega y(t)\| < c\ h \ , \quad t_{m-1} \leq t \leq t_m \tag{11.74}$$

Proof 1. The solution of system (11.70) is bounded and there exists a real $\alpha_1 > 0$ such that $\|y(t)\| \leq \alpha_1$, this leads to $|v(t)| \leq \alpha_2$, where $\alpha_2 = \alpha_1^2|||P_\omega B|||$.

2. We show that

$$\|\nabla_\omega y(t)\| \le M\|\nabla_\omega y_0\|e^{-(\lambda - Mb|||B|||\alpha_2)t}$$

Indeed, let $j(t) = \nabla_\omega T(t)$, we have

$$\begin{aligned}\|\nabla_\omega y(t)\| &\le \|j(t)y_0\| + \int_0^t |u(s)|\|j(t-s)By(s)\|ds \\ &\le Me^{-\lambda t}\|\nabla_\omega y_0\|_n + Mb|||B|||\alpha_2 \int_0^t e^{-\alpha(t-s)}\|\nabla_\omega y(s)\|_n ds\end{aligned}$$

More again

$$e^{\lambda t}\|\nabla_\omega y(t)\| \le M\|\nabla_\omega y_0\| + Mb|||B|||\alpha_2 \int_0^t e^{\alpha s}\|\nabla_\omega y(s)\|ds$$

and Gronwall lemma gives

$$e^{\lambda t}\|\nabla_\omega y(t)\| \le M\|\nabla_\omega y_0\|e^{Mb|||B|||\alpha_2 t}$$

It follows $\|\nabla_\omega y(t)\| \le M\|\nabla_\omega y_0\|e^{-(\lambda - Mb|||B|||\alpha_2)t}$.
3. We show that for $m > 1$, there exist two positive constants M_1 and $\tilde{\lambda}$ such that

$$\|\nabla_\omega z_m^t\| \le M\|\nabla_\omega z_{m-1}^{t_{m-1}}\| + M_1 h e^{\tilde{\lambda}(m-1)h} \tag{11.75}$$

where $z_m^t = y_m(t) - y(t)$. Indeed,

$$\begin{aligned}\|\nabla_\omega z_m^t\| &\le Me^{-\lambda(t-t_{m-1})}\|\nabla_\omega z_{m-1}^{t_{m-1}}\| \\ &\quad + bM|||B||| \int_{t_{m-1}}^t e^{-\lambda(t-s)}|u(t_{m-1}) - u(s)|\|\nabla_\omega y(s)\|ds \\ &\quad + bM|||B||| \int_{t_{m-1}}^t e^{-\lambda(t-s)}|u(t_{m-1})|\|\nabla_\omega z_m^s\|ds \\ &\le Me^{-\lambda(t-t_{m-1})}\|\nabla_\omega z_{m-1}^{t_{m-1}}\| \\ &\quad + 2\alpha_2 M^2 b\|\nabla_\omega y_0\||||B||| \int_{t_{m-1}}^t e^{-\lambda(t-s)}e^{-(\lambda - Mb|||B|||\alpha_2)s}ds \\ &\quad + \alpha_2 Mb|||B||| \int_{t_{m-1}}^t e^{-\lambda(t-s)}\|\nabla_\omega z_m^s\|ds\end{aligned}$$

Which gives

$$\begin{aligned}e^{\lambda t}\|\nabla_\omega z_m^t\| &\le Me^{\lambda t_{m-1}}\|\nabla_\omega z_{m-1}^{t_{m-1}}\| + 2\alpha_2 M^2 b\|\nabla_\omega y_0\||||B|||he^{Mb|||B|||\alpha_2 t} \\ &\quad + \alpha_2 Mb|||B||| \int_{t_{m-1}}^t e^{\lambda s}\|\nabla_\omega z_m^s\|ds\end{aligned}$$

Once again Gronwall lemma gives

$$e^{\lambda t}\|\nabla_\omega z_m^t\| \le Me^{\lambda t_{m-1}}\|\nabla_\omega z_{m-1}^{t_{m-1}}\| + 2\alpha_2 M^2 b\|\nabla_\omega y_0\| |||B|||he^{Mb|||B|||\alpha_2 t}$$

$$+\alpha_2 M^2 b|||B|||e^{\lambda t_{m-1}}\|\nabla_\omega z_{m-1}^{t_{m-1}}\| \int_{t_{m-1}}^{t} e^{Mb|||B|||\alpha_2(t-s)}ds$$

$$+2\alpha_2^2 M^3 b^2\|\nabla_\omega y_0\|\|B\|^2 h^2 e^{Mb|||B|||\alpha_2 t}$$

which leads to

$$e^{\lambda t}\|\nabla_\omega z_m^t\| \le Me^{\lambda t_{m-1}}\|\nabla_\omega z_{m-1}^{t_{m-1}}\|e^{Mb|||B|||\alpha_2(t-t_{m-1})}$$

$$+2\lambda M\|\nabla_\omega y_0\|he^{Mb|||B|||\alpha_2 t}$$

$$+2M\|\nabla_\omega y_0\|\lambda^2 h^2 e^{Mb|||B|||\alpha_2 t}$$

We note $\tilde{\alpha} = -\lambda + bM\|B\|\alpha_2$, it follows

$$\|\nabla_\omega z_m^t\| \le M\|\nabla_\omega z_{m-1}^{t_{m-1}}\|_n e^{\tilde{\alpha}(t-t_{m-1})} + 2\lambda M\|\nabla_\omega y_0\|he^{\tilde{\alpha}t}$$

$$+2M\|\nabla_\omega y_0\|\lambda^2 h^2 e^{\tilde{\alpha}t}$$

$$\le M\|\nabla_\omega z_{m-1}^{t_{m-1}}\| + 2\lambda M\|\nabla_\omega y_0\|he^{\tilde{\alpha}(m-1)h}$$

$$+2M\|\nabla_\omega y_0\|\lambda^2 h^2 e^{\tilde{\alpha}(m-1)h}$$

We obtain

$$\|\nabla_\omega z_m^t\| \le M\|\nabla_\omega z_{m-1}^{t_{m-1}}\| + M_1 he^{\tilde{\alpha}(m-1)h} \tag{11.76}$$

with $M_1 = 2\lambda M\|\nabla_\omega y_0\|(1 + h\lambda)$.

4. We show the estimation (11.74). Following the same above techniques, we get

$$\|\nabla_\omega z_1^t\|e^{\lambda t} \le 2\lambda M\|\nabla_\omega y_0\|he^{Mb|||B|||\alpha_2 t} + Mb|||B|||\alpha_2 \int_0^t \|\nabla_\omega z_1^t\|e^{\lambda s}ds$$

and with Gronwall lemma we have

$$\|\nabla_\omega z_1^t\|e^{\lambda t} \le 2\lambda M\|\nabla_\omega y_0\|_n he^{Mb|||B|||\alpha_2 t} + 2Mh^2\lambda^2\|\nabla_\omega y_0\|e^{Mb|||B|||\alpha_2 t}$$

hence $\|\nabla_\omega z_1^t\| \le M_1 h$. As $-\lambda + \alpha_2 Mb|||B||| < 0$, then

$$\lim_{m \longrightarrow +\infty} (m+1)^2 e^{(-\lambda+\alpha_2 Mb|||B|||)mh} = 0$$

so there exists $N_1 > 0$ such that for all $m > N_1$,

$$e^{(-\lambda+\alpha_2 Mb|||B|||)mh)} < \frac{1}{(1+m)^2}$$

Using (11.76), for all $m > N_1$, we have

$$\|\nabla_\omega z^t_m\| \le M\|\nabla_\omega z^{t_{m-1}}_{m-1}\| + \frac{M_1 h}{m^2}$$

thus

$$\|\nabla_\omega z^t_m\| \le M^2\|\nabla_\omega z^{t_{m-2}}_{m-2}\| + M\frac{M_1 h}{(m-1)^2} + \frac{M_1 h}{m^2}$$

so

$$\begin{aligned}\|\nabla_\omega z^t_m\| \le M^{m-N_1}\|\nabla_\omega z^{t_{N_1}}_{N_1}\| + M^{m-N_1-1}\frac{M_1 h}{(N_1+1)^2}\\ +M^{m-N_1-2}\frac{M_1 h}{(N_1+2)^2} + \cdots + \frac{M_1 h}{m^2}\end{aligned} \tag{11.77}$$

and by (11.76) for $m' \le N_1$, we have

$$\begin{aligned}\|\nabla_\omega z^t_{m'}\| &\le M\|\nabla_\omega z^{t_{m'-1}}_{m'-1}\| + M_1 h\\ &\le M^2\|\nabla_\omega z^{t_{m'-2}}_{m'-2}\| + MM_1 h + M_1 h\\ &\le \sum_{i=0}^{m'} M^i M_1 h\end{aligned}$$

then for all $m' \le N_1$, $\|\nabla_\omega z^{t_{m'}}_{m'}\| \le hM_1 \sum_{i=0}^{m'} M^i$. Replacing in (11.77), we obtain

$$\begin{aligned}\|\nabla_\omega z^t_m\| &\le M^{m-N_1}\sum_{i=0}^{N_1} M^i M_1 h + \sum_{i=0}^{m-N_1-1}\frac{M^i M_1}{(m-i)^2}h\\ &\le (M^{m-N_1}\sum_{i=0}^{N_1} M^i M_1 + \sum_{i=0}^{m-N_1-1}\frac{M^i M_1}{(m-i)^2})h\\ &\le (M^{m-N_1}\sum_{i=0}^{N_1} M^i + \sum_{i=N_1+1}^{m}\frac{M^{m-i}}{i^2})M_1 h,\end{aligned}$$

As the last serial is convergent, then the estimation (11.74) is verified for $c > 0$. □

Now we solve the algebraic Riccati equation (11.55). We set $H^N = \text{span}\{\phi_i \ \ i = 1, \cdots, N\}$ is a subspace of $H^1(\Omega)$ where $\{\phi_i \ |i \in N^*\}$ is an orthogonal basis of $H^1(\Omega)$.

H^N is endowed with the restriction of the inner product of $H^1(\Omega)$.

We define the projection operator by:

$$\begin{aligned}\Pi_N : H^1(\Omega) &\longrightarrow H^N \\ y &\mapsto \sum_{i=1}^{N} \langle y, \phi_i \rangle \phi_i(x)\end{aligned} \tag{11.78}$$

Solving (11.55), turns to solving the algebraic Riccati equation:

$$\langle P_\omega{}^N A^N y^N, y^N \rangle + \langle y^N, P_\omega^N A^N y^N \rangle + \langle R^N y^N, y^N \rangle = 0 \tag{11.79}$$

where A^N, P_ω^N, and R^N are respectively the projections of operators A, P_ω and R.

As for all $y \in H^1(\Omega)$, $\lim_{N\to\infty} \|\Pi^N(y) - y\| = 0$ then $\lim_{N\to\infty} \|P_\omega^N \Pi^N y - Py\| = 0$ or again $P_\omega^N \Pi^N$ converges to P_ω, (see [3]).

For $m \geq 1$ and $t_{m-1} \leq t \leq t_m$, we consider the system

$$\begin{cases} \dfrac{\partial y_m(t)}{\partial t} = Ay_m(t) + u(t_{m-1})By_m(t) \\ y_m(t_{m-1}) = y_{m-1}(t_{m-1}) \\ u(t_m) = -\langle P^N B^N y_m^N(t_m), y_m^N(t_m) \rangle \\ u(0) = -\langle P^N B^N y_0^N, y_0^N \rangle \end{cases} \tag{11.80}$$

where the operator $(A + u^N(t_{m-1})B)$ generates a C_0-semigroup

$$T^{m,N} = \sum_{i \geq 0} T_i^{m,N}$$

with

$$\begin{cases} T_i^{m,N}(t)y = T^{m,N}y + u(t_{m-1}) \displaystyle\int_0^t T^{m,N}(t-s)BT_{i-1}^{m,N} y ds \\ T_0^{m,N} = T \end{cases}$$

Then $\lim_{N\to+\infty} \|T^{m,N} - T^m\| = 0$, where T^m is the C_0-semigroup of generator $(A + u(t_{m-1})B)$. So the weak solution of system (11.80) strongly converges to the weak solution of system (11.72) when $N \to +\infty$.

This leads the the following algorithm.
Step 1 Initialisation of data:
A tolerance $\varepsilon > 0$, a target subregion ω, N the dimension of the projection space and y_0^N, the projection of the initial state.
Step 2:
Choice of a sequence $(t_m)_{m\in N}$, $t_{m+1} = t_m + h$, with $h > 0$ small enough for numerical considerations.

Step 3
Solving (11.79) with the algorithm given in [4] gives P_ω^N
Step 4:
1. $u(0) = -\langle P_\omega^N B^N y_0^N, y_0^N \rangle$
Repeat.
2. $u(t) = -\langle P_\omega^N B^N y_m^N(t_{m-1}), y_m^N(t_{m-1})\rangle$ on $[t_{m-1}, t_m]$.
3. Solving (11.80) gives $y_m^N(t_m)$
4. If $\| \nabla_\omega y_m(t_m) \| \leq \epsilon$, otherwise $m = m+1$ and go to 2.

We give illustrations through some examples.

Example 11.5 On $\Omega =]0, 1[$, we consider the system

$$\begin{cases} \dfrac{\partial y}{\partial t}(x,t) = 0.01\Delta y(x,t) + 0.68 y(x,t) + u(t) y(x,t) \text{ on } Q \\ \dfrac{\partial y(0,t)}{\partial x} = \dfrac{\partial y(1,t)}{\partial x} = 0 \text{ on } \Omega \\ y(x,0) \;\; = x^2(1-x^3)^2 \end{cases} \tag{11.81}$$

and a target subregion $\omega =]0, 0.4[$. We consider the problem (11.50) with $R = 5\nabla_\omega^* \nabla_\omega$.

The eigenfunctions of the operator $A = 0.01\Delta y(x,t) + 0.68 y(x,t)$ are $\phi_i(x) = a_i \cos(i\pi x)$, with $a_i = \sqrt{\dfrac{2}{1+(i\pi)^2}}$ for all $i \geq 0$, and constitute an orthogonal basis in $H^1(\Omega)$.

We consider the space H^N generated $\phi_i(x)$, $i = 1, N$, and for simulations we take $N = 6$. The projection of the operator $A = 0.01\Delta y(x,t) + 0.68 y(x,t)$ is a matrix given by

$$A^6 = (\langle (0.01\Delta + 0.68)\phi_{i-1}, \phi_{j-1} \rangle)_{1 \leq i,j \leq 6}$$

with

$$diag(A^6) = (0.5813, 0.2852, -0.2083, -0.8991, -1.7874, -2.8731)$$

Applying the previous algorithm, we obtain $P_\omega^6 = (p_{ij})_{1 \leq i,j \leq 6}$, a symmetric matrix with:

$$P_\omega^6 = \begin{pmatrix} 1.6725 & 0.6623 & 0.2405 & -0.1397 & -0.1499 & -0.0089 \\ 0.6623 & 1.5564 & 0.5736 & 0.0236 & -0.1218 & -0.0444 \\ 0.2405 & 0.5736 & 0.8415 & 0.5082 & 0.1496 & -0.0458 \\ -0.1397 & 0.0236 & 0.5082 & 0.6683 & 0.4124 & 0.0814 \\ -0.1499 & -0.1218 & 0.1496 & 0.4124 & 0.4252 & 0.2482 \\ -0.0089 & -0.0444 & -0.0458 & 0.0814 & 0.2482 & 0.3048 \end{pmatrix}$$

and this gives the results represented by the following figures (Fig. 11.1).

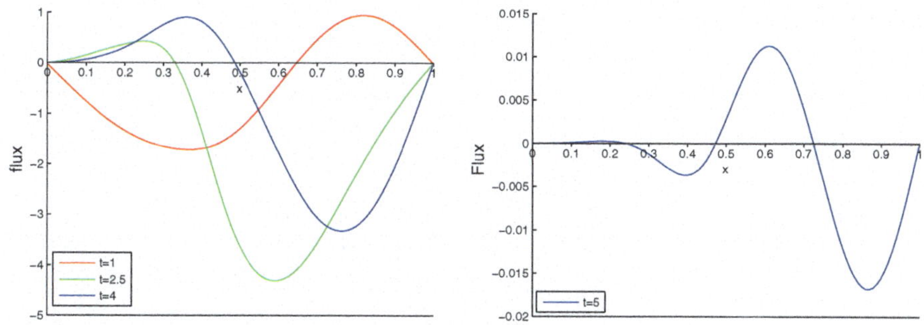

Fig. 11.1 Evolution du gradient

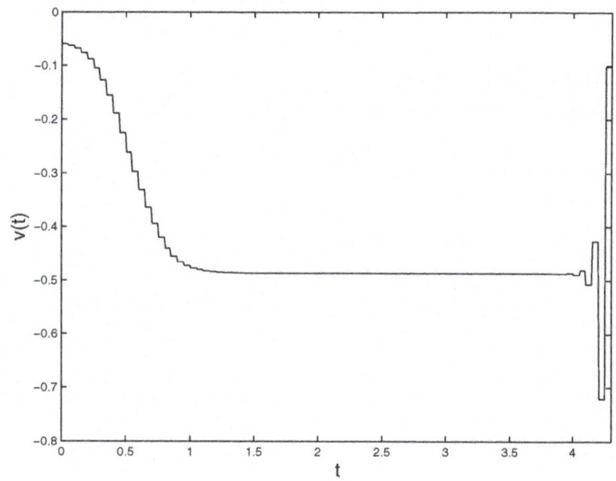

Fig. 11.2 Evolution of control

The above figures show that the system (11.81) is G-stabilizable on $\omega =]0, 0.4[$ with error of $5.9678\ 10^{-7}$ and cost equals to $3.1\ 10^{-4}$.

The next figure presents the evolution of the stabilizing control (Fig. 11.2).

We study the evolution of cost and error G-stabilization with respect to the area of the subregion target ω. We obtain the following array.

ω	]0,0.3[	]0,0.4[	]0,0.65[	]0,0.7[	]0,0.9[	]0,1[
Error $\times 10^6$	0.0059	3.8522	3.5058	4.7438	26.048	27.163
Cost $\times 10^2$	5.43	5.83	6.58	6.65	6.79	6.8

If we fix $\omega =]0.6, 1[$, the error and cost of the G-stabilization on ω are respectively $1.6037\ 10^{-4}$ and $2.83\ 10^{-2}$.

The following figures show that at $t = 4.5$, the system (11.81) is G-stabilizable on ω (Fig. 11.3).

The following figure shows that the evolution of the G-stabilizing control (Fig. 11.4).

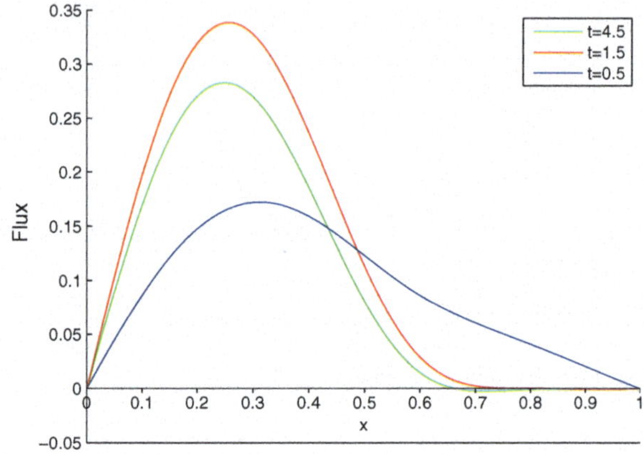

Fig. 11.3 Evolution of the gradient

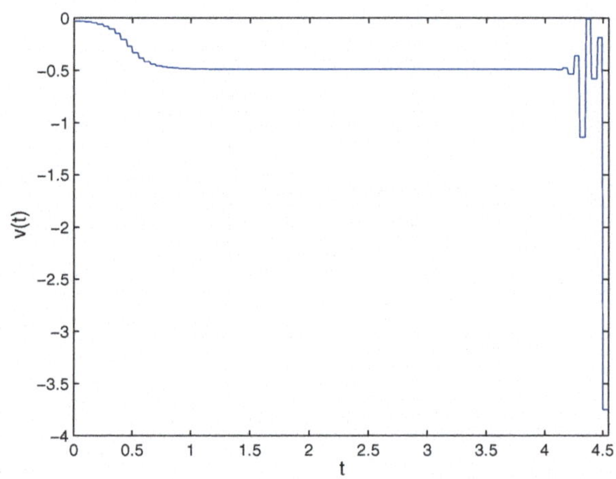

Fig. 11.4 Evolution of control

Example 11.6 On $\Omega =]0, 1[$, we consider the system

$$\begin{cases} \dfrac{\partial y}{\partial t}(x,t) = 0.01\Delta y(x,t) + 0.5y(x,t) + 0.7u(t)y(x,t) \text{ on } Q \\ y(0,t) \quad = y(1,t) = 0 \\ y(x,0) \quad = 0.5x(1-x) \text{ on } \Omega \end{cases} \tag{11.82}$$

We take $\omega =]0, 0.4[$ and we consider the problem (11.50) with $R = 3\nabla_\omega^*\nabla_\omega$.

For $N = 5$, $A^5$is the diagonal matrix projection of $A = 0.01\Delta + 0.5Id$, given by

$$diag(A^5) = (0.4013, 0.10521, -0.3882, -1.07913, -1.9674)$$

and

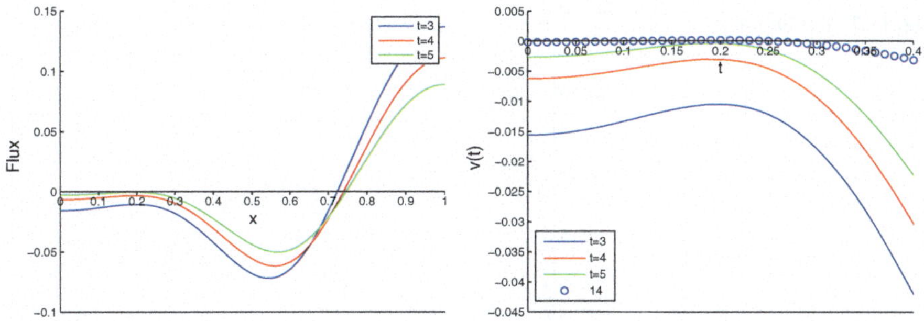

Fig. 11.5 Evolution of the gradient

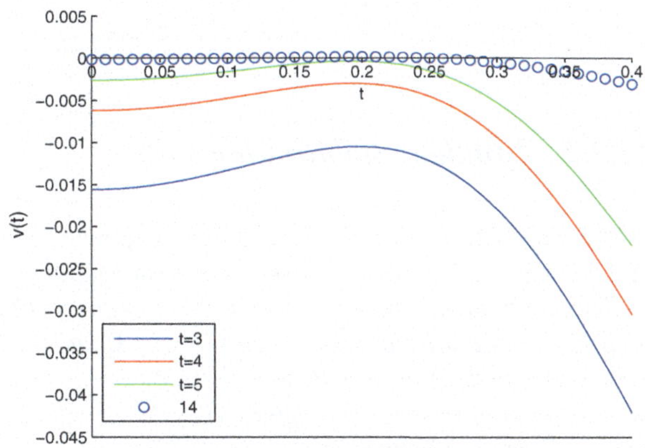

Fig. 11.6 Evolution of the gradient on $\omega =]0, 0.4[$

$$P^5_\omega = \begin{pmatrix} 1.8654 & 0.4911 & -0.0599 & -0.1141 & -0.01013 \\ 0.4911 & 1.0613 & 0.63035 & 0.1827 & -0.0805 \\ -0.0599 & 0.6303 & 0.9209 & 0.5089 & 0.0623 \\ -0.1141 & 0.1827 & 0.5089 & 0.5401 & 0.3287 \\ -0.0101 & -0.0805 & 0.0623 & 0.3287 & 0.4292 \end{pmatrix}$$

The simulations results are given by the following figures (Fig. 11.5).

The next figure indicates the gradient evolution of system (11.82) only on the target subregion $\omega =]0, 0.4[$ (Fig. 11.6).

We remark that at $t = 14$, the gradient evolves close to 0, on $\omega =]0, 0.4[$, hence the system (11.82) is G-stabilizable on ω, with error equals $9.9235\ 10^{-4}$ and cost equals to $12.39\ 10^{-2}$ (Fig. 11.7).

The following shows the relation between the area of the target subregion, the cost and the error of the G-stabilization.

ω	]0,0.1[	]0,0.3[	]0,0.5[	]0,0.8[	]0,1[
Error $\times 10^4$	8.2691	9.5172	9.9658	9.9878	9.9954
Cost $\times 10^2$	11.9	12.22	13.01	13.5	14.29

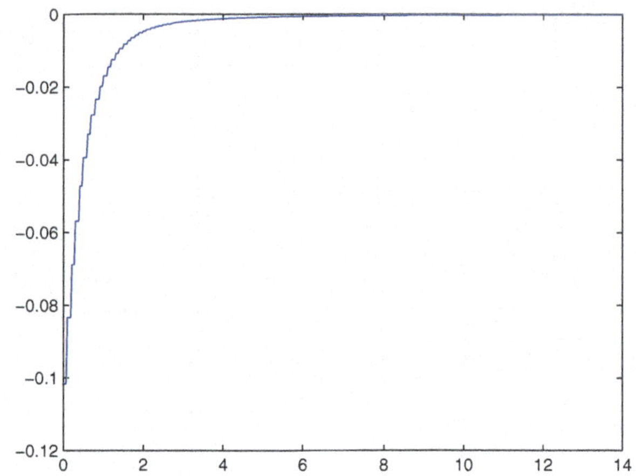

Fig. 11.7 Evolution of control

11.3.2 Gradient Stabilization Case

We have seen that control (11.4) stabilizes exponentially the gradient of system (11.1) and control (11.22) ensures strong and weak stabilization.

The gradient stabilization of system (11.1) may follow the steps:
Step 1: Initial data: threshold accuracy $\epsilon > 0$ *and initial condition* y_0;
Step 2: Apply $u(t_i) = -\langle \nabla By(t_i), \nabla y(t_i)\rangle_n$;
Step 3: Solving system (11.1) *gives* $y(t_{i+1})$;
Step 4: Calculate $\nabla y(t_{i+1})$;
Step 5: If $\|\nabla y(t_{i+1})\|_n < \epsilon$ *stop, else ;*
Step 6: $i = i + 1$ *and go to 2.*

Example 11.7 On $\Omega =]0, 1[$, we consider the system

$$\begin{cases} \dfrac{\partial y(x,t)}{\partial t} = \dfrac{\partial^2 y(x,t)}{\partial x^2} + y(x,t) + u(t)By(x,t), & \Omega\times]0,+\infty[\\ y(x,0) = x(1-x), & \Omega \\ y(0,t) = y(1,t) = 0, &]0,+\infty[\end{cases} \tag{11.83}$$

where $By = \displaystyle\sum_{j=1}^{+\infty} \frac{1}{j^2}\langle y, \varphi_j\rangle\varphi_j$, with $\varphi_j(x) = \sqrt{\dfrac{2}{1+(j\pi)^2}}\sin(j\pi x)$,
$\forall j \in \mathbb{N}^*$.

Fig. 11.8 Gradient evolution

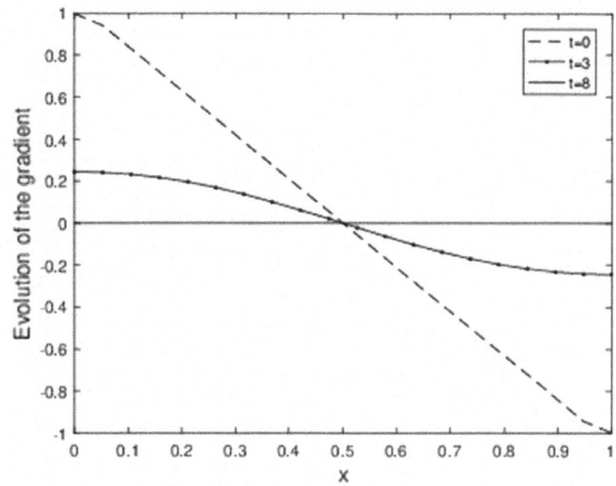

Fig. 11.9 State evolution

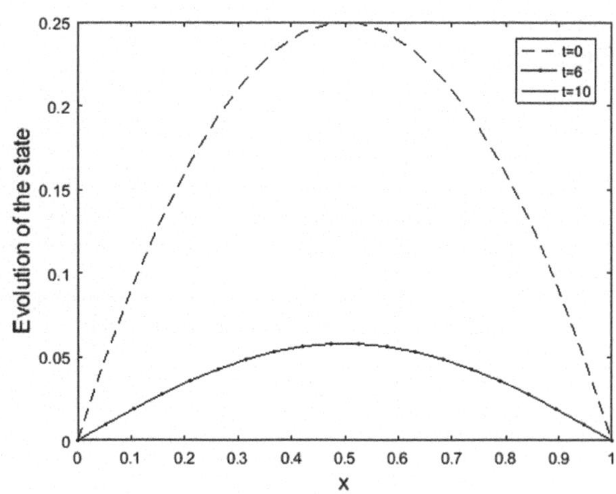

Applying the above algorithm and solving (11.83) using Petrov-Galerkin method given in (Skeel and Berzins 1990), Fig. 11.8 shows that the gradient of system (11.83) is weakly stabilized at time $t = 8$ by control (11.22), with a stabilization error equals to 4.73×10^{-5} and Fig. 11.9 shows that the state of the system is stabilized at time $t = 10$ with stabilization error equals to 8.69×10^{-5} (Fig. 11.10).

We remark that the state of the system is stabilized in time greater than the one needed for the stablization of the gradient of the system.

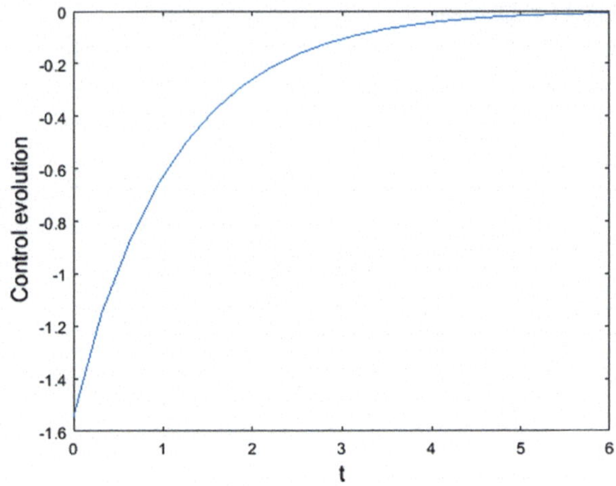

Fig. 11.10 Control evolution

References

1. Pazy, A.: Semigroups of Linear Operators and Applications to Partial Differential Equations. Springer, New York (1983)
2. Ball, J.M.: On the asymptotic behavior of generalized processes with application to nonlinear evolutions equations. J. Differ. Equ. **27**, 224–265 (1978)
3. Banks, H., Kunisch, B.: The linear regulator problem for parabolic systems. SIAM J. Control Optim. **22**, 684–696 (1984)
4. Arnold, W.F., Laub, A.J.: Generalized eigenproblem: algorithms and software for algebraic Riccati equations. Proc. IEEE **72**, 1746–1754 (1984)

Chapter 12
Regional Gradient Stabilization of Infinite Dimensional Semilinear Systems

This chapter deals with regional gradient stabilization of a semilinear system defined on a bounded and regular spatial domain $\Omega \subset \mathbb{R}^n$ by

$$\frac{dy(t)}{dt} = Ay(t) + u(t)By(t), \qquad y(0) = y_0 \tag{12.1}$$

where the state space is considered to be complex-valued Hilbert space $H^1(\Omega)$ endowed with its usual inner product, denoted by $\langle .,. \rangle$ and $\|.\|$ the associated norm, A is a linear operator with domain $\mathcal{D}(A) \subset H^1(\Omega)$ and generates a semigroup of contractions $S(t)$, $u(t)$ is a scalar valued control and B is a nonlinear and bounded operator mapping $H^1(\Omega)$ into itself and verifies $B(0) = 0$ (0 is an equilibrium state of system (12.1)).

12.1 Controls for Regional Gradient Stabilization

In this section we study regional strong and weak gradient stabilization for distributed semilinear system (12.1) by considering the class of feedback controls of the form

$$u(t) = -\frac{\langle y(t), By(t) \rangle}{g(y(t))}, \tag{12.2}$$

where $g : H^1(\Omega) \longrightarrow (0, +\infty)$ is an appropriate function.

Let ω be a subregion of Ω of positive Lebesgue measure and we define the operator

$$\begin{aligned} \nabla_\omega : H^1(\Omega) &\longrightarrow \qquad\qquad (L^2(\omega))^n \\ y \quad &\longmapsto (\chi_\omega \frac{\partial y(x)}{\partial x_1}, \chi_\omega \frac{\partial y(x)}{\partial x_2}, \ldots, \chi_\omega \frac{\partial y(x)}{\partial x_n}) \end{aligned}$$

E. H. Zerrik and O. Castillo, *Stabilization of Infinite Dimensional Systems*, Studies in Systems, Decision and Control 355,
https://doi.org/10.1007/978-3-030-68600-0_12

where χ_ω is the restriction operator to ω, defined by

$$\begin{array}{rcl}\chi_\omega : L^2(\Omega) & \longrightarrow & L^2(\omega) \\ y & \longmapsto & y_{/\omega}\end{array}$$

12.1.1 Regional Gradient Strong Stabilization

In this subsection, we develop sufficient conditions for regional gradient strong stabilization of system (12.1).

Theorem 12.1 *Assume that g is locally Lipschitz such that* $g(.) \geq m > 0$ *on* $H^1(\Omega)$, *B is locally Lipschitz and*

$$\int_0^T |\langle S(s)y, BS(s)y\rangle| ds \geq \delta \|\nabla_\omega y\|^2_{(L^2(\omega))^n}, \ \forall y \in H^1(\Omega), \ (T, \delta > 0). \quad (12.3)$$

Then the control (12.2) regionally strongly stabilizes the gradient of system (12.1) on ω.

Proof Let us first show that the operator $N(y) = \dfrac{\langle y, By\rangle}{g(y)} By$ is locally Lipshitz.

Let $x, y, z \in H^1(\Omega)$ and $R > 0$ such that $\|x - y\|, \|x - z\| \leq R$.

Without loss of generality, we can take $x = 0$ and we have

$$\|g(y) - g(z)\| \leq L_R(g)\|y - z\| \text{ and } \|By - Bz\| \leq L_R(B)\|y - z\|$$

Thus

$$\begin{aligned}\|Nz - Ny\| &= \frac{\|g(y)\langle z, Bz\rangle Bz - g(z)\langle y, By\rangle By\|}{g(z)g(y)} \\ &\leq \frac{1}{m^2}(\|g(y)(\langle z, Bz\rangle Bz - \langle y, By\rangle By)\| \\ &+ \|(g(y) - g(z))\langle y, By\rangle By\|) \\ &\leq \frac{1}{m^2}(\|g(y)\langle z, Bz\rangle(Bz - By) + g(y)(\langle z, Bz\rangle - \langle y, By\rangle)By\| \\ &+ \|(g(y) - g(z))\langle y, By\rangle By\|) \\ &\leq \frac{K_R(g)R^2L_R^2(B)}{m^2}(3 + R)\|z - y\|.\end{aligned}$$

where $K_R(g) = \sup(M_R(g), L_R(g))$, $M_R(g) = \sup_{\|x\| \leq R} g(x)$ and $L_R(g) = \sup_{(y,z)\in B_R^2, y\neq z} \frac{|g(y) - g(z)|}{\|y - z\|}$ which implies that N is locally Lipschitz and

$$
\begin{aligned}
L_R(N) := & \sup_{(y,z)\in B_R^2, y\neq z} \frac{|N(y) - N(z)|}{\|y - z\|} \\
& \leq \frac{K_R(g) R^2 L_R^2(B)}{m^2}(3 + R).
\end{aligned}
$$

Since N is locally Lipschitz and $\langle Ny, y\rangle \geq 0$, $\forall y \in H^1(\Omega)$, system (12.1) has a unique global mild solution $y(t)$.

Since $S(t)$ is a semigroup of contractions (so that A is dissipative), then

$$
\frac{d}{dt}\|y(t)\|^2 \leq -2\langle Ny(t), y(t)\rangle, \ \forall t \geq 0.
$$

Integrating this last inequality, we get

$$
\|y(t)\|^2 - \|y(\tau)\|^2 \leq -2\int_\tau^t \langle Ny(s), y(s)\rangle ds, \ \forall t \geq \tau \geq 0. \tag{12.4}
$$

It follows that

$$
\|y(t)\| \leq \|y_0\|, \ \forall t \geq 0. \tag{12.5}
$$

It remains to show that $\nabla_\omega y(t) \longrightarrow 0$, as $t \longrightarrow +\infty$.

From the relation

$$
\begin{aligned}
\langle NS(t)y_0, S(t)y_0\rangle = & \langle NS(t)y_0 - Ny(t), S(t)y_0\rangle - \langle Ny(t), \Phi(t)\rangle \\
& + \langle Ny(t), y(t)\rangle
\end{aligned}
$$

we obtain the following inequality

$$
\begin{aligned}
|\langle NS(t)y_0, S(t)y_0\rangle| \leq & L_{\|y_0\|}(N)\|\Phi(t)\|(\|S(t)y_0\| + \|y(t)\|) \\
& + |\langle Ny(t), y(t)\rangle|, \forall t \in [0, T].
\end{aligned}
$$

where $\Phi(t) = \int_0^t S(t-s)Ny(s)ds$.

Moreover, since $S(t)$ is a semigroup of contractions, then, for a fixed $T > 0$, Schwartz's inequality yields

$$
\|\Phi(t)\| \leq \frac{\sqrt{T}}{m}\|B\|\|y_0\|\Big(\int_0^T |\langle y(s), By(s)\rangle|^2 ds\Big)^{\frac{1}{2}}.
$$

Furthermore, we have

$$|\langle NS(t)y_0, S(t)y_0\rangle| \leq 2\|y_0\| L_{\|y_0\|}(N)\|\Phi(t)\| + |\langle Ny(t), y(t)\rangle|, \forall t \in [0, T].$$

Then, we deduce the following estimate

$$\begin{aligned}\int_0^T |\langle NS(s)y_0, S(s)y_0\rangle| ds \leq & \frac{2T^{\frac{3}{2}}}{m}\|B\|\|y_0\|^2 L_{\|y_0\|}(N) \\ & \times \left(\int_0^T |\langle y(s), By(s)\rangle|^2 ds\right)^{\frac{1}{2}} \\ & + \frac{1}{m}\int_0^T |\langle y(s), By(s)\rangle|^2 ds.\end{aligned}$$

By remarking that

$$|\langle NS(s)y_0, S(s)y_0\rangle| \geq \frac{|\langle S(s)y_0, BS(s)y_0\rangle|}{M_{\|y_0\|}(g)},$$

we obtain

$$\begin{aligned}\int_0^T |\langle S(s)y_0, BS(s)y_0\rangle| ds \leq & M_{\|y_0\|}(g)\frac{2T^{\frac{3}{2}}}{m}\|B\|\|y_0\|^2 L_{\|y_0\|}(N) \\ & \times \left(\int_0^T |\langle y(s), By(s)\rangle|^2 ds\right)^{\frac{1}{2}} \\ & + \frac{M_{\|y_0\|}(g)}{m}\int_0^T |\langle y(s), By(s)\rangle|^2 ds.\end{aligned}$$

Then, replacing y_0 by $y(t)$ and using (12.5), we get

$$\begin{aligned}\int_0^T |\langle S(s)y(t), BS(s)y(t)\rangle| ds \leq & K_{\|y_0\|}\left(\int_t^{t+T} |\langle y(s), By(s)\rangle|^2 ds\right)^{\frac{1}{2}} \\ & + \frac{M_{\|y_0\|}(g)}{m}\int_t^{t+T} |\langle y(s), By(s)\rangle|^2 ds\end{aligned} \quad (12.6)$$

where $K_{\|y_0\|} = \frac{2T^{\frac{3}{2}}}{m}\|B\|\|y_0\|^2 L_{\|y_0\|}(N)$.

Setting $\tau = 0$ in (12.4), we obtain

$$2\int_0^t \langle Ny(s), y(s)\rangle ds \leq \|y_0\|^2 - \|y(t)\|^2, \; \forall t \geq 0.$$

then

$$2\int_0^t \langle Ny(s), y(s)\rangle ds \le \|y_0\|^2, \ \forall t \ge 0.$$

Since $\dfrac{|\langle y(s), By(s)\rangle|^2}{M_{\|y_0\|}(g)} \le \dfrac{|\langle y(s), By(s)\rangle|^2}{g(y(s))}$, we get

$$\int_0^t |\langle y(s), By(s)\rangle|^2 ds \le \frac{M_{\|y_0\|}(g)}{2}\|y_0\|^2, \ \forall t \ge 0.$$

which implies that

$$\int_0^{+\infty} |\langle y(s), By(s)\rangle|^2 ds < +\infty.$$

Hence

$$\int_t^{t+T} |\langle y(s), By(s)\rangle|^2 ds \longrightarrow 0, \ as\ t \longrightarrow +\infty.$$

It follows from (12.6) that

$$\int_0^T |\langle S(s)y(t), BS(s)y(t)\rangle| ds \longrightarrow 0, \ as\ t \longrightarrow +\infty. \tag{12.7}$$

From (12.3) and (12.7), we deduce that $\nabla_\omega y(t) \longrightarrow 0, \ as\ t \longrightarrow +\infty$. ■

The following Corollary is a direct consequence of Theorem 12.1.

Corollary 12.1 *Assume that B is locally Lipschitz and (12.3) holds. Then both controls*

$$u_1(t) = -\frac{\langle y(t), By(t)\rangle}{1 + |\langle y(t), By(t)\rangle|} \tag{12.8}$$

and

$$u_2(t) = -\frac{\langle y(t), By(t)\rangle}{\sup(1, |\langle y(t), By(t)\rangle|)}, \tag{12.9}$$

regionally strongly stabilize the gradient of system (12.1) on ω.

Example 12.1 Let $\Omega =]0, 1[$, we consider the following beam equation

$$\begin{cases} \dfrac{\partial^2 y}{\partial t^2}(x,t) = -\dfrac{\partial^4 y}{\partial x^4}(x,t) + u(t) i_\omega \dfrac{\partial y}{\partial t}(x,t) & in\ \Omega \times]0, +\infty[\\ y(\xi, t) = \dfrac{\partial^2 y}{\partial x^2}(\xi, t) = 0, \ \xi = 0, 1 & on\ ]0, +\infty[\end{cases} \tag{12.10}$$

where $i_\omega = \chi_\omega^* \chi_\omega$, with χ_ω is the restriction operator to ω.

Let $P = \dfrac{\partial^4 y}{\partial x^4}$ with domain $\mathcal{D}(P) = \{y \in L^2(0,1) / \dfrac{\partial^4 y}{\partial x^4} \in L^2(0,1),\ y(\xi,t) = \dfrac{\partial^2 y}{\partial x^2}(\xi,t) = 0,\ \xi = 0, 1\}$.

The set $\mathcal{D}(P^{\frac{1}{2}})$ forms a Hilbert space under the inner product $\langle y_1, y_2\rangle_P = \langle P^{\frac{1}{2}} y_1, P^{\frac{1}{2}} y_2\rangle_{L^2(0,1)}$. The eigenvalues of P are $\lambda_j = (j\pi)^4$, corresponding to eigenfunctions $\varphi_j(x) = \sqrt{2}\sin(j\pi x)\ \forall j \in \mathbb{N}^*$.

To write system (12.10) in the form (12.1) we set $X = (H^2(0,1) \cap H_0^1(0,1)) \times L^2(0,1)$ endowed with the inner product $\langle (y_1, z_1), (y_2, z_2)\rangle = \langle y_1, y_2\rangle_P + \langle z_1, z_2\rangle_{L^2(0,1)}$

$$A = \begin{pmatrix} 0 & I \\ -P & 0 \end{pmatrix} \ and\ B = \begin{pmatrix} 0 & 0 \\ 0 & i_\omega \end{pmatrix}.$$

The operator A is skew-adjoint, and the operator B satisfies the condition (12.3), indeed, for $y \in X$, we have

$$y = \sum_{j=1}^{\infty} \begin{pmatrix} \alpha_j \\ \lambda_j^{\frac{1}{2}} \beta_j \end{pmatrix} \varphi_j$$

where $(\alpha_j, \beta_j) \in \mathbb{R}^2\ \forall j \geq 1$.

Then

$$S(s)y = \sum_{j=1}^{\infty} \begin{pmatrix} \alpha_j \cos(\lambda_j^{\frac{1}{2}} s) + \beta_j \sin(\lambda_j^{\frac{1}{2}} s) \\ \beta_j \lambda_j^{\frac{1}{2}} \cos(\lambda_j^{\frac{1}{2}} s) - \alpha_j \lambda_j^{\frac{1}{2}} \sin(\lambda_j^{\frac{1}{2}} s) \end{pmatrix} \varphi_j, \quad \forall s \geq 0.$$

It follows that

$$\begin{aligned} \langle S(s)y, BS(s)y\rangle &= \sum_{j=1}^{\infty} \lambda_j \left\{\alpha_j \sin(\lambda_j^{\frac{1}{2}} s) - \beta_j \cos(\lambda_j^{\frac{1}{2}} s)\right\} \\ &= \sum_{j=1}^{\infty} \lambda_j \left\{\alpha_j^2 \sin^2(j\pi s) - \beta_j^2 \cos^2(j\pi s) - \sin(2j\pi s)\beta_j \alpha_j\right\}. \end{aligned}$$

Integrating this relation over the time interval $[0, 2]$, we obtain

$$\begin{aligned} \int_0^2 |\langle S(s)y, BS(s)y\rangle| ds &= \sum_{j=1}^{\infty} \lambda_j (\alpha_j^2 + \beta_j^2) \\ &\geq \delta \|\nabla_\omega y\|^2_{(L^2(\omega))^n} \end{aligned}$$

with $\delta = \min\{\lambda_j^{-\frac{1}{2}} | j \in \mathbb{N}^*\}$, so that (12.3) holds.

We conclude that the feedbacks

$$u_1(t) = -\frac{\|\chi_\omega \partial_t y(.,t)\|^2}{\sup(1, \|\chi_\omega \partial_t y(.,t)\|^2)}$$

and

$$u_2(t) = -\frac{\|\chi_\omega \partial_t y(.,t)\|^2}{1 + \|\chi_\omega \partial_t y(.,t)\|^2}$$

regionally strongly stabilize the gradient of system (12.10). ■

Remark 12.1 Note that this example has also been considered in [1] where a weak stabilization result using quadratic feedback has been established.

To prove our next result, we will need the following lemma.

Lemma 12.1 ([2]) *Let p be a positive increasing function such that $p(0) = 0$ and set $\tilde{p}(t) = t - (I + p)^{-1}(t)$, where I denotes the identity function.*

Let $(z_k)_{k\geq 0}$ be a sequence of positive numbers such that

$$p(z_{k+1}) + z_{k+1} \leq z_k, \ k \geq 0$$

then $z_k \leq S(k)z_0$ where $S(t)z_0$ is the solution of

$$\begin{cases} \dfrac{dx(t)}{dt} = -\tilde{p}(x(t)) \\ x(0) = x_0 \geq 0. \end{cases} \tag{12.11}$$

Let $G_\omega = \nabla_\omega^* \nabla_\omega$ be a bounded operator applying $H^1(\Omega)$ into itself, where ∇_ω^* denotes the adjoint operator of ∇_ω.

The following theorem characterizes the asymptotic behavior of the gradient and gives sufficient conditions for regional strong stabilization of the gradient on ω with explicit decay rate.

Theorem 12.2 *Assume that B is locally Lipschitz and satisfies the following inequality*

$$\langle y, By\rangle^2 \leq \delta\langle y, By\rangle\langle y, G_\omega By\rangle, \ \ (for\ some\ \delta > 0) \tag{12.12}$$

moreover, if the following assumption holds

$$\|\nabla_\omega y\|^2_{(L^2(\omega))^n} \leq \int_0^T |\langle S(s)y, BS(s)y\rangle| ds, \ (for\ some\ T > 0). \tag{12.13}$$

Then the control $u(t) = -\langle y(t), By(t)\rangle$ regionally strongly stabilizes the gradient of system (12.1) on ω, and for all y_0 we have the following decay estimate

$$\|\nabla_\omega y(t)\|^2_{(L^2(\omega))^n} = O(\frac{1}{t}) \ as\ t \longrightarrow +\infty.$$

Proof For all $y_0 \in H^1(\Omega)$, the system (12.1) has a unique mild solution $y(.)$ defined on a maximal interval $[0, t_{max}[$ (see [1]), given by the variation of constants formula

$$y(t) = S(t)y_0 + \int_0^t S(t-s)v(s)By(s)ds.$$

Since $S(t)$ is a semigroup of contractions (so that A is dissipative), then a standard approximation argument shows that

$$\|y(t)\|^2 \leq \|y_0\|^2 - 2\int_0^t \langle y(s), By(s)\rangle^2 ds, \ \forall t \in [0, t_{max}[\tag{12.14}$$

so for all $y_0 \in H^1(\Omega)$, we have

$$\|y(t)\| \leq \|y_0\|. \tag{12.15}$$

Hence for all $y_0 \in H^1(\Omega)$, the solution $y(t)$ is global.

For $y_0 \in H^1(\Omega)$, we define the function $z(t) = y(t) - S(t)y_0 \ \forall t \geq 0$. Once more using the variation of constants formula, we have

$$y(t) = -\int_0^t \langle y(s), By(s)\rangle S(t-s)By(s)ds, \ \forall t \geq 0.$$

Using Schwartz's inequality and the fact that $S(t)$ is a semigroup of contractions (so that $\|S(s)\| \leq 1, \forall s \geq 0$), we get

$$\|y(t)\| \leq \|B\|\|y_0\|\sqrt{T}\Big(\int_0^T |\langle y(s), By(s)\rangle|^2 ds\Big)^{\frac{1}{2}}, \ \forall t \in [0, T]$$

Then

$$\|y(t)\|^2 \leq C_1 \int_0^T |\langle y(s), By(s)\rangle|^2 ds, \ \forall t \in [0, T] \tag{12.16}$$

where $C_1 = \|B\|^2\|y_0\|^2 T$.

From the relation

$$|\langle S(t)y_0, BS(t)y_0\rangle| \leq |\langle y(t) - z(t), BS(t)y_0 - By(t)\rangle| + |\langle y(t), By(t)\rangle|$$
$$+ |\langle z(t), By(t) - Bz(t)\rangle| + |\langle z(t), Bz(t)\rangle|$$

and using (12.15) and the fact that B is locally Lipschitz, we deduce that

$$|\langle S(t)y_0, BS(t)y_0\rangle| \leq k_1\|S(t)y_0 - y(t)\|\|y(t) - z(t)\| + |\langle y(t), By(t)\rangle|$$
$$+ k_2\|y(t) - z(t)\|\|z(t)\| + \alpha\|z(t)\|^2$$
$$\leq 2k\|z(t)\| + |\langle y(t), By(t)\rangle| + \alpha\|z(t)\|^2$$

with $k = \max\{k_1, k_2\}$.

From (12.16), for all $t \in [0, T]$ we have

$$|\langle S(t)y_0, BS(t)y_0\rangle| \leq 2k\sqrt{C_1C_2} + |\langle y(t), By(t)\rangle| + \alpha C_1C_2$$

with $C_2 = \int_0^T |\langle y(s), By(s)\rangle|^2 ds$.

Integrating this last inequality over the interval $[0, T]$ and using (12.13) we obtain

$$\|\nabla_\omega y(T)\|^2_{(L^2(\omega))^n} \leq C_3(\sqrt{C_2} + C_2)$$

where $C_3 = \max\{(2k\sqrt{C_1}T + 1), \alpha TC_1\}$.

It follows from (12.12) that

$$\|\nabla_\omega y(t)\|^2_{(L^2(\omega))^n} \leq \|\nabla_\omega y_0\|^2_{(L^2(\omega))^n} - 2\delta\int_0^t \langle y(s), By(s)\rangle^2 ds, \ \forall t \in [0, T] \tag{12.17}$$

thus $2\delta C_2 \leq \|\nabla_\omega y_0\|^2_{(L^2(\omega))^n} - \|\nabla_\omega y(T)\|^2_{(L^2(\omega))^n}$, hence

$$\|\nabla_\omega y(T)\|^2_{(L^2(\omega))^n} \leq \frac{1}{\sqrt{2\delta}}C_3h(\|\nabla_\omega y_0\|^2_{(L^2(\omega))^n} - \|\nabla_\omega y(T)\|^2_{(L^2(\omega))^n}) \tag{12.18}$$

where $h(t) = t + \sqrt{t}, \forall t \geq 0$.

Let $p_0(.)$ denotes the inverse of $\frac{1}{\sqrt{2\delta}}C_3h(.)$, then (12.18) can be written

$$p_0(\|\nabla_\omega y(T)\|^2_{(L^2(\omega))^n}) + \|\nabla_\omega y(T)\|^2_{(L^2(\omega))^n} \leq \|\nabla_\omega y_0\|^2_{(L^2(\omega))^n}.$$

On the other hand, From (12.17), we obtain $\|\nabla_\omega y(t)\|^2_{(L^2(\omega))^n} \leq \|\nabla_\omega y_0\|^2_{(L^2(\omega))^n}$ $\forall t \in [0, T]$, instead of the interval $[0, T]$ we could just as well work on the interval $[kT, (k+1)T]$, $k \in \mathbb{N}^*$, thus $\|\nabla_\omega y(t)\|^2_{(L^2(\omega))^n} \leq \|\nabla_\omega y_0\|^2_{(L^2(\omega))^n}$ $\forall t \in [kT, (k+1)T]$, it follows that $\|\nabla_\omega y(kT)\|^2_{(L^2(\omega))^n} \leq \|\nabla_\omega y_0\|^2_{(L^2(\omega))^n}$, then

$$p_0(W((k+1)T)) + W((k+1)T) \leq W(kT), \ \forall k \in \mathbb{N}^*$$

with $W(t) = \|\nabla_\omega y(t)\|^2_{(L^2(\omega))^n}$.

Let $z_k = W(kT)$ and applying Lemma 12.1, we obtain

$$W(kT) \le S(k)W(0),\ k \in \mathbb{N}$$

For any $t > 0$ we have $t = kT + \tau$ for some integer $k \ge 0$ and $\tau \in [0, T[$.

Since $W(t)$ and $S(t)z_0$ are nonincreasing, we have

$$\begin{aligned} W(t) &\le W(kT) \\ &\le S(\tfrac{t-\tau}{T})W(0) \\ &\le S(\tfrac{t}{T} - 1)W(0) \end{aligned}$$

On the other hand, the function $\tilde{p}(t) = t - (I + p)^{-1}(t)$ satisfies $\tilde{p}_0(t + p_0(t)) = p_0(t)$, so $\tilde{p}_0(t)$ has the same asymptotic behavior as $p_0(t)$ as $t \longrightarrow 0$.

The function

$$g(t) = \int_t^{\xi} \frac{ds}{\tilde{p}_0(s)},\ t > 0,$$

is decreasing with $g(\xi) = 0$ and $g(0^+) = +\infty$.

Thus $[0, +\infty[$ is in the range of g and the solution of (12.11) is given by $z(t) = g^{-1}(t)$.

Since $g(0^+) = +\infty$, $\lim_{t \to +\infty} z(t) = \lim_{t \to +\infty} g^{-1}(t) = 0$.

For $0 < \epsilon < 1$, there exists $\eta_\epsilon > 0$ such that $|\tilde{p}_0(s) - C_3 s^2| \le \epsilon C_3 s^2$ for $0 < s < \eta_\epsilon$.

Moreover, there exists $t_\epsilon > 0$ such that $0 < z(t) < \eta_\epsilon$ for $t \ge t_\epsilon$.

Therefore, if $t \ge t_\epsilon$, we have $-\tilde{p}_0(z(t)) \le C_3(\epsilon - 1)z(t))^2$.

Then,

$$\frac{dz(t)}{dt} + C_3(1 - \epsilon)(z(t))^2 \le 0,\ t \ge t_\epsilon.$$

It follows that $z(t) = O(\frac{1}{t})$ as $t \longrightarrow \infty$, hence $\|\nabla_\omega y(t)\|^2_{(L^2(\omega))^n} = O(\frac{1}{t})$ as $t \longrightarrow +\infty$, which completes the proof. ■

12.1.2 Regional Gradient Weak Stabilization

In this subsection we give sufficient conditions for weak regional gradient stabilization of system (12.1).

Theorem 12.3 *Assume that B is Locally Lipschitz and weakly sequentially continuous, g is Locally Lipschitz such that $g(.) \ge m > 0$ on $H^1(\Omega)$ and*

$$\langle S(t)\Psi, BS(t)\Psi\rangle = 0, \forall t \ge 0 \Longrightarrow \nabla_\omega \Psi = 0. \qquad (12.19)$$

Then the control (12.2) regionally weakly stabilizes the gradient of (12.1) on ω.

Proof Let $\psi \in H^1(\Omega)$ be fixed, we shall show that $\langle \nabla_\omega y(t), \psi \rangle \longrightarrow 0,\ as\ t \longrightarrow +\infty$.

Let $\{t_n\}$ be a sequence of positive numbers such that $t_n \longrightarrow +\infty$, as $n \longrightarrow +\infty$. The sequence $x_n = \langle \nabla_\omega y(t_n), \psi \rangle$ is, by virtue of (12.5) bounded.

Now, let $x_{\gamma(n)}$ be an arbitrary convergent subsequence of x_n.

Since $y(t_{\gamma(n)})$ is bounded in $H^1(\Omega)$, we can extract a subsequence also denoted $y(t_{\gamma(n)})$ such that $y(t_{\gamma(n)}) \rightharpoonup \Psi \in H^1(\Omega)$, as $n \longrightarrow +\infty$.

By similar arguments using in the proof of Theorem 12.1, we obtain the following inequality

$$\begin{aligned} &\int_0^T |\langle S(s)y(t_{\gamma(n)}), BS(s)y(t_{\gamma(n)})\rangle| ds \\ &\leq K_{\|y_0\|}\Big(\int_{t_{\gamma(n)}}^{t_{\gamma(n)}+T} |\langle y(s), By(s)\rangle|^2 ds\Big)^{\frac{1}{2}} \\ &+ \frac{M_{\|y_0\|}(g)}{m} \int_{t_{\gamma(n)}}^{t_{\gamma(n)}+T} |\langle y(s), By(s)\rangle|^2 ds \end{aligned}$$

It follows that

$$\int_0^T |\langle S(s)y(t_{\gamma(n)}), BS(s)y(t_{\gamma(n)})\rangle| ds \longrightarrow 0,\ as\ n \longrightarrow +\infty.$$

Using the dominated convergence theorem and the fact that B is weakly sequentially continuous, we deduce that $\langle S(s)\Psi, BS(s)\Psi \rangle = 0$ for all $t \geq 0$, which implies, by virtue of (12.19), that $\nabla_\omega \Psi = 0$. Then $\nabla_\omega y(t_{\gamma(n)}) \rightharpoonup 0$.

It follows that $x_n \longrightarrow 0,\ as\ t \longrightarrow +\infty$.

We deduce that $\langle \nabla_\omega y(t), \psi \rangle \longrightarrow 0,\ as\ t \longrightarrow +\infty$. In other words $\nabla_\omega y(t)$ converges weakly to 0 as $t \longrightarrow +\infty$, which proves the theorem. ■

The next result is a direct consequence of the above theorem.

Corollary 12.2 *Assume that B is locally Lipschitz and (12.19) holds. Then both the control (12.8) and the control (12.9) regionally weakly stabilize the gradient of system (12.1) on* ω.

The following example illustrates the established results.

Example 12.2 . Let Ω be an open regular subset of $\mathbb{R}^n$ such that $\omega \in \Omega$ and consider the following Schrödinger equation

$$\begin{cases} i\dfrac{\partial y(x,t)}{\partial t} = \Delta y(x,t) + u(t)By(x,t), & \Omega \times]0,\infty[\\ y(.,t) = 0 \quad \partial\Omega \times]0,\infty[\end{cases} \tag{12.20}$$

where the operator B is given by $By = \dfrac{1}{1+|\langle y,\phi\rangle|}\sum_{j=1}^{+\infty}\alpha_j\langle y,\varphi_j|_\omega\rangle\varphi_j|_\omega$, where $\phi \in H^1(\Omega)$, $\alpha_j > 0$, $\forall j \geq 1$, $\sum_{j=1}^{+\infty}\alpha_j^2 < \infty$ and (φ_j) is an orthonormal basis of the state space $H^1(\Omega)$ endowed with its natural complex inner product.

It is well known that the operator $A = -i\Delta$ with domain $\mathcal{D}(A) = H^2(\Omega)\cap H_0^1(\Omega)$ generates a semigroup of isometries ($\|S(t)y\| = \|y\|$).

Also, it is easy to see that B is weakly sequentially continuous and locally Lipschitz and $B(0) = 0$, (0 is an equilibrium state for system(12.20)).

For $\omega \subset \Omega$, the condition (12.19) holds.

Then the gradient of system (12.20) is regionally weakly stabilizable on the region ω by the controls

$$u_1(t) = \frac{-1}{\sup\left(1, \dfrac{\sum_{j=1}^{+\infty}|\alpha_j\langle y(.,t),\varphi_j|_\omega\rangle|^2}{1+|\langle y,\phi\rangle|}\right)} \times \frac{\sum_{j=1}^{+\infty}|\alpha_j\langle y(.,t),\varphi_j|_\omega\rangle|^2}{1+|\langle y,\phi\rangle|}$$

and

$$u_2(t) = \frac{-1}{1+\dfrac{\sum_{j=1}^{+\infty}|\alpha_j\langle y(.,t),\varphi_j|_\omega\rangle|^2}{1+|\langle y,\phi\rangle|}} \times \frac{\sum_{j=1}^{+\infty}|\alpha_j\langle y(.,t),\varphi_j|_\omega\rangle|^2}{1+|\langle y,\phi\rangle|}.$$

■

Now if we consider the control law

$$v(t) = -\langle y(t), Ky(t)\rangle \tag{12.21}$$

where $K : H^1(\Omega) \longrightarrow H^1(\Omega)$ is a continuous operator such that the following assumption holds

$$\langle y, Ky\rangle\langle y, By\rangle \geq 0, \forall y \in H^1(\Omega) \tag{12.22}$$

then we have the following result.

Proposition 12.1 *1. Suppose that there exists $\eta > 0$ such that $\forall y \in \mathcal{D}(A)$*

$$\langle G_\omega Ay, y\rangle - \langle y, By\rangle\langle y, G_\omega By\rangle \geq \eta(\langle Ay, y\rangle - \langle y, By\rangle^2), \text{ where } G_\omega = \nabla_\omega^*\nabla_\omega$$

If the control (12.21) regionally weakly stabilizes the gradient of system (12.1) on ω, then the corresponding state remains bounded on the whole domain Ω.

2. *Suppose that B is locally Lipschitz and sequentially continuous. If (12.22) holds and*

$$\langle S(t)\psi, KS(t)\psi\rangle\langle S(t)\psi, BS(t)\psi\rangle = 0,\ t \geq 0 \Longrightarrow \nabla_\omega \psi = 0. \qquad (12.23)$$

Then the control (12.21) regionally weakly stabilizes the gradient of (12.1) on ω, and the state remains bounded on Ω.

Proof 1. For all $y_0 \in \mathcal{D}(A)$ we have

$$\frac{d}{dt}\|\nabla_\omega y(t)\|^2_{(L^2(\omega))^n} \geq \eta \frac{d}{dt}\|y(t)\|^2$$

which implies that

$$\|\nabla_\omega y(t)\|^2_{(L^2(\omega))^n} - \|\nabla_\omega y_0\|^2_{(L^2(\omega))^n} \geq \eta\|y(t)\|^2 - \eta\|y_0\|^2,\ t \geq 0.$$

This inequality holds for all initial states, as its terms are continuous in $y_0 \in H^1(\Omega)$. The regional weak gradient stabilization of (12.1) on ω implies that $\|\nabla_\omega y(t)\|_{(L^2(\omega))^n}$ is bounded in t, then the above inequality ensures the boundedness of $\|y(t)\|$.

2. First, we note that from the work of Ball and Slemrod [1], system (12.1) has a unique global solution $y(t)$, which is bounded in t and there exists $\psi \in H^1(\Omega)$ such that $y(t) \rightharpoonup \psi$ as $t \longrightarrow +\infty$ and that

$$\langle S(t)\psi, BS(t)\psi\rangle = 0, t \geq 0$$

From (12.23), it follows that $\nabla_\omega \psi = 0$, and as the operator ∇_ω is continuous, then

$$\forall \varphi \in H^1(\Omega),\ \langle \nabla_\omega y(t), \varphi\rangle \longrightarrow \langle \nabla_\omega \psi, \varphi\rangle,\ as\ t \longrightarrow +\infty$$

which achieves the proof. ■

12.2 Regional Gradient Stabilization Problem

The purpose of this section is to give the minimum energy control that achieves regional gradient stabilization of system (12.1) on a subregion $\omega \subset \Omega$ which may be formulated as

$$\begin{cases} \min J(u) = \displaystyle\int_0^{+\infty} \langle y(t), Ry(t)\rangle dt + \int_0^{+\infty} |\langle y(t), P_\omega By(t)\rangle|^2 dt \\ \qquad + \displaystyle\int_0^{+\infty} |u(t)|^2 dt \\ u \in \mathcal{U}_{ad} = \{u | y(t)\ is\ a\ global\ solution\ and\ J(u) < +\infty\}. \end{cases} \qquad (12.24)$$

where $R \in \mathcal{L}(H^1(\Omega))$ is a self-adjoint and positive operator and P is a self-adjoint and positive operator verifying the Lyapunov equation

$$\langle P_\omega Ay, y\rangle + \langle y, P_\omega Ay\rangle + \langle y, Ry\rangle = 0, \ \ y \in \mathcal{D}(A) \tag{12.25}$$

with $P_\omega = G_\omega PG_\omega$ and $G_\omega = \nabla^*_\omega \nabla_\omega$.

Proposition 12.2 *Suppose that system (12.1) has a unique mild global solution $y^*(t)$ which satisfies*

$$\langle P_\omega y^*(t), y^*(t)\rangle \longrightarrow 0 \text{ as } t \longrightarrow \infty.$$

Then the control

$$u^*(t) = -\langle y^*(t), P_\omega By^*(t)\rangle \tag{12.26}$$

is the unique control solution of problem (12.24).

Proof Let us define the function $F(y) = \langle P_\omega y, y\rangle$, $y \in H^1(\Omega)$.

For all $y_0 \in \mathcal{D}(A)$ and $t \geq 0$, we have

$$\frac{dF(y^*(t))}{dt} = -\langle y^*(t), Ry^*(t)\rangle - 2\langle y^*(t), P_\omega By^*(t)\rangle^2$$

Integrating this relation, we obtain

$$\int_0^t \langle y^*(s), Ry^*(s)\rangle ds + 2\int_0^t \langle y^*(s), P_\omega By^*(s)\rangle^2 ds \leq F(y_0), \ t \geq 0 \tag{12.27}$$

The solution $y(.)$ is continuous with respect to the initial condition (see [3]), and since $F(y)$ and $\langle y, Ry\rangle$ are continuous then (12.27) holds for all $y_0 \in H^1(\Omega)$.

It follows that $J(u^*)$ is finite for all initial condition $y_0 \in H^1(\Omega)$ and $\mathcal{U}_{ad} \neq \emptyset$. Let $u \in \mathcal{U}_{ad}$, for $y_0 \in \mathcal{D}(A)$, integrating the relation

$$\frac{dF(y(t))}{dt} = [\langle y(t), P_\omega By(t)\rangle + u(t)]^2 - \langle y(t), Ry(t)\rangle - \langle y(t), P_\omega By(t)\rangle^2 - u^2(t)$$

we have

$$J(u) = F(y_0) + \int_0^{+\infty} [\langle y(t), P_\omega By(t)\rangle + u(t)]^2 dt$$

then

$$J(u) \geq F(y_0), \ \forall u \in \mathcal{U}_{ad}.$$

Setting $u = u^*$, we obtain $J(u^*) = F(y_0)$.

Let $y_0 \in H^1(\Omega)$, and a sequence $(y_{0n})_n \subset \mathcal{D}(A)$ such that $y_{0n} \longrightarrow y_0$ as $n \longrightarrow +\infty$.

we have

$$J(u) = F(y_{0n}) + \int_0^{+\infty} [\langle y_n(t), P_\omega B y_n(t)\rangle + v(t)]^2 dt$$

thus $J(u) \geq F(y_{0n})$.

By the continuity of F we deduce that $J(u) \geq F(y_0) = J(u^*)$. In other words (12.26) is the optimal control solution of the problem (12.24).

The uniqueness of u^* follows from the uniqueness of $y(t)$ solution of system (12.1), which completes the proof. ■

The following result gives sufficient conditions for the optimal control (12.26) to be a stabilizing one.

Proposition 12.3 *Suppose that $S(t)$ is a semigroup of contractions and P is compact such that $K = P_\omega B$ verifies*

$$\langle KS(t)\psi, S(t)\psi\rangle\langle S(t)\psi, BS(t)\psi\rangle = 0, \ t \geq 0 \Longrightarrow \nabla_\omega \psi = 0. \qquad (12.28)$$

Then (12.26) is the unique feedback control solution of (12.24) and stabilizes regionally weakly the gradient of system (12.1) on ω.

Proof From Proposition 12.1 we deduce that the control (12.26) stabilizes regionally weakly the gradient of (12.1) on ω.

Now as G_ω is continuous and P is supposed compact, then P_ω is also compact, so for any sequence of positive real number (t_n) verifying $t_n \longrightarrow +\infty$, the sequence $P_\omega y(t_n)$ tends to 0 strongly and the real sequence $F(y(t_n))$ converges to 0.

Then from the above proposition, we deduce that (12.26) is the unique solution of problem (12.24). ■

12.3 Simulation Results

We have shown that the gradient of system (12.1) may be stabilized on a subregion by the controls (12.8) and (12.9) or by the control (12.26) solution of problem (12.24). In order to illustrate the previous results using numerical computations, we perform the following steps:

Step 1: *Solve the Lyapunov equation* (12.25) *using Bartels-Stewart method* [4];
Step 2: *Apply a feedback control given by* (12.8), (12.9) *or* (12.26);
Step 3: *Solve system using Petrov-Galerkin method given in* [5];
Step 4: *Calculate* $\nabla_\omega y(t_i)$;
Step 5: *Calculate the stabilization cost* J.

Example 12.3 . On $\Omega =]0, 1[$, we consider the question of regional gradient stabilization of the following system

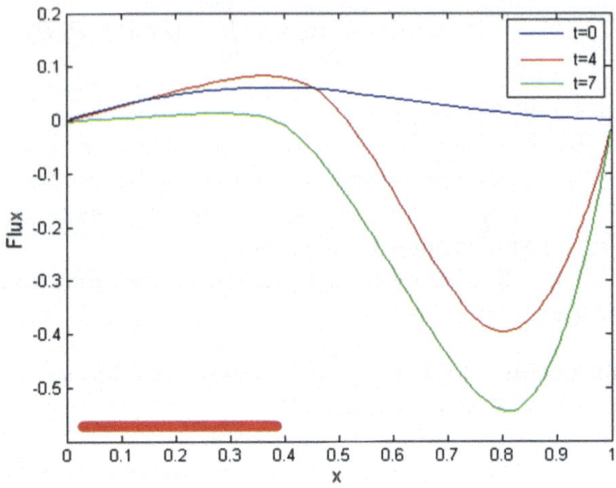

Fig. 12.1 The gradient evolution

$$
\begin{cases}
\dfrac{\partial y}{\partial t}(x,t) = 0.01\Delta y(x,t) + y(x,t) + \dfrac{1}{2}u(t)|y(x,t)| & \Omega\times]0,+\infty[\\
\dfrac{\partial y}{\partial x}(0,t) = \dfrac{\partial y}{\partial x}(1,t) = 0 & \partial\Omega\times]0,+\infty[\\
y(x,0) = x^2(1-x)^3
\end{cases}
\tag{12.29}
$$

We consider the problem (12.24) with $R = 4\nabla_\omega^*\nabla_\omega$ and the subregion $\omega =]0, 0.4[$.

1. Optimal stabilizing control.
The solution of problem (12.24) is given by

$$u(t) = -\frac{1}{2}\langle P_\omega|y|, y\rangle, \tag{12.30}$$

the Fig. 12.1 shows how the gradient of system (12.29) is stabilized by the control (12.30) on ω with a stabilization error equals to 2.758×10^{-4} and minimizes the performance cost J that equals to 3.133×10^2 (Fig. 12.2).

The following table shows that there exists a relation between the area of target subregion ω, the cost and the error of gradient stabilization.

ω	$]0, 0.3[$	$]0, 0.4[$	$]0, 0.6[$	$]0, 0.8[$	$]0, 1[$
Error	2.163×10^{-5}	2.758×10^{-5}	4.567×10^{-5}	7.326×10^{-5}	9.648×10^{-5}
Cost	2.626×10^{2}	3.133×10^{2}	4.14×10^{2}	6.22×10^{2}	7.95×10^{2}

More the area of the target region increases, more both the gradient stabilization error and the cost increase.

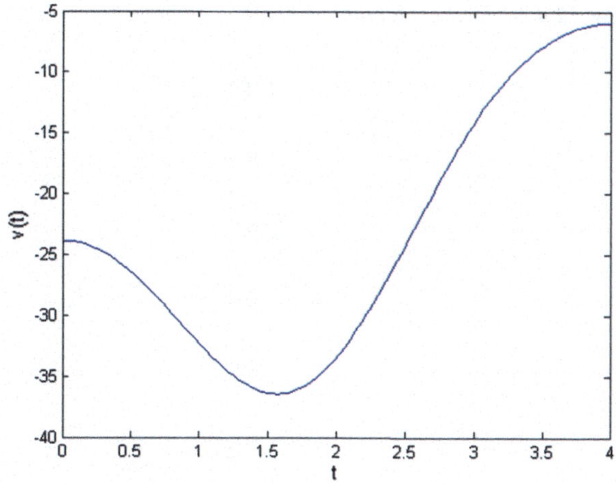

Fig. 12.2 Control evolution

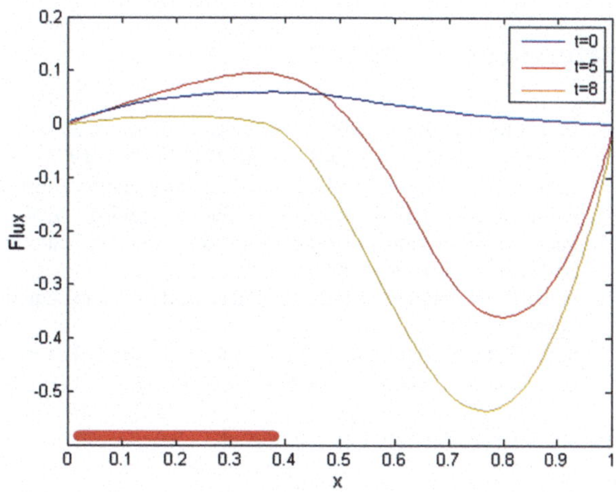

Fig. 12.3 The gradient evolution

2. Stabilizing control.
Following steps: 2, 3 and 4 we get the Fig. 12.3 which shows that the following control

$$u(t) = -\frac{\langle \frac{1}{2}|y|, y\rangle}{\sup(1, |\langle \frac{1}{2}|y|, y\rangle|)}, \tag{12.31}$$

stabilizes the gradient of system (12.29) on ω with a stabilization error equals to 4.201×10^{-4} and cost J equals to 1.007×10^3 (Fig. 12.4).

Remark 12.2 With respect to the error and cost, the optimal control (12.30) is better than the stabilizing one (12.31).

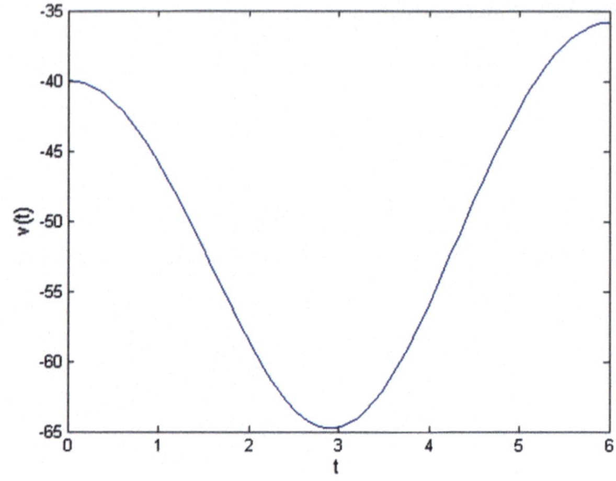

Fig. 12.4 Control evolution

References

1. Ball, J.M.: On the asymptotic behavior of generalized processes with application to nonlinear evolutions equations. J. Differ. Equ. **27**, 224–265 (1978)
2. Lasiecka, I., Tataru, D.: Uniform boundary stabilisation of semilinear wave equation with nonlinear boundary damping. J. Differ. Integral Equ. **6**, 507–533 (1993)
3. Pazy, A.: Semigroups of Linear Operators and Applications to Partial Differential Equations. Springer, New York (1983)
4. Penzl, T.: Numerical solution of generalized Lyapunov equations. Adv. Comput. Math **8**, 33–48 (1998)
5. Skeel, R.D., Berzins, M.: A Method for the spatial discretization of parabolic equations in one space variable. SIAM J. ScI. Stat. Comput. **11**, 1–32 (1990)

Chapter 13
Conclusion and Perspectives

In this work, we have considered the problem of the stability of the stabilization of infinite dimensional systems. We first recall the main results on the global stability and stabilization. Then we explore the concept of regional stability and stabilization for various classes of infinite dimensional systems. The gradient stabilization of such systems is also discussed and various approaches are developed. stabilization of such systems and various approaches are developed. Then we study the stabilization of a class of second order systems. This work shows that many technical difficulties have to be overcome. But the richness of this subject leads to many questions still open and may be the subjects of future explorations. This is the special case of the few problems posed below.

1. Exponential stabilization of a class of second order systems
Let us consider the following system

$$\begin{cases} y_{tt}(t) = Ay(t) + v(t)By(t), \\ y(0) = y_0, \ y_t(0) = y_1, \end{cases} \tag{13.1}$$

where A is a dissipative operator defined on a Hilbert space X endowed with inner product $\langle ., . \rangle$ and the corresponding norm $|.|$, B is a linear and bounded operator on X and $v \in L^\infty(0, +\infty)$.

The question is to study the exponential stabilization of (13.1), This problem could be resolved using a control verifying the dissipativity energy inequality and the observability assumption.

2. Boundary stabilization of infinite dimensional semilinear systems
Let $\Omega \subset \mathbb{R}^n$ be an open and bounded domain, we consider the following semilinear system

E. H. Zerrik and O. Castillo, *Stabilization of Infinite Dimensional Systems*,
Studies in Systems, Decision and Control 355,
https://doi.org/10.1007/978-3-030-68600-0_13

$$\begin{cases} \frac{dy(t)}{dt} = Ay(t) + g(y(t)) \\ y = f(v(t), y(t)) \\ y(0) = y_0, \end{cases} \tag{13.2}$$

where A generates a semigroup of contraction $S(t)$ on a Hilbert space X, $v(t)$ is a real valued control, g is a nonlinear function on X and f is a linear function.

The aim is to characterize the controls $v(t)$ which stabilize exponentially, strongly or weakly the system (13.2).

3. On the stabilization of the gradient of linear systems and strategic regions
We can study the notion of stabilization of the gradient through the admissibility notion.

On Ω an open regular subset of $\mathbb{R}^n$, let a system described by the equation

$$\begin{cases} \frac{\partial y}{\partial t}(t) = Ay(t) \\ y(0, x) = y_0(x) \in X \end{cases} \tag{13.3}$$

where A generates a semi group $S(t)$ on $X = L^2(\Omega)$.

The system is augmented with the output $z(t) = \nabla_\omega y(t)$.

Let $\omega \subset \overline{\Omega}$ be a target subregion. and we recall the definition of the operator gradient

$$\nabla_\omega : X \longrightarrow (L^2(\omega))^n$$

$$y \longrightarrow (\chi_\omega \frac{\partial y_1}{\partial t}, \chi_\omega \frac{\partial y_2}{\partial t}, \cdots, \chi_\omega \frac{\partial y_n}{\partial t})$$

The space $(L^2(\Omega))^n$ is endowed with its usual inner product $\langle ., . \rangle_n$, and $(L^2(\omega))^n$ is endowed with the restriction of $\langle ., . \rangle_n$. To make sense to the output $z(t)$, we suppose that $\mathcal{D}(A)$ is endowed with the graph norm $||.||_A$. We can consider the following definitions.

A region ω is said admissible for $S(t)$ if
(1) $\nabla_\omega : (\mathcal{D}(A), ||.||_A) \longrightarrow (L^2(\omega))^n$ is continue.
(2) there exists $\alpha > 0$ such that

$$||z(.)||_{L^2(0,\infty;(L^2(\omega))^n)} = ||\nabla_\omega S(.)y_0||_{L^2(0,\infty;L^2(\omega))} \leq \alpha ||y_0|| \,, \; y_0 \in \mathcal{D}(A) \tag{13.4}$$

The question that remains is to characterize the stability of the output $z(t)$ on admissible regions.

Note that ω is said to be admissible, if the operator ∇_ω is admissible in the meaning of Weiss (1989). This allows to extend the application $y_0 \longmapsto \nabla_\omega S(.)y_0$ in a continuous map from X to $(L^2(0, \infty))^n$, thus the output $z(t)$ is well defined for all $y_0 \in X$.

The notion of admissibility allows to work in a more general functional framework, $(X = L^2(\Omega))$, consequently the existence of ∇y will be discussed only on ω.

4. Exponential stabilization and stabilization with bounded controls of the gradient of bilinear systems.
On Ω we consider a bilinear system described by the equation

$$\begin{cases} \dfrac{dy}{dt} = Ay(t) + v(t)By(t) \\ y(0) = y_0 \end{cases} \tag{13.5}$$

Can we characterize the control $v(t)$? which exponentially stabilizes the gradient of the system (13.5)?

We can also consider the stabilization of the gradient of the system (13.5) and minimizing a functional cost with bounded controls, ($|v(t)| \leq M$ for $M > 0$.) That is to consider the questions
1. Under what condition the system (13.5) with control $u(t)$ admits a global and unique weak solution?
2. Characterize the stabilizing control at different degrees, the gradient of the system (13.5).

5. Regional detectability of the gradient.
We consider the system (13.5) augmented with the output equation

$$z(t) = Cy(t) \tag{13.6}$$

where $C \in \mathcal{L}(X, O)$, with O the observation space supposed to be Hilbert space.

We recall that the system (13.5), is said to be detectable, if there exists an operator $F \in \mathcal{L}(X, O)$ such that the operator $A + FC$ generates a C_0 -semigroup exponentially stable. It is said G-detectable on ω if there exists an operator $F \in \mathcal{L}(X, O)$ such that the system

$$\begin{cases} \dfrac{\partial y}{\partial t}(t) = (A + FC)y(t) \\ y(0, x) = y_0(x) \end{cases} \tag{13.7}$$

be exponentially G-stable on ω. Can we then characterize the detectability of the gradient?

Relation of Stability with Inputs and Outputs Structures
Interesting work consists of taking up various notions of analysis of infinite dimensional systems and connecting them to actuators and sensors structure.

Several results would be interesting to explore, among others, the characterization of stability through the choice of the number and the spatial distribution of the actuators.

It would also be interesting to take up the various criteria defining the cost of stabilization and to study their dependence with the number of actuators; whether for linear or non-linear systems or in internal and boundary cases.

6. Stabilization of a bilinear wave equation using distributed control
We consider the wave equation in a smooth bounded domain Ω of $\mathbb{R}^n$ described by the equation

$$\begin{cases} y_{tt}(x,t) - \Delta y(x,t) + u(x,t)y_t(x,t) = 0 & Q = \Omega \times]0, +\infty[\\ y(x,t) = 0 & \partial\Omega \times]0, +\infty[\\ y(x,0) = y_0, \;\; y_t(x,t) = y_1(x) & \Omega, \end{cases} \tag{13.8}$$

where $u \in L^\infty(Q)$ are distributed controls.

The goal is to:
1. characterize a control u and give sufficient conditions for weak, strong and exponential stabilisation of system (13.8),
2. study the strong stabilization of system (13.8) using a minimization problem

$$\begin{cases} \min J(u) = \displaystyle\int_0^{+\infty} \|(y(t), y_t(t)\|^2 dt + \int_Q u^2(x,t)dxdt \\ u \in \mathcal{U}_{ad} := \{u \in L^\infty(Q) \mid (y(t), y_t(t)) \text{ is a global solution and } J(u) < +\infty\}, \end{cases}$$

3. develop sufficient conditions for regional and gradient stabilization of system (13.8).

References

1. Afifi, L.: Analyse régionale: Systèmes discrets, sources et détection. These d'état, Kenitra, Maroc. Juillet (1994)
2. Badraoui, L., Seidman, T.: Some 'Regional controllabilty' issues for the heat equation. Rev. Matemaátcas A plicadas (Chile) **22**, 1–23 (2001)
3. Balakrishnan, A.V.: Applied Functional Analysis. Springer, New York (1976)
4. Ball, J.M., Marsden, J.E., Slemrod, M.: Controllability for distributed bilinear systems, SIAM. J. Control Optim. **20**, 575–597 (1982)
5. Bartle, R.G.: The Elements of Integration and Lebesgue Measure. Wiley, New York (2014)
6. Belfekih, A.: Répartition spatiale optimale du contrôle dans les systèmes distribués. Université Med V, Diplôme de spécialité de troisième cycle (1987)
7. Belfekih, A.: Une nouvelle approche pour la contrôlabilité et l'observabilité des systèmes distribués. Université Ibn Tofail, Thèse d'état (1990)
8. Bensoussan, A.: Representation and Control of Infinite Dimensional Systems, vol. 1. Boston (1992)
9. Berrahmoune, L.: Stabilization and decay estimate for distributed bilinear systems. Syst. Control Lett. **36**, 167–171 (1999)
10. Berrahmoune, L.: Stabilization of bilinear control systems in Hilbert space with nonquadratic feedback. Rendiconti Matematico di palermo **58**, 275–282 (2009)
11. Benaddi, A., Rao, B.: Energy decay rate of wave equations with indefinite damping. J. Differential Equ. **161**, 337–357 (2000)
12. Bounabat, A., Gauthier, J.: Weak stabilizability of infinite dimensional nonlinear systems. Appl. Math. Lett **4**, 95–98 (1991)
13. Bounabat, A, Bouslouss, and Hammouri, H.: Stabilisation faible d'un système bilinéaire dissipatif en dimension finie. Extracta Mathematicae **7**(2), 95–98 (1993)
14. Bounit, H.: Contribution à la stabilisation et à la construction d'observateurs pour une classe de systèmes à paramètres distribués. Universitè de Lyon I, France, Thèse de doctorat (1996)
15. Butkovskii, A.G.: Theory of optimal control of distributed parameter systems. Elsevier, Amsterdam (1969)
16. Chazarain, J., Piriou, A.: Introduction à la théorie des équations aux dérivées partielles linéaires. Bordas, Paris (1981)
17. Chilov, G.: Analyse mathématique: fonctions de plusieurs variables réelles, Mir-Moscou (1975)
18. Chen, G., Fulling, A., Narcowich, F.J., Sun, S.: Exponential Decay of Energy of Evolution Equations with Locally Distributed Damping. SIAM J. Appl. Math. **51**, 266–301 (1991)

E. H. Zerrik and O. Castillo, *Stabilization of Infinite Dimensional Systems*, Studies in Systems, Decision and Control 355, https://doi.org/10.1007/978-3-030-68600-0

19. Curtain, R.F., Pritchard, A.J.: A semi-group approach to infinite dimensional systems theory. IMA Symp. On recent dev. in control, Leicester (1976)
20. Curtain, R.F., Pritchard, A.J.: An abstract theory for unbounded control action for distributed parameter systems. SIAM J. Contr. **15**, 566–611 (1977)
21. Couchoroun, J.M.: Stabilization of controlled vibrating systems. ESAIM: COCV **17**, 1144–1157 (2011)
22. Dautry, R., Lions, J.L.: Analyse mathématique et calcul numérique pour les sciences et les techniques. INSTN, CEA collection enseignement, Masson (1988)
23. Delfour, M.C., Mitter, S.K.: Controlability and observability for infinite dimensional systems. SIAM J. Control. **10**, 329–333 (1972)
24. Dunford, N., Schwartz.: Linear Operator. Part I. Interscience, New York (1958)
25. El Alami, N.: Analyse et commande optimale des systèmes bilinéaires à paramètres distribués-Application aux procédés énérgetique. Université de Perpignan, Thèse d'état (1986)
26. El Jai, A., Pritchard, A.J.: Capteurs et actionneurs dans l'analyse des systèmes distribués. Masson (1986)
27. El Jai, A., Amouroux, M.: Sensors and observers in distributed parameter systems. Int. J. Control. **47**(1), 333–347 (1988)
28. El Jai, A., Afifi, L.: Spy-sensors and detection. Int. J. Syst. Sci. **26**(8), 1446–1447 (1995)
29. El Jai, A., Zerrik, E., Simon, M.C.: Regional observability of a thermal process. IEEE, Trans. Autom. Control **40**(3), 518–521 (1995)
30. Fattorini, H.O.: On complete controllability of linear systems. J. Differ. Equ. **3**, 391–402 (1967)
31. Fattorini, H.O.: Controllability of Higher Order Linear Systems in Mathematical Theory of Control, Balakrishnan, Neustad (eds.). Academic, New York (1967)
32. Haraux, A.: Une remarque sur la stabilisation de certains systémes du deuxiéme ordre en temps. Port. Math **46**, 245–258 (1989)
33. Haraux, A.: Remarks on weak stabilization of semilinear wave equations. ESAIM Control Optimisation and Calculus of Variations **6**, 553–560 (2001)
34. Ghisi, M., Gobbino, M., Haraux, A.: The remarkable effectiveness of time-dependent damping terms for second order evolution equations. SIAM J. Control Optim. **54**(3), 1266–1294 (2016)
35. Khapalov, A.Y.: Controllability of Partial Differential Equations Governed by Multiplicative Controls. Springer, Berlin (2010)
36. Yosida, K.: Functional Analysis. Springer, Berlin (1980)
37. Martinez, P., Vancostenoble. J.: Exponential Stability for the Wave Equation with Weak Non-monotone Damping. Portugaliae Mathematica, 57 Fasc. 3 (2000)
38. Ouzahra, M.: Partial stabilisation of semilinear systems using bounded controls. Int. J. Control **86**, 2253–2262 (2013)
39. Ouzhra, M.: Feedback stabilization of parabolic systems with bilinear controls. Electron. J. Deffer. Equ. **38**, 1–10 (2011)
40. Pardalos, P.M., Yatsenko, V.A.: Optimization and Control of Bilinear Systems: Theory, Algorithms and Applications, Springer Science and Business Media (2010)
41. Prichard, A.J., Triggiani, R.: Stabilizability in Banach spaces. University of Warwick. CTC report N 35 (1974)
42. Kangsheng, L., Bopeng, R., Xu, Z.: Stabilization of the wave equations with potential and indefinite damping. J. Math. Anal. Appl. **269**, 747–769 (2002)
43. Ryan, E., Buckingham, N.J.: On asymptotically stabilizing feedback control of bilinear systems. IEEE. Trans. Automa Contr.AC-28 **8**(1), 862–865 (1983)
44. Slemrod, M.: The linear stabilization problem in Hilbert space. J. Funct. Anal **11**, 334–345 (1974)
45. Staffans, O.: Admissible factorizations of Hankel operators induce well-posed linear systems. Syst. Control Lett. **37**, 301–307 (1999)
46. Staffans, O.: Lax-Phillips, scattering and well-posed linear systems. In: Proceedings of the 7th Mediterranean Conference on Control and Automation (MED99), pp. 1359–1370. Israel, Haifa (1999)

47. Tebou, L.: Equivalence between observability and stabilization for a class of second order semilinear evolution equation. Discret. Contin. Dyn. Syst. 744–752 (2009)
48. Triggiani, R.: Extension of rank conditions for controllability and observability to Banach spaces and unbounded operators. Siam. J. Contr. Opt. **14**, 313–338 (1976)
49. Wang, P.K.C.: Feedback Stabilization of Distributed Systems with Application to Plasma Stabilization. Instability of Continuous Systems, Leipholz, H (ed.). Springer, Berlin (1971)
50. Weiss, G.: Admissibility of unbounded control operators. SIAM J. Contr. Optim. **27**, 527–545 (1989)
51. Yoshida, K.: Functional Analysis, 6th edn. Springer, Berlin (1980)
52. Zabczyk, J.: Complete stabilizability implies exact controllability. Seminarul de Ecuatti Functionale. Universitatea din Timosoara, Rumania (1976)
53. Zerrik, E., Ouzahra, M.: Output stabilization for infinite bilinear systems. Appl. Math. Comput. Sci. **15**(2), 187–195 (2005)
54. Zerrik, E., Ouzahra, M.: Output stabilization for distributed semilinear systems. IET. Control Theory Appl. **1**(3), 838–843 (2007)
55. Zerrik, E., Boutoulout, A., Kamal, A.: Regional gradient controllability of parabolic systems. Int. J. Appl. Math. Comput. Sci. **9**, 767–787 (1999)
56. Zerrik, E., Laghrissi, R.: Regional target control of the wave equation. Int. J. Syst. Sci. **32**(10), 1233–1242 (2001)
57. Zerrik, E., Ghafrani, F.: Minimum energy control subject to output constraints. Numerical approach. IEE - Proc.-D, Control Theory Appl. **149**(1), 105–110 (2002)
58. Zerrik, E., Boutoulout, A., El Jai, A.: Actuators and regional boundary controllability for parabolic systems. Int. J. Syst. Sci. **31**(1), 73–82 (2000)
59. Zerrik, E., Boutoulout, A., Kamal, A.: Regional gradient controllability of parabolic systems. Appl. Math. Comput. Sci. **9**, 767–787 (1999)
60. Zerrik, E., Kamal, A., Boutoulout, A.: Regional gradient controllability and actuators. Int. J. Syst. Sci. **33**(4), 239–246 (2001)
61. Zerrik, E., Bourray, H., El Jai, A.: Regional flux reconstruction for parabolic systems. Int. J. Syst. Sci. **34**, 641–650 (2003)
62. Zerrik, E., Bourray, H., El Jai, A.: Regional observability for semilinear distributed parabolic systems. Dyn. Control Syst. **10**, 413–430 (2004)
63. Zerrik, E., Benslimane, Y., El Jai, A.: An output stabilization of bilinear distributed systems. Int. J. Math. Anal. **7**, 195–211 (2013)
64. Ztot, K.: Analyse et contrôle des systèmes bilinéaires distribués. Thèse d'état, Ecole Mohammadia d'ingénieurs, Rabat (1996)
65. Zuazua, E.: Exponential decay for the semilinear wave equation with locally distributed damping. Commun. Partial Differ. Equ. **15**(2), 205–235 (1990)

Index

E. H. Zerrik and O. Castillo, *Stabilization of Infinite Dimensional Systems*,
Studies in Systems, Decision and Control 355,
https://doi.org/10.1007/978-3-030-68600-0

The manufacturer's authorised representative in the EU is Springer
Nature Customer Service Centre GmbH, Europaplatz 3, 69115 Heidelberg,
Germany. If you have any concerns regarding our products, please
contact ProductSafety@springernature.com

Printed and bound by CPI Group (UK) Ltd, Croydon, CR0 4YY
30/09/2026
02225448-0010